Maths for Chemists
2nd Edition

Tutorial Chemistry Texts

Editor-in-Chief: Prof EW Abel
Executive Editors: Prof AG Davies, Prof D Phillips and Prof JD Woollins
Educational Consultant: Mr M Berry

This series of books consist of short, single-topic or modular texts, concentrating on the fundamental areas of chemistry taught in undergraduate science courses. Each book provides a concise account of the basic principles underlying a given subject, embodying an independent-learning philosophy and including worked examples. The "one topic, one book" approach, ensures that the series is adaptable to chemistry courses across a variety of institutions.

Titles in the Series:

For further information please contact:
Book Sales Department, Royal Society of Chemistry, Thomas Graham House, Science Park, Milton Road, Cambridge, CB4 0WF, UK
Telephone: +44 (0)1223 420066, Fax: +44 (0)1223 420247,
Email: books@rsc.org
Visit our website at http://www.rsc.org/Shop/Books/

Maths for Chemists
2nd Edition

Martin Cockett
Department of Chemistry, University of York, Heslington, York, UK

Graham Doggett
University of York, Heslington, York, UK (retired)

RSC Publishing

Tutorial Chemistry Texts No. 25

ISBN: 978-1-84973-359-5
ISSN: 2045-6611

A catalogue record for this book is available from the British Library

Published by The Royal Society of Chemistry,
Thomas Graham House, Science Park, Milton Road,
Cambridge CB4 0WF, UK

Registered Charity Number 207890

For further information see our web site at www.rsc.org

Preface
Maths for Chemists Second Edition

This revised edition of the Maths for Chemists introductory text provides a foundation in the key mathematical topics required for most degree level chemistry courses. While it is aimed primarily at students with limited backgrounds in mathematics, the text should prove accessible and useful to all chemistry undergraduates. We have chosen from the outset to place the mathematics in a chemical context, a challenging approach because the context can often make the problem appear more difficult than it actually is. However, it is equally important to convince students of the relevance of mathematics in all branches of chemistry. Our approach links mathematical principles with the chemical context, by introducing the basic concepts first and then demonstrates how they translate into a chemical setting.

Historically, physical chemistry has been the target for much of the undergraduate level mathematical support. However, in all branches of chemistry – be they the more traditional areas of inorganic, organic and physical, or the rapidly developing areas of biochemistry, materials, analytical and environmental chemistry – mathematical tools are required to build models of varying degrees of complexity, in order to develop a language for providing insight and understanding together with, ideally, some predictive capability.

Since the target student readership possesses a wide range of mathematical experience, we have created a course of study in which selected key topics are treated without going too far into the finer mathematical details. The first two chapters focus on numbers, algebra and functions in some detail, as these topics form an important foundation for further mathematical developments in calculus and for working with quantitative models in chemistry. There then follow chapters on limits, differential calculus, differentials and integral calculus. Later chapters go on to cover power series, complex numbers, and the properties and applications of determinants, matrices and vectors. A final chapter on simple statistics and error analysis is a new addition to this second edition of the book, and covers the key principles required for a successful appraisal and treatment of experimental errors – an area increasingly neglected in formal undergraduate teaching, but which remains a vital component in all areas of experimental chemistry.

Contents

1
Numbers and Algebra

Numbers of one kind or another permeate all branches of chemistry (and science generally), simply because any measuring device we use to record a characteristic of a system can only yield a number as output. For example, we might measure or determine the:

- Weight of a sample.
- Intensity or frequency of light absorption of a solution.
- Vibration frequency for the HCl molecule.
- Relative molecular mass of a carbohydrate molecule.

Or we might:

- Confirm the identity of an organic species by measuring its boiling point.
- Measure, or deduce, the equilibrium constant of a reversible reaction.
- Wish to count the number of isomeric hydrocarbon species with the formula C_4H_{10}.

In some of these examples, we also need to:

- Specify units.
- Estimate the error in the measured property.

Clearly then, the manner in which we interact with the world around us leads us quite naturally to use numbers to interpret our experiences.

In many situations, we routinely handle very large and very small numbers, so disparate in size that it is difficult to have an intuitive feel for order of magnitude. For example:

- The number of coulombs (the basic unit of electrical charge) associated with a single electron is approximately 0.000 000 000 000 000 000 160 2177.
- The equilibrium constant for the electrochemical process

$$Au^{3+}(aq) + Al(s) \rightleftharpoons Au(s) + Al^{3+}(aq)$$

is of the order of 1 followed by 4343 zeros.[1] In chemical terms, we have no problem with this answer, as it indicates that the equilibrium is

Decimal numbers are commonly written with a space between every group of three digits after the decimal point (sometimes omitted if there are only four such digits).

totally towards the right side (which means that the aluminium electrode will be completely consumed and the gold electrode untouched).

These two widely different examples, of a type commonly experienced in chemistry, illustrate why it is so important to feel at ease using numbers of all types and sizes. A familiarity and confidence with numbers is of such fundamental importance in solving quantitative chemical problems that we devote the first two chapters of this book to underpinning these foundations. Our main objective is to supply the necessary tools for constructing models to help in interpreting numerical data, as well as in achieving an understanding of the significance of such data.

Aims:

In this introductory chapter, we provide the necessary tools for working with numbers and algebraic symbols, as a necessary prelude to understanding functions and their properties – a key topic of mathematics that impinges directly on all areas of chemistry. By the end of the chapter you should be able to:

- Understand the different types of numbers and the rules for their combination.
- Work with the scientific notation for dealing with very large and very small numbers.
- Work with numerical and algebraic expressions.
- Simplify algebraic expressions by eliminating common factors.
- Combine rational expressions by using a common denominator.
- Treat units as algebraic entities.

1.1 Real Numbers

1.1.1 Integers

Counting numbers have been in use for a very long time, but the recognition of zero as a numeral originated in India over two millennia ago, and only became widely accepted in the West with the advent of the printed book in the 13th century.

One of the earliest skills we learn from childhood is the concept of counting: at first we learn to deal with natural numbers (positive, whole numbers), including zero, but we tend to ignore the concept of negative numbers, because they are not generally used to count objects. However, we soon run into difficulties when we have to subtract two numbers, as this process sometimes yields a negative result. The concept of a negative counting number applied to an object can lead us into all sorts of trouble, although it does allow us to account for the notion of debt (you owe me 2 apples is the equivalent of saying "I own –2 apples"). We therefore extend natural numbers to a wider category of number called integers, which consist of all positive and negative whole numbers, as well as zero:

$$\ldots, -3, -2, -1, 0, 1, 2, 3, \ldots$$

We use integers in chemistry to specify:

- The atomic number, Z, defined as the number of protons in the nucleus: Z is a positive integer, less than or equal to 112.
- The number of atoms of a given type (positive) in the formula of a chemical species.
- The number of electrons (a positive integer) involved in a redox reaction occurring in an electrochemical cell.
- The quantum numbers required in the mathematical specification of individual atomic orbitals. These can take positive or negative integer values or zero depending on the choice of orbital.

1.1.2 Rational Numbers

When we divide one integer by another, we sometimes obtain another integer; for example, $6/-3 = -2$. At other times, however, we obtain a fraction, or rational number, of the form a/b, where the integers a and b are known as the numerator and denominator, respectively; for example, $2/3$. The denominator, b, cannot take the value zero because $a/0$ is of indeterminate value.

Rational numbers occur in chemistry:

- In defining the spin quantum number of an electron ($s = 1/2$), and the nuclear spin quantum number, I, of an atomic nucleus; for example, ^{45}Sc has $I = 7/2$.
- In specifying the coordinates $(0,0,0)$ and $\left(\frac{a}{2}, \frac{a}{2}, \frac{a}{2}\right)$, which define the locations of two of the nuclei that generate a body-centred unit cell of side a.

1.1.3 Irrational Numbers

Rational numbers can always be expressed as ratios of integers, but sometimes we encounter numbers which cannot be written in this form. These numbers are known as irrational numbers and include:

At the time of writing, the heaviest (named) element to have been isolated is the highly radioactive element copernicium ($Z = 112$). In June 2011, elements 114 and 116 officially joined the periodic table. Element 116 was made by bombarding targets made of the radioactive element curium ($Z = 96$) with calcium nuclei ($Z = 20$). The nuclei of element 116 lasted only a few milliseconds before decaying into element 114, which itself lasted less than half a second before decaying to copernicium. The heaviest naturally occurring element is uranium, $Z = 92$.

$\sqrt{2}$ is obtained as a solution of the equation $x^2 - 2 = 0$; likewise, $\sqrt[3]{2}$ is obtained as a solution of $x^3 - 2 = 0$.

- **Surds**, of the form $\sqrt{2}$, $\sqrt[3]{2}$, which are obtained from the solution of a quadratic or higher order equation.
- **Transcendental numbers**, which, in contrast to surds, do not derive from the solution to algebraic equations. Examples include π, which we know as the ratio of the circumference to diameter of a circle, and e, which is the base of natural logarithms.

1.1.4 Decimal Numbers

Decimal numbers occur in:

- Measuring chemical properties, and interpreting chemical data.
- Defining relative atomic masses.
- Specifying the values of fundamental constants.

Decimal numbers are so-called because they use base 10 for counting.

Decimal numbers consist of two parts separated by a **decimal point**:

- Digits to the left of the decimal point give the integral part of the number in units, tens, hundreds, thousands, *etc.*
- A series of digits to the right of the decimal point specify the fractional (or decimal) part of the number (tenths, hundredths, thousandths, *etc.*).

We can now more easily discuss the distinction between rational and irrational numbers by considering how they are represented using decimal numbers.

Rational numbers, expressed in decimal form, may have either of the following representations:

- A finite number of digits after the decimal point: for example, 3/8 becomes 0.375.
- A never-ending number of digits after the decimal point, but with a repeating pattern: for example, 70/33 becomes 2.121 212..., with an infinite repeat pattern of '12'.

Irrational numbers, expressed in decimal form have a never-ending number of decimal places in which there is *no* repeat pattern: for example, π is expressed as 3.141 592 653... and e as 2.718 281 82... . As irrational numbers like π and e cannot be represented exactly by a finite number of digits, there will always be an error associated with their decimal representation, no matter how many decimal places we include. For example, the important irrational number e, which is the base for natural logarithms (not to be confused with the electron charge), appears widely in chemistry. This number is defined by the infinite sum of terms:

$$e = 1 + \frac{1}{1!} + \frac{1}{2!} + \frac{1}{3!} + \frac{1}{4!} + \cdots + \frac{1}{n!} + \cdots \qquad (1.1)$$

where $n!$ is the **factorial** (pronounced 'n factorial') of the number n, defined as $n! = 1 \times 2 \times 3 \times 4 \times \ldots \times n$: for example, $4! = 1 \times 2 \times 3 \times 4$. The form of eqn (1.1) indicates that the value for e keeps getting larger (but by increasingly smaller amounts), as we include progressively more and more terms in the sum; a feature clearly seen in Table 1.1, in which the value for e has been truncated to 18 decimal places.

We can represent a sum of terms using a shorthand notation involving the summation symbol Σ: for example, the sum of terms $e = 1 + \frac{1}{1!} + \frac{1}{2!} + \frac{1}{3!} + \frac{1}{4!} + \cdots + \frac{1}{n!} + \cdots$ may be written as $\sum_{r=0}^{\infty} \frac{1}{r!}$, where the **counting index**, which we have arbitrarily named r, runs from 0 to ∞. A sum of terms which extends indefinitely, is known as an **infinite series**, whereas one which extends to a finite number of terms is known as a **finite series**. We shall discuss series in more detail in Chapter 8.

Table 1.1. An illustration of the effect of successive truncations to the estimated value of e derived from the infinite sum of terms given in eqn (1.1).

n	Successive estimated values for e
1	2.000 000 000 000 000 000
5	2.716 666 666 666 666 666
10	2.718 281 801 146 384 797
15	2.718 281 828 458 994 464
20	2.718 281 828 459 045 235
25	2.718 281 828 459 045 235
30	2.718 281 828 459 045 235

Although the value of e has converged to 18 decimal places, it is still not exact; the addition of more terms causes the calculated value to change beyond the 18th decimal place. Likewise, attempts to calculate π are all based on the use of formulae with an infinite number of terms:

- Perhaps the most astonishing method uses only the number 2 and surds involving sums of 2:

$$\pi = 2 \times \frac{2}{\sqrt{2}} \times \frac{2}{\sqrt{2+2}} \times \frac{2}{\sqrt{2+2+2}} \times \cdots$$

- Another method involves an infinite sum of terms:

$$\frac{\pi}{2} = \frac{1}{1} + \frac{1 \times 1}{1 \times 3} + \frac{1 \times 1 \times 2}{1 \times 3 \times 5} + \frac{1 \times 1 \times 2 \times 3}{1 \times 3 \times 5 \times 7} + \cdots,$$

- A particularly elegant method uses a formula that relates the square of π to the sum of the inverses of the squares of all positive whole numbers:

$$\frac{\pi^2}{6} = 1 + \frac{1}{2^2} + \frac{1}{3^2} + \frac{1}{4^2} + \frac{1}{5^2} \cdots$$

However, this requires an enormous number of terms to achieve a satisfactory level of precision (see Chapter 8 for more information regarding infinite series and convergence).

1.1.4.1 Working with Decimal Numbers

As we have seen above, numbers in decimal form may have a finite or infinite number of digits after the decimal point: thus, for example, we say that the number 1.4623 has 4 decimal places. However, since the decimal representations of irrational numbers, such as π or the surd $\sqrt{2}$, all have an infinite number of digits, it is necessary, when working with such decimal numbers, to reduce the number of digits to those that are significant (often indicated by the shorthand 'sig. fig.'). In specifying the number of significant figures of a number displayed in decimal form, all zeros to the left of the first non-zero digit are taken as not significant and are therefore ignored. Thus, for example, both 0.1456 and 0.000 097 44 have 4 significant figures.

There are basically two approaches for reducing the number of digits to those deemed significant:

- Truncation of the decimal part of the number to an appropriate number of decimal places or significant digits: for example, we could truncate π, 3.141 592 653... to 7 significant figures (6 decimal places), by dropping all digits after the 2, to yield 3.141 592. For future reference, we refer to the sequence of digits removed as the 'tail' which, in this example, is 653...
- Rounding up or rounding down the decimal part of a number to a given number of decimal places is achieved by some generally accepted rules. The number is first truncated to the required number of decimal places in the manner described above; attention is then focused on the tail (see above).

(i)   If the leading digit of the tail is greater than 5, then the last digit of the truncated decimal number is increased by unity (rounded up), *e.g.* rounding π to 6 decimal places (d.p.) yields 3.141 593;

(ii) If the leading digit of the tail is less than 5, then the last digit of the truncated decimal number is left unchanged (the number is rounded down); *e.g.* rounding π to 5 d.p. yields 3.141 59;

(iii) If the leading digit of the tail is 5, then:

(a)   If this is the only digit, or there are also trailing zeros, *e.g.* 3.7500, then the last digit of the truncated decimal number is rounded up if it is odd or down if it is even. Thus 3.75 is rounded up to 3.8 because the last digit of the truncated number is 7 and therefore odd, but 3.45 is rounded down to 3.4 because the last digit of the truncated number is 4 and therefore even. This somewhat complicated rule ensures that there is no bias in rounding up or down in cases where the leading digit of the tail is 5;

(b) If any other non-zero digits appear in the tail, then the last digit of the truncated decimal number is rounded up, *e.g.* 3.751 is rounded up to 3.8.

Worked Problem 1.1

Q. Compare the results obtained by sequentially rounding 7.455 to an integer with the result obtained using a single act of rounding.
A. On applying the rules for rounding, the numbers produced in sequence are 7.46, 7.5, 8. Rounding directly from 7.455, we obtain 7.

Problem 1.1

Give the values of (a) 2.554 455, (b) 1.723 205 08, (c) π and (d) e to:

(i)   5, 4 and 3 decimal places, by a single act of rounding in each case,
(ii) 3 significant figures,

using $\pi = 3.141\ 592\ 653$ and e $= 2.718\ 281\ 828$.

1.1.4.2 Observations on Rounding

Worked Problem 1.1 illustrates that different answers may be produced if the rules are not applied in the accepted way. In particular, sequential rounding is not acceptable, as potential errors may be introduced because more than one rounding is carried out. In general, it is accepted practice to present the result of a chemical calculation by rounding the result to the number of significant figures that are known to be reliable (zeros to the left of the first non-zero digit are not included). Thus, although π is given as 3.142 to 4 significant figures (3 decimal places), $\pi/1000$ is given to 4 significant figures (and 6 decimal places) as 0.003142.

1.1.4.3 Rounding Errors

It should always be born in mind that, in rounding a number up or down, we are introducing an error: the number thus represented is merely an approximation of the actual number. The conventions discussed above for truncating and rounding a number imply that a number obtained by rounding actually represents a range of numbers spanned by the implied error bound. Thus, π expressed to 4 decimal places, 3.1416, represents all numbers between 3.14155 and 3.14165, a feature that we can indicate by writing this rounded form of π as 3.14160 $\pm$ 0.00005. Whenever we use rounded numbers, it is prudent to aim to minimise the rounding error by expressing the number to a sufficient number of decimal places. However, we must also be aware that if we subsequently combine our number with other rounded numbers through addition, subtraction, multiplication and division, the errors associated with each number also combine, propagate and, generally, grow in size through the calculation.

Problem 1.2

(a) Specify whether each of the following numbers are rational or irrational and, where appropriate, give their values to 4 significant figures. You should assume that any repeat pattern will manifest itself within the given number of decimal places:

(i) 1.378 423 7842; (ii) 1.378 423 7842...; (iii) $\dfrac{1}{70}$;

(iv) $\dfrac{\pi}{4}$; (v) 0.005068; (vi) $\dfrac{e}{10}$.

Note: The number e expressed to 9 decimal places, 2.718 281 828, appears to have a repeating pattern, which might wrongly suggest it is a rational number; however, if we extend it to a further 2 decimal places, 2.718 281 828 46, we see that there is no repeating pattern and the number is irrational.

(b) In a titration experiment, the volume delivered by a burette is recorded as 23.3 cm^3. Give the number of significant figures; the number of decimal places; and estimates for the maximum and minimum titres.

1.1.5 Combining Numbers

Numbers may be combined using the arithmetic operations of addition ($+$), subtraction ($-$), multiplication ($\times$) and division (/ or $\div$). The type of number (integer, rational or irrational) is not necessarily maintained under combination; thus, for example, addition of the fractions ¼ and ¾ yields an integer, but division of 3 by 4 (both integers) yields the rational number (fraction) ¾. When a number (for example, 8) is multiplied by a fraction (for example, ¾), we say in words that we want the number which is three quarters *of* 8, which, in this case, is 6.

For addition and multiplication the order of operation is unimportant, regardless of how many numbers are being combined. Thus,

$$2 + 3 = 3 + 2$$

and

$$2 \times 3 = 3 \times 2$$

and we say both addition and multiplication are commutative. However, for subtraction and division, the order of operation *is* important, and we say that both are non-commutative:

$$2 - 3 \neq 3 - 2$$

and

$$\frac{2}{3} \neq \frac{3}{2}.$$

One consequence of combining operations in an arithmetic expression is that ambiguity may arise in expressing the outcome. In such cases, it is imperative to include brackets (in the generic sense), where appropriate, to indicate which arithmetic operations should be evaluated first. The order in which arithmetic operations may be combined is described, by convention, by the BODMAS rules of precedence. These state that the order of preference is as follows:

Brackets
Of (multiplication by a fraction)
Division
Multiplication
Addition/Subtraction

For example:

- If we wish to evaluate $2 \times 3 + 5$, the result depends upon whether we perform the addition prior to multiplication or *vice versa*. The BODMAS rules tell us that multiplication takes precedence over addition and so the result should be $6 + 5 = 11$ and not $2 \times 8 = 16$. Using parentheses ( ) in this case removes any ambiguity, as we would then write the expression as $(2 \times 3) + 5$.

- If we wish to divide the sum of 15 and 21 by 3, then the expression $15 + 21 / 3$ yields the unintended result $15 + 7 = 22$, instead of 12, as division takes precedence over addition. Thus, in order to obtain the intended result, we introduce parentheses to ensure that summation of 15 and 21 takes place before division:

$$(15 + 21) / 3 = 36 / 3 = 12.$$

Alternatively, this ambiguity is avoided by expressing the quotient in the form:

$$\frac{15 + 12}{3}$$

However, as the solidus sign, /, for division is in widespread use, it is important to be aware of possible ambiguity.

1.1.5.1 Powers or Indices

When a number is repeatedly multiplied by itself, in an arithmetic expression, such as $3 \times 3 \times 3$, or $\frac{3}{2} \times \frac{3}{2} \times \frac{3}{2} \times \frac{3}{2}$, the power or index notation (also often called the exponent) is used to write such products in the forms 3^3 and $\left(\frac{3}{2}\right)^4$, respectively. Both numbers are in the general form a^n, where n is the index. If the index, n, is a positive integer, we define the number a^n as a raised to the nth power.

We can define a number of laws for combining numbers written in this form simply by inspecting expressions such as those given above: For example, we can rewrite the expression:

$$\frac{3}{2} \times \frac{3}{2} \times \frac{3}{2} \times \frac{3}{2} = \left(\frac{3}{2}\right)^4$$

as

$$\frac{3}{2} \times \frac{3}{2} \times \frac{3}{2} \times \frac{3}{2} = \left(\frac{3}{2}\right)^3 \left(\frac{3}{2}\right)^1 = \left(\frac{3}{2}\right)^4$$

and we see that the result is obtained simply by adding the indices of the numbers being combined. This rule is expressed in a general form as:

$$a^n a^m = a^{n+m} \qquad (1.2)$$

For rational numbers, of the form $\frac{a}{b}$, raised to a power n, we can rewrite the number as a product of the numerator with a positive index and the denominator with a negative index:

$$\left(\frac{a}{b}\right)^n = \frac{a^n}{b^n} = a^n \times b^{-n} = a^n b^{-n} \qquad (1.3)$$

which, in the case of the above example, yields:

$$\left(\frac{3}{2}\right)^4 = \frac{3^4}{2^4} = 3^4 \times 2^{-4}.$$

On the other hand, if $b=a$, and their respective powers are different, then the rule gives:

$$\frac{a^n}{a^m} = a^n a^{-m} = a^{n-m}. \qquad (1.4)$$

The same rules apply for rational indices, as is seen in the following example:

$$\left(\frac{3}{2}\right)^{3/2} = \frac{3^{3/2}}{2^{3/2}} = 3^{3/2} \times 2^{-3/2}.$$

1.1.5.2 Rational Powers

Numbers raised to powers $\frac{1}{2}, \frac{1}{3}, \frac{1}{4}, \ldots, \frac{1}{n}$ define the square root, cube root, fourth root,..., n^{th} root, respectively. Numbers raised to the power m/n are interpreted either as the mth power of the nth root or as the nth root of the mth power: for example, $3^{m/n} = (3^{1/n})^m = (3^m)^{1/n}$. Numbers raised to a rational power may either simplify to an integer, for example $(27)^{1/3} = 3$, or may yield an irrational number; for example $(27)^{1/2} = 3 \times 3^{1/2} = 3^{3/2}$.

1.1.5.3 Further Properties of Indices

Consider the simplification of the expression $(3^2 \times 10^3)^2$:

$$(3^2 \times 10^3)^2 = 3^2 \times 10^3 \times 3^2 \times 10^3 = 3^4 \times 10^6.$$

The above example illustrates the further property of indices that $(a^n)^m = a^{n \times m}$. Thus, we can summarise the rules for handling indices in eqn (1.5).

$$a^n \times a^m = a^{n+m}; \quad \frac{1}{a^n} = a^{-n}; \quad \frac{a^n}{a^m} = a^{n-m};$$

$$(a^n)^m = a^{n \times m} = a^{nm}; \quad a^0 = 1$$

(1.5)

Note that, when multiplying symbols representing numbers, the multiplication sign ($\times$) may be dropped. For example, in the penultimate expression in eqn (1.5), $a^{n \times m}$ becomes a^{nm}. In these kinds of expression, n and m can be integer or rational. Finally, if the product of two different numbers is raised to the power n, then the result is given by:

$$(ab)^n = a^n b^n$$

(1.6)

Problem 1.3

Simplify the following expressions:

(a) $\dfrac{10^2 \times 10^{-4}}{10^6}$; (b) $\dfrac{9 \times 2^4 \times 3^{-2}}{4^2}$; (c) $\left(\dfrac{10}{3^2+4^2+5^2}\right)^{-1/2}$; (d) $\dfrac{\left(2^4\right)^3}{4^4}$.

Worked Problem 1.2 and Problem 1.4 further illustrate how the (BODMAS) rules of precedence operate.

Worked Problem 1.2

Q. Simplify the following expressions.

(a) $\dfrac{1}{2} \times (5-2) + 3 - 12/4 + 3 \times 2^2$

(b) $\dfrac{1}{25}\left(3 \times 10^2\right)^2 + 3200/4 - 6 \times 120$.

A. (a) $\dfrac{1}{2} \times (5-2) + 3 - \dfrac{12}{4} + 3 \times 2^2 = \dfrac{3}{2} + 3 - 3 + 12$

$$= \dfrac{3}{2} + 12 = \dfrac{3}{2} + \dfrac{24}{2} = \dfrac{27}{2}$$

The penultimate step involves creating fractions with a common denominator, to make the addition easier.

$$\text{(b)} \quad \frac{1}{25}\left(3 \times 10^2\right)^2 + 3200/4 - 6 \times 120$$

$$= \frac{1}{25} \times (300 \times 300) + 800 - 720 = 3600 + 80 = 3680$$

In both parts, the rules of precedence are used to evaluate each bracket, and the results are combined using further applications of the rules.

Problem 1.4

Evaluate the following expressions, without using a calculator:

(a) $\left(2.5 \times 10^2 - 0.5 \times 10^2\right)^2/4 \times 10^4$, (b) $\left(\dfrac{2}{2 \times 4}\right)^{1/3} - 4 \times \dfrac{1}{16}$.

1.1.6 Scientific Notation

As has been noted earlier, many numbers occurring in chemical calculations are either extremely small or extremely large. Clearly, it becomes increasingly inconvenient to express such numbers using decimal notation as the order of magnitude becomes increasingly large or small. For example, as seen in the introduction to this chapter, the charge on an electron (in coulombs), expressed as a decimal number, is given by:

$$0.000\ 000\ 000\ 000\ 000\ 000\ 160\ 2177$$

To get around this problem we can use scientific notation to write such numbers as a signed decimal number, usually with magnitude greater than or equal to 1 and less than 10, multiplied by an appropriate power of 10. Thus, we write the fundamental unit of charge to 9 significant figures as:

$$1.602\ 177\ 33 \times 10^{-19}\ \text{C}.$$

Likewise, for very large numbers, such as the speed of light, we write $c = 299\ 792\ 458$ ms^{-1} which, in scientific notation, becomes $2.997\ 924\ 58 \times 10^8$ ms^{-1} (9 sig. fig.). Often we use $c = 3 \times 10^8$ ms^{-1}, using only 1 sig. fig., if we are carrying out a rough calculation.

The conventional representation of numbers using scientific notation is not always followed. For example, we might represent a bond length as 0.14×10^{-9} m rather than 1.4×10^{-10} m, if we were referencing it to another measurement of length given in integer unit multiples of 10^{-9} m.

The negative of the fundamental unit of charge is the charge carried by an electron.

Sometimes, an alternative notation is used for expressing a number in scientific form: instead of specifying a power of 10 explicitly, it is common practice (particularly in computer programming) to give expressions for the speed of light and the fundamental unit of charge as 2.998e8 ms^{-1} and 1.6e–19 C, respectively. In this notation, the number after the e is the power of 10 multiplying the decimal number prefix.

1.1.6.1 Combining Numbers Given in Scientific Form

Consider the two numbers 4.2×10^{-8} and 3.5×10^{-6}; their product and quotient are given respectively by:

$$4.2 \times 10^{-8} \times 3.5 \times 10^{-6} = 14.7 \times 10^{-14} = 1.47 \times 10^{-13}$$

and

$$\frac{4.2 \times 10^{-8}}{3.5 \times 10^{-6}} = 1.2 \times 10^{-2}.$$

However, in order to calculate the sum of the two numbers (by hand!), it may be necessary to adjust one of the powers of ten to ensure equality of powers of 10 in the two numbers. Thus, for example:

$$4.2 \times 10^{-8} + 3.5 \times 10^{-6} = 0.042 \times 10^{-6} + 3.5 \times 10^{-6}$$

$$= 3.542 \times 10^{-6}$$

1.1.6.2 Names and Abbreviations for Powers of Ten

As we have seen in some of the examples described above, an added complication in performing chemical calculations often involves the presence of units. More often than not, these numbers may be expressed in scientific form and so, in order to rationalise and simplify their specification, it is conventional to use the names and abbreviations given in Table 1.2, adjusting the decimal number given as prefix as appropriate.

Table 1.2 Names and abbreviations used to specify the order of magnitude of numbers expressed in scientific notation.

10^{15}	10^{12}	10^9	10^6	10^3	10^{-1}	10^{-2}	10^{-3}	10^{-6}	10^{-9}	10^{-12}	10^{-15}	10^{-18}
peta	tera	giga	mega	kilo	deci	centi	milli	micro	nano	pico	femto	atto
P	T	G	M	k	d	c	m	μ	n	p	f	a

Thus, for example:

- The charge on the electron is given as 0.16 aC, to 2 significant figures.
- The binding energy of the electron in the hydrogen atom is given by 2.179×10^{-18} J, which is specified as 2.179 aJ.
- the bond vibration frequency for HF is 1.2404×10^{14} s^{-1}, which is given as 124.04 Ts^{-1} or 0.12404 Ps^{-1}.

Some of these data are used in the Problem 1.5.

Problem 1.5

(a) Given that 1 eV of energy is equivalent to 0.1602 aJ, use the information given above to calculate the ionisation energy of the hydrogen atom to 3 sig. fig. in electron volts (a common macroscopic energy unit).

(b) Given that the Planck constant, h, has the value 6.626×10^{-34} Js, and that vibrational energy, ε_{vib}, is related to vibration frequency, v, according to $\varepsilon_{vib} = hv$, calculate the value of ε_{vib} for HF to 3 sig. fig.

The results you obtain for Problem 1.5 should show that, since the joule (J) is a macroscopic base unit of energy, property values on the microscopic scale have extremely small magnitudes. We now explore this idea further in Worked Problem 1.3 and in Problem 1.6 to give more practice in manipulating numbers in scientific form; but, more importantly, to provide further insight into size differences in the microscopic and macroscopic worlds.

Worked Problem 1.3

Q. Calculate the whole number of football pitches that would be covered by 1 mol of benzene, spread out in a molecular monolayer, assuming that:

(a) a benzene molecule can be considered as a disc of radius $r = 300$ pm, and
(b) The area of a soccer pitch is 6900 m^2.

A. The number of molecules in 1 mol of benzene is equal to 6.022×10^{23} (to 4 sig. fig.). If we assume that a reasonable estimate for the area, A, covered can be calculated by multiplying the effective area of each molecule, a, by the number of molecules in 1 mol, then;

$$A = 6.022 \times 10^{23} \times \pi \times \left(300 \times 10^{-12} \text{ m}\right)^2 = 1.703 \times 10^5 \text{ m}^2,$$

where $a = \pi \times r^2$ for the area of a disc of radius r. Thus, if we now divide A by the area of a football pitch, FP, we obtain an estimate for the area covered in terms of the number of football pitches, n, covered by 1 mol of benzene:

$$n = \frac{A}{FP} = \frac{1.703 \times 10^5}{6900} = 24.7,$$

where the units m^2 in the numerator and denominator cancel. There is evidently enough benzene to cover 25 football pitches.

Problem 1.6

Given that the density, ρ, of benzene is 879 kg m^{-3} at 298 K, show that 1 cm^3 of benzene contains 0.0113 mol (assume a molar mass for benzene of 0.078 kg mol^{-1}).

Problem 1.7

Assuming that the earth's radius is 6378 km, and that a molecule of benzene may be treated as a disc of radius 300 pm, calculate the mass of benzene needed to create a chain of molecules around the equator of the earth.

Worked Problem 1.4

Q. Gold crystallises in a cubic close packed structure, based on a cube with side a. Given that the density of metallic gold is 19.3×10^3 kg m^{-3} (density is mass divided by volume) and that $a = 4.08 \times 10^{-10}$ m, find the number of Au atoms per unit cell using the molar mass of ^{197}Au (100%) = 197 g mol^{-1}. Give your answer to 3 sig. fig.

A. The mass, m, of the unit cell is given by volume times density; thus:

$$m = \left(4.08 \times 10^{-10} \text{ m}\right)^3 \times 19.3 \times 10^3 \text{ kg m}^{-3}$$
$$= 1.31 \times 10^{-24} \text{ kg} = 1.31 \times 10^{-21} \text{ g}.$$

If M(Au) = 197 g mol^{-1}, then the mass of one Au atom is:

$$\frac{197 \text{ g mol}^{-1}}{6.02 \times 10^{23} \text{ mol}^{-1}} = 3.27 \times 10^{-22} \text{ g}.$$

Thus, the number of atoms per unit cell is:

$$n = \frac{1.31 \times 10^{-21} \text{ g}}{3.27 \times 10^{-22} \text{ g}} = 4.01.$$

Cubic close packing implies 4 atoms per unit cell so this result would seem to conform to our expectation in this case. The deviation from the theoretically correct result of 4 derives simply from the precision with which we specify the density, the value for a, the molar mass of Au and the Avogadro's constant. If we had chosen to specify some or all of these quantities to 4 or 5 significant figures rather than 3, then we might reasonably expect to achieve a better match. It is worth remarking that in carrying out this type of calculation, we need also consider whether the metal in question is isotopically pure or not. For example, metallic copper, which also crystallises in a cubic close-packed structure, is made up of a mixture of ^{63}Cu and ^{65}Cu. In this case, the calculated number of Cu atoms per unit cell will be affected by the difference in isotopic masses in the sample because we assume that each unit cell has the same isotopic composition (since the definition of a unit cell describes it as the basic repeating unit).

Problem 1.8

Calculate the number of Au atoms per unit cell to 3 d.p., given that the molar mass of ^{197}Au (100%) = 196.97 g mol^{-1}, the density of metallic gold is 19.321 $\times$ 10^3 kg m^{-3}, that a = 4.0783 $\times$ 10^{-10} m and given Avogadro's constant = 6.0221 $\times$ 10^{23} mol^{-1}.

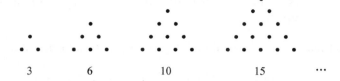

3   6   10   15   ...

1.1.7 Relationships Between Numbers

Frequently in chemistry we find ourselves considering the significance of a numerical quantity, associated with some property of a system, in terms of its relationship to some accepted standard. For example, we might measure a rate constant which tells us whether a particular reaction is fast or slow but we can only draw a conclusion in this respect by comparing it to some standard which we know to imply one extreme or the other or somewhere in between. Of course, this activity is important in all areas of life, and highlights the value of being able to assess how numbers relate to one another. Historically, this relationship has been made easier by associating numbers with patterns, as shown in Figure 1.1.

These so-called figurate numbers (in this case, triangular numbers) are more easily presented in order of increasing magnitude, simply because it is easy to see that there are more dots to the right than to the left. By following this convention, it is then straightforward to deduce the next number in the sequence (here 21, by constructing a triangle with a row of six dots at the base. Intuitively, we can see that 6 is of greater ($>$) magnitude than 3, and of lesser ($<$) magnitude than 15 simply by counting dots. The mathematical notation for describing these two relations is $6>3$ and $6<15$. Such relations are termed **inequalities**. Note that it is equally true that $3<6$ and $15>6$. We can also combine these two relations into one: either $3<6<15$ or $15>6>3$.

The association of numbers with patterns was common in ancient Greece. The so-called figurate numbers 3, 6, 10 and 15 given in the text are examples of triangular numbers; other examples include square (1, 4, 9, 16, 25...) and pentagonal (1, 5, 12, 22, 35...) numbers.

Problem 1.9

Use the inequality symbols ($<$ and $>$) to express two relationships between the following pairs of numbers:

(a) 2, 6;
(b) 1.467, 1.469;
(c) π, e.

1.1.7.1 Negative Numbers

The question of negative numbers must now be addressed. All negative numbers are less than zero, and hence we can say immediately that $-6 < 3$. Furthermore, as $6 > -3$, we can obtain the latter inequality from the former simply by changing the sign of the two numbers (multiplying through by -1) *and* reversing the inequality sign.

Problem 1.10

(a) Use inequalities to express the fact that a number given as 2.456 lies between 2.4555 and 2.4565.
(b) Express each of the following inequalities in alternative ways:

$$-5.35 < -5.34, \quad -5.35 < 5.34.$$

1.1.7.2 Very Large and Very Small Numbers

The numerical value of the Avogadro's constant is 6.022×10^{23}; a very large number. An expression of the disparity in the size between this number and unity may be expressed in the form $6.022 \times 10^{23} \gg 1$; likewise, for the magnitude of the charge on the electron, we can express its smallness with respect to unity as $1.602 \times 10^{-19} \ll 1$.

1.1.7.3 Infinity

The concept of an unquantifiably enormous number is of considerable importance to us in many contexts, but probably is most familiar to us when we think about the size of the universe or the concept of time as never ending. For example, the sums of the first 100, 1000 and 1000 000 positive integers are 5050, 500 500 and 500 000 500 000, respectively. If the upper limit is extended to 1000 000 000 and so on, we see that the total sum increases without limit. Such summations of numbers – be they integers, rationals or decimal numbers – which display this behaviour, are said to tend to infinity. The use of the symbol ∞ to designate infinity should not be taken to suggest that infinity is a number: it is not! The symbol ∞ simply represents the concept of indefinable, un-ending enormity. It also arises in situations where a constant is divided by an increasing small number. Thus, the sequence of values $\frac{1}{10^{-6}}, \frac{1}{10^{-20}}, \frac{1}{10^{-1000}}, \cdots$ (that is 10^6, 10^{20}, 10^{1000}) clearly tends to infinity, whilst the same sequence of negative terms tends to $-\infty$. Once again, there is no limiting value for the growing negative number $-\frac{1}{10^{-n}}$ as the value of n increases (the denominator decreases towards zero). Although it is tempting to write $\frac{1}{0} = \infty$, this statement is devoid of mathematical meaning because we could then just as easily write $\frac{2}{0} = \infty$, which would imply that $1=2$, which is clearly not the case. We shall see in Chapter 3 how to evaluate limiting values of expressions in which the denominator approaches zero.

1.1.7.4 The Magnitude

The magnitude of a number is always positive, and is obtained by removing any sign: thus, the magnitude of -4.2 is given using the modulus notation as $|-4.2| = 4.2$.

Problem 1.11

(a) Give the values of $|4-9|$, $|-3-6|$ and $|9-4|$.
(b) Give the limiting values of the numbers 10^{-m}, 10^m, -10^{-m} and -10^m, as m tends to infinity.

1.2 Algebra

Much of the preceding discussion has concerned numbers and some of the laws of arithmetic used for their manipulation. In practice, however, we do not generally undertake arithmetic operations on numbers obtained from some experimental measurement at the outset – we need a set of instructions telling us how to process the number(s) to obtain some useful property of the system. This set of instructions takes the form of a formula involving constants, of fixed value, and variables represented by a symbol or letter: the symbols designate quantities that, at some future stage, we might give specific numerical values determined by measurements on the system. Formulae of all kinds are important, and their construction and use are based on the rules of algebra. The quantity associated with the symbol is usually called a variable because it can take its value from some given set of values: these variables may be continuous variables if they can take any value from within some interval of numbers (for example, tempera-ture or concentration), or they may be **discrete variables** if their value is restricted to a discrete set of values, such as a subset of positive integers (for example, atomic number). One further complicating issue is that, in processing a number associated with some physical property of the system, we also have to consider the units associated with that property. In practice the units are also processed by the formula, but some care is needed in how to present units within a formula – an issue discussed later on in Chapter 2. However, the most important point is that algebra provides us with a tool for advancing from single one-off calculations to a general formula which provides us with the means to understand the chemistry. Without formulae, mathematics and theory, we are in the dark!

Symbols used to represent variables or constants in a formula may additionally have subscripts and/or superscripts attached. For example, x_1, k_2 and A^{12}.

1.2.1 Generating a Formula for the Sum of the First n Positive Integers

Consider first the simple problem of summing the integers 1, 2, 3, 4 and 5. The result by arithmetic (mental or otherwise) is 15. However, what if we want to sum the integers from 1 to 20? We can accomplish this easily enough by typing the numbers into a calculator or adding them in our head, to obtain the result 210, but the process becomes somewhat more tedious. Now, if we want to sum the sequence of integers from 1 to some, as yet unspecified, upper limit, denoted by the letter n, we need a formula that allows us to evaluate this sum, without actually having to add each of the numbers individually. We can accomplish this as follows:

- Write down the sum of the first five integers, 1 to 5, from highest to lowest, and introduce the symbol S_s to represent this sum:

$$S_5 = 5 + 4 + 3 + 2 + 1$$

- Repeat the exercise by summing the same five integers from lowest to highest:

$$S_5 = 1 + 2 + 3 + 4 + 5$$

- Add the two expressions to obtain:

$$2S_5 = 6 + 6 + 6 + 6 + 6 = 5 \times 6 = 30 \Rightarrow S_5 = 15,$$

where the symbol $\Rightarrow$ means 'implies'.

If we repeat this procedure for the first six integers, rather than the first five, we obtain:

$$2S_6 = 7 + 7 + 7 + 7 + 7 + 7 = 6 \times 7 = 42 \Rightarrow S_6 = 21.$$

We can see that in each case, the respective sum is obtained by multiplying the number of integers, n, in the sum by the same number incremented by 1, and dividing the result by 2: that is:

$$S_5 = \frac{5 \times 6}{2} = 15$$

$$S_6 = \frac{6 \times 7}{2} = 21.$$

Problem 1.12

Use the result given above to deduce expressions that yield the sum of the first:

(a) 100 positive integers;
(b) 68 negative integers.

The pattern should now be apparent, and we can generalise the expression for the sum of the first n positive integers by multiplying n by $n+1$, and dividing the result by 2:

$$S_n = \frac{n \times (n+1)}{2}$$

It is usual practice, when symbols are involved, to drop explicit use of the multiplication sign $\times$, thus enabling the formula for S_n to be given in the form:

$$S_n = \frac{n(n+1)}{2} \tag{1.7}$$

We can test our new formula, by using it to determine the sum of the positive integers from 1 to 20:

$$S_{20} = \frac{20 \times 21}{2} = 210$$

1.2.2 Algebraic Manipulation

The rules for manipulating algebraic symbols are the same as those for numbers: thus we can formally add, subtract, multiply and divide combinations of symbols, just as if they were numbers. In the example given above, we have used parentheses to avoid ambiguity in how to evaluate the sum. The general rules for expanding expressions in parentheses ( ), brackets [ ] or braces { } take the following forms:

$$a(b + c) = ab + ac = (b + c)a \qquad (1.8)$$

and

$$(b + c)/d = b / d + c / d. \qquad (1.9)$$

When we want to multiply two expressions in parentheses together, the first rule is simply applied twice. Thus, if we are given the expression $(a + b)(c + d)$, we can expand by letting $X = (a + b)$, and then:

$$(a + b)(c + d) = X(c + d) = Xc + Xd$$
$$= (a + b)c + (a + b)d$$
$$= ac + bc + ad + bd. \qquad (1.10)$$

We can use these rules to expand our expression for the sum of n integers above to obtain either

$$S_n = n^2 / 2 + n / 2 \text{ or } S_n = n / 2 + n^2 / 2.$$

However, it would be usual in this case to stick to our original expression because it is more compact and aesthetically pleasing.

The ordering of symbols representing numerical quantities in product and summation forms is of no consequence – the symbols **commute** under both addition and multiplication.

Worked Problem 1.5

A Formula of Chemical Importance
The spin of a proton has two orientations with respect to the direction of a homogeneous magnetic field: either 'spin up' or 'spin down', often represented by the arrows ↑ or ↓, respectively. In an NMR experiment, the two orientations have different energies. Let us now consider how many possible spin combinations there are for two and for three equivalent protons, and then derive the result for n equivalent protons.

For two equivalent protons, the first proton can have one of two possible spin states, spin up or spin down. Each of these can combine with either one of the spin states possible for the second proton. Thus the total number of two-spin state combinations is $2 \times 2 = 2^2$ (a result which is more useful than the arithmetic result, 4). The four two-spin states are as follows:

First spin state 'up':   ↑ ↑ ↑ ↓
First spin state 'down': ↓ ↑ ↓ ↓

If we now include a third equivalent proton, as would be appropriate for the methyl group, each parent two-spin state gives rise to two three-spin states – thus doubling the overall number of spin states to yield a total of $2 \times 2^2 = 2^3$ which is, of course 8:

First two-spin state: ↑ ↑ ↑ ↑ ↑ ↓
Second two-spin state: ↑ ↓ ↑ ↑ ↓ ↓
Third two-spin state: ↓ ↑ ↑ ↓ ↑ ↓
Fourth two-spin state: ↓ ↓ ↑ ↓ ↓ ↓

It should now be apparent that, for each additional equivalent proton, the number of spin states is doubled: thus, for 4 protons, there are 2^4 spin states and hence, for n equivalent protons, there are 2^n spin states.

Problem 1.13

The nuclear spin of the deuteron, which has a spin quantum number I of 1, can have three orientations, represented symbolically by ↑, →, ↓. The expression derived above for two-spin state systems can readily be extended to any number of spin states – thus for n equivalent nuclei having m possible spin orientations, the number of possible spin states is given by m^n.

(a) Deduce the number of three-spin states for the CD_2 radical (where D is the chemical symbol for deuterium). Now do the same for a fully deuterated methyl radical.
(b) Given that there are $2I + 1$ spin states for a nuclear spin quantum number, I, give the number of spin states associated with n equivalent nuclei with spin I.
(c) State the number of nuclear spin states for a single atom of ^{51}V, which has a nuclear spin quantum number of 7/2 (this is useful for understanding the electronic characteristics of complexes of vanadium).

The deuteron is the name given to the nucleus of deuterium which contains one proton and one neutron.

1.2.2.1 Dealing with Negative and Positive Signs

In the algebraic expressions considered so far, all the constituent terms carried a positive sign. In general, however, we have to work with expressions involving terms carrying positive or negative signs. Dealing with signed terms is straightforward when we appreciate that a negative or positive sign associated with a number or symbol simply implies the **operation**, *multiply by* –1 or +1, respectively. For example, the operation:

$$(-a) \times (-b)$$

is equivalent to writing

$$-1 \times a \times -1 \times b = (-1 \times -1) \times a \times b = ab.$$

A simple set of rules can be constructed, using this reasoning, to help us to carry out multiplication and division of signed numbers or symbols:

Multiplication:	Division:
$[(+a) \times (+b)] = ab$	$[(+a) / (+b)] = a / b$
$[(+a) \times (-b)] = -ab$	$[(+a) / (-b)] = -a / b$
$[(-a) \times (+b)] = -ab$	$[(-a) / (+b)] = -a / b$
$[(-a) \times (-b)] = ab$	$[(-a) / (-b)] = a / b$

These rules are valid if a, b are numbers, symbols, or algebraic expressions.

Worked Problem 1.6

Q. Given that $x = 6$, $y = -3$ and $z = 2$, find the value of each of the following algebraic expressions.

(a) $xy + 2y - 3z/y$,

(b) $(x + 2z)(z - y)$.

A.

(a) $xy + 2y - 3z/y = -18 - 6 + 6/3 = -22$;

(b) $(x + 2z)(z - y) = 10 \times 5 = 50$

Problem 1.14

Given that $u = (x + y)$ and $v = (x - y)$, find an expression for:

(a) $(u^2 + v^2) / (v - u)$.

(b) $uv / (2u - v)$

(c) $10^{u+v} / 10^{u-v}$

Problem 1.15

Simplify the following expressions:

(a) (i) $4p - q - (2q + 3p)$; (ii) $3p^2 - p(4p - 7)$;

(b) (i) $(1 + x)^2 - (1 - x)^2$; (ii) $x(2x + 1) - (1 + x - x^2)$.

1.2.2.2 Working with Rational Expressions

A rational expression (often called a **quotient**) takes the form $\frac{a}{b}$, where a and b may be simple or complicated expressions. In many instances, it is necessary to simplify the appearance of such expressions by searching for **common factors** (symbols or numbers common to each term) and, if necessary, by deleting such factors in both the numerator and denominator. For example, in:

$$\frac{3x^2 - 12xy}{3},$$

the numerator has 3 and x as common factors, whereas the denominator has 3. Since the denominator and numerator both have the common factor 3, this may be cancelled from each term to give:

$$\frac{x^2 - 4xy}{1} = x^2 - 4xy,$$

which simplifies further to $x(x - 4y)$, once the common factor x has been removed from each term. In this case, the rational expression reduces to a simple expression. We should also be aware that, whenever we are faced with a rational expression involving symbols, it is necessary to specify that any symbol appearing as a common factor, in both numerator and denominator, cannot take the value zero, because otherwise the resulting expression would become $\dfrac{0}{0}$, which is indeterminate (*i.e.* meaningless!).

Problem 1.16

Simplify the following expressions, and indicate any restrictions on the symbol values.

(a) $\dfrac{p^4 q^2}{p^2 q^3}$;   (b) $\dfrac{p^8 q^{-3}}{p^{-5} q^2}$;   (c) $\dfrac{4x}{6x^2 - 2x}$;   (d) $\dfrac{3x^2 - 12xy}{3}$.

1.2.3 Polynomials

A **polynomial** is represented by a sum of symbols raised to different powers, each with a different coefficient; for example, $3x^3 - 2x + 1$ involves a sum of x raised to the third, first and zeroth powers (remember that $x^0 = 1$) with coefficients $3, -2$ and 1, respectively. The highest power indicates the **degree** of the polynomial and so, for this example, the expression is a polynomial of the third degree.

1.2.3.1 Factorising a Polynomial

Since x does not appear in all three terms in the polynomial $3x^3 - 2x + 1$, it cannot be common factor. However, if we can find a number a, such that $3a^3 - 2a + 1 = 0$, then $x - a$ *is* a common factor of the polynomial. Thus, in order to factorise the example given, we need first to solve the expression:

$$3a^3 - 2a + 1 = 0.$$

Trial and error shows that $a = -1$ is a solution of this equation, which means that $x - (-1) = x + 1$ is a factor of $3x^3 - 2x + 1$. It is now possible to express $3x^3 - 2x + 1$ in the form $(x + 1)(3x^2 - 3x - 1)$. Note that the second degree polynomial in parentheses does, in fact, factorise further, but the resulting expression is not very simple in appearance.

Problem 1.17

Express the following polynomials in factored form, indicating any limitations on the values of x.

(a) (i) $x^2 - 3x + 2$;   (ii) $x^3 - 7x + 6$;

(b) (i) $\dfrac{x^3 - 7x + 6}{x - 2}$;   (ii) $\dfrac{x^2 - 1}{x - 1}$.

1.2.3.2 Forming a Common Denominator

An expression of the form $\dfrac{x}{a} + \dfrac{y}{b}$ may be written as one rational expression with a common denominator ab as follows:

$$\frac{x}{a} + \frac{y}{b} = \frac{xb + ya}{ab} \tag{1.11}$$

If there are three terms to combine, we reduce the first two terms to a rational expression, and then repeat the process with the new and the third terms.

Problem 1.18

Express the following in common denominator form, in which there are no common factors in numerator and denominator:

(a) (i) $\dfrac{3x}{4} - \dfrac{x}{2}$;   (ii) $\dfrac{2}{x} - \dfrac{1}{x^2}$;   (iii) $1 - \dfrac{1}{x} + \dfrac{2}{x^2}$

(b) (i) $\dfrac{1}{1+x} - \dfrac{1}{1-x}$;   (ii) $\dfrac{2x}{x^2+1} - \dfrac{2}{x}$.

1.2.4 Coping with Units

In chemistry, we work with algebraic expressions involving symbols representing particular properties or quantities, such as temperature, concentration, wavelength and so on. Any physical quantity is described in terms, not only of its magnitude but also of its dimensions; the latter giving rise to units, the natures of which are determined by the chosen system of units. In chemistry, we use the S.I. system of units: for example, if we specify a temperature of 273 K, then the dimension is temperature, usually given the symbol T, the magnitude is 273 and the base unit of temperature is Kelvin, with name K. Similarly, a distance between nuclei of 150 pm in a molecule has dimensions of length, given the symbol l, a magnitude of 150 and a unit of pm (10^{-12} m). All such physical quantities must be thought of as the product of the magnitude, given by a number, and the appropriate unit(s), specified by one or more names. Since each symbol representing a physical quantity is understood to involve a number and appropriate units (unless we are dealing with a pure number such as percentage absorbance in spectroscopy), we treat the property symbols and unit names as algebraic quantities. All the usual rules apply and, for example, in the case of:

- The molar energy property, E, we may wish to use the rules of indices to write $E = 200$ kJ/mol as 200 kJ mol^{-1}.
- Concentration, c, we are concerned with amount of substance (name is n, unit is mol) per unit volume (name is V, unit is m^3).

If necessary, as seen earlier, we can manipulate the various prefixes for the S.I. base units as required. Further practice is given in the following problem.

Problem 1.19

(a) The expression $\dfrac{RT}{F}$ occurs widely in electrochemistry. The gas constant, R, has units J K^{-1} mol^{-1} ; the Faraday constant, F, has units C mol^{-1}, and temperature, T, is measured in K. Given that 1 coulomb volt (CV) is equivalent to 1 joule (J), find the units of the given expression.

(b) The Rydberg constant, $R_\infty = \dfrac{m_e e^4}{8h^3 c \varepsilon_0^2}$, occurs in models used for interpreting atomic spectral data; m_e is the mass of the electron (kg), e is the elementary charge (C), c is the speed of light (m s^{-1}), ε_0 is the vacuum permitivity (J^{-1} C^2 m^{-1}), and h the Planck constant (J s). Given that 1 J is equivalent to 1 kg m^2 s^{-2}, find the units of R_∞.

Summary of Key Points

This chapter has revisited the elementary but important mathematical concepts of numbers and algebra as a foundation to the following chapter on functions and equations. The key points discussed include:

1. The different types of number: integers, rational, irrational and decimal.
2. The rules for rounding decimal numbers.
3. The rules for combining numbers: powers and indices.
4. Scientific notation for very large and very small numbers.
5. The relationships between numbers: how we reference numbers with respect to magnitude and sign.
6. The principles of algebra: generating a formula and algebraic manipulation.
7. Working with polynomials.
8. An introduction to units.

References

1. R.N. Perutz, *Univ. Chem. Educ.*, 1999, **3**, 37.

Further Reading

D. W., *The Penguin Dictionary of Curious and Interesting Numbers*, Penguin, London, 1987.

2

Functions and Equations: Their Form and Use

As we saw in Chapter 1, the importance of numbers in chemistry derives from the fact that experimental measurement of a particular chemical or physical property will always yield a numerical value to which we attach some significance. This might involve direct measurement of an intrinsic property of an atom or molecule, such as ionisation energy or conductivity, but, more frequently, we find it necessary to use theory to relate the measured property to other properties of the system. For example, the rotational constant, B, for the diatomic molecule CO can be obtained directly from a measurement of the separation of adjacent rotational lines in the infrared spectrum. Theory provides the link between the measured rotational constant and the moment of inertia, I, of the molecule by the formula:

$$B = \frac{h}{8\pi^2 Ic},$$

where h is Planck's constant and c is the speed of light. The moment of inertia itself is related to the square of the bond length of the molecule by:

$$I = \mu r^2,$$

where μ is the reduced mass. The relationship between B and r was originally derived, in part, from the application of quantum mechanics to the problem of the rigid rotor. In general, relationships between one chemical or physical property of a system and another are described by mathematical functions. Such functions are especially important for building the mathematical models we need to predict changes in given property values that result from changes in the parameters defining the system – if we can predict such changes then we are well on the way to understanding our system better! However, before we can explore these applications further, we have to define function in its mathematical sense. This is a necessary step

For a diatomic molecule, AB, the reduced mass is given by $\mu = \frac{m_A m_B}{m_A + m_B}$ where m_A and m_B are the masses of A and B, respectively.

because, in chemistry, the all-pervasive presence of units complicates the issue.

Aims

This chapter aims to demonstrate the importance of mathematical functions and equations in a chemical context. By the end of the chapter you should be able to:

- Work with functions in the form of a table, formula or prescription and, for each type, specify the independent and dependent variables, an appropriate domain, and construct a suitable graphical plot using Cartesian coordinates.
- Recognise periodic, symmetric or antisymmetric character in a function.
- Find the factors and roots of simple polynomial equations using either algebraic or graphical procedures.
- Use the laws of algebra to simplify expressions of all kinds.
- Work with units using algebra.
- Use the formulae for the logarithm of a product or quotient of expressions.
- Understand the difference between degrees and radians for measurement of angles.
- Be familiar with the use of trigonometric identities and addition formulae.

2.1 Defining Functions

Our aim in this section is to show what features need to be understood in order to define a function properly as a mathematical object. First of all, let us consider the association between an arbitrary *number*, x, and the *number* $2x + 1$. We can thus associate the number 6 with 13, π with $2\pi + 1$, 1.414 with 3.828, and so on. It is conventional practice to express this association as a formula, or equation:

$$y = 2x + 1,$$

where the unspecified number, x, is the input number for the formula, and y the output number. Before we can say that this association expresses y as a function of x, we need to:

- Specify the set of numbers for which the formula applies (the **domain**)
- Check that each value of x is associated with only *one* value of y.

In the present example, we could specify the domain as either the collection of all real numbers (conventionally described as the **set R**), or the set of all integers, **I**, or a subset of either or both, thereby satisfying the first requirement.

We can also see in this case that any number chosen as input generates a single number, y, as output and so the second requirement is also satisfied. It is very important to keep in mind that the function is defined not only by the formula but also by the domain: consequently, if we specify that the formula $y = 2x + 1$ applies to any real number as input, x, then we can define the function $y = f(x)$ where $f(x) = 2x + 1$ with domain **R**. In this case f is the name of the function that describes both the formula *and* the permitted values of x. If we had specified that the same formula $y = 2x + 1$ is used with the domain consisting of all *integers*, **I**, then we would be dealing with a *different* function, which we might wish to call $y = g(x)$. For both functions, y is the number produced by the formula for a given x defined within the specified domain of the function. The association between x and y defines different functions for different subsets of numbers – although the formula of association is the same. Where a function is defined by a formula, and the domain is not explicitly stated, then it is assumed that the domain consists of all real numbers for which the function has a real value. This is called the **natural domain** of the function.

A subset of **R** or **I**, consists of a selection of real numbers or integers, respectively. It follows that **I** itself is by definition a subset of **R**.

Worked Problem 2.1

Q. Define two possible domains of the function

$$f(x) = \frac{1}{(x-4)(x+3)}.$$

A. The value of $f(x)$ is indeterminate for $x = 4$ and $x = -3$ because division by zero yields an indeterminate result. For all other values of x, the function is defined, and consequently the domain of $f(x)$ could be either all real numbers, excluding $x = 4$ and $x = -3$, or all integers excluding $x = 4$ and $x = -3$. We could write this explicitly as either:

$$f(x) = \frac{1}{(x-4)(x+3)}, \ x \in \mathbf{R}, \ x \neq 4, \ x \neq -3,$$

$$\text{or } f(x) = \frac{1}{(x-4)(x+3)}, \; x \in \mathbf{I}, \; x \neq 4, \; x \neq -3.$$

It is often not necessary to specify the domain explicitly, as any restrictions are evident from the formula: this is especially true in the chemical context.

The symbol $\in$ means "is an element of" or alternatively "belongs to". Thus $x \in \mathbf{R}$ means that "x is an element of the set of real numbers".

The symbols x and y, used in the function formulae, are conventionally termed the **independent variable** and **dependent variable**, respectively. This terminology conveys the idea that we are free to assign values to the independent variable but that, once we have done so, a unique value for the dependent variable results. A function may have more than one independent variable, in which case a domain needs to be specified for each variable. For example, the formula:

$$y = pq \, / \, r$$

expresses an association between the three independent variables p, q and r and the dependent variable y. The domain is defined by specifying the permitted values associated with each of the independent variables. Having checked that only one value of the dependent variable results for a given set of values of p, q and r, we may then define the function:

$$y = f(p, q, r) = pq \, / \, r. \tag{2.1}$$

In practice, although many functions that we meet in a chemical context have more than one independent variable, the function may be reduced to a single variable by specifying that one or more of the other variables remain constant.

Frequently, we specify functions by formulae that do not explicitly involve a dependent variable, but express the function simply in terms of the formula and the label used to denote the function. For example, the formula $f(x) = 2x + 1$ defines the function f that associates the number $2x + 1$ with the number x. Thus, $f(-5) = -9$ implies that f associates –9 with –5, while $f(3) = 7$ implies that f associates 7 with 3. Although this way of presenting the function does not involve a dependent variable, we can introduce one by letting $y = f(x)$, and rewriting the function as $y = 2x + 1$. The most frequently encountered labels used for the independent and dependent variables are x and y, respectively, but these labels are entirely arbitrary. We might just as easily use the labels p and r or ϕ and ρ; similarly, when labelling the function, instead of f we might use g, h, F or ψ, or indeed any label which we think appropriate: for example, if we wanted to collectively label the group of 1s, 2s, 3s,... atomic orbital functions, we might use the name ϕ and then distinguish each function using a numbered suffix, say, ϕ_1, ϕ_2 and ϕ_3. Our choice here is entirely arbitrary, but is designed to allow us, in this case, to group similar types of functions under a common name. You could rightly argue that the labels ϕ_1, ϕ_2 and ϕ_3 provide little that the labels 1s, 2s, 3s,... do not: the point here is that either will do, and it is really just a matter of taste, context, convenience or convention that dictates what labels and names we use. It is very important that we do not allow unfamiliar labels to give the impression that an otherwise straightforward association or function is more complicated than it actually is!

Worked Problem 2.2

Q. The energies of the electron orbits in Bohr's model of the hydrogen atom take discrete values according to the expression, $E_n = \dfrac{-m_e e^4}{8\varepsilon_0^2 h^2 n^2}$, where m_e and e are the mass and charge of the electron, respectively, h is Planck's constant, ε_0 is the vacuum permittivity and n is the principal quantum number. Write down an equivalent expression using the labels x and y for the independent and dependent variables, respectively, and any labels you feel appropriate for the constants.

A. The formula for E_n has exactly the same mathematical form as the formula $y = -ax^{-2}$. The energy E_n is the dependent variable, equivalent to y in the second formula and carries units of J. The principal quantum number, n, is the independent variable, equivalent to x, and can take only integer values greater than or equal to 1. The constants m_e, e, h and ε_0, can be collected together into a single constant, equivalent to a in the second formula. The domain of the original formula is restricted to all positive integers. However, this restriction may not necessarily apply to the second formula, which might have as domain all real numbers, for example. Consequently, although the formulae are identical in form, we define two distinct functions, which are distinguished from each other by their domains.

The requirement that a function be single-valued, for a given input value for the independent variable, will hold for the majority of associations between one number and another. However, the association between any real number and its square root always yields both a positive and a negative result. For example, the two square roots of 9 are ± 3, and so we say that 9 is associated with both -3 and 3. Thus, if we write this association as $y = x^{1/2}$, then we cannot define the function $y = f(x) = x^{1/2}$. However, if we explicitly limit the values of y to the positive (or negative) roots only, then we can redefine the association as a single-valued function. Alternatively, if we square both sides to yield $y^2 = x$, we can take the association between x now as *dependent* variable and y as *independent* variable and define the function $x = g(x) = y^2$ for which there is only one value of x for any value of y.

An everyday example might be the association between price and item in a supermarket. As there are in all likelihood many items that have the same price, we cannot describe this association as a function; however, the association of item with price *does* define a function, as each article has only one price. In this latter case, the domain of the function is simply a list of all items for sale in the supermarket.

The nth root of a number x may be written using the radical notation $\sqrt[n]{x}$, where n is the index, and the $\sqrt{}$ sign is known as the radical. The square root of x is thus given by $\sqrt[2]{x}$ but more commonly this index is omitted and we simply write $\sqrt{x}$. By convention, use of the radical sign implies the principal or positive root. If we wish to specify explicitly the negative root then we must write $-\sqrt{x}$. However, we may alternatively write $x^{1/2}$ which represents both positive and negative roots.

Problem 2.1

State whether each of the following associations defines a function; if so, give its domain.

(a) The set of car registration numbers and registered keepers.
(b) The set of registered keepers and car registration numbers.
(c) Periodic Table Group number and element name.
(d) Element name and Periodic Table Group number.

2.1.1 Functions in a Chemical Context

As we have seen in the previous section, functions involve associations between numbers. However, when we work with functions in a chemical context, we have to recognise that any association involving chemical properties necessarily involves units. Consider, for example, the relation between atomic number, Z, and atomic first ionisation energy, IE. While there are clearly no units associated with atomic number, the ionisation energy, IE, has units of kJ mol^{-1} (although we could just as easily have chosen any other unit of energy such as eV, cm^{-1}, J, kcal mol^{-1} and so on). In this example, there is clearly a relation between the subset of the positive integers (1, 2, 3, 4, ... 116), corresponding to Z, and the subset of the 116 decimal numbers corresponding to the values of IE / kJ mol^{-1}. Note that the association remains between two numbers devoid of units because, in the case of the ionisation energy, we have divided IE by the units of energy chosen. For example, in the case of atomic nitrogen, where $Z = 7$, and IE / kJ mol^{-1} = 1402.3, there is a relation between the positive integer 7, and the decimal number 1402.3. This relation has physical meaning only for positive integers greater than or equal to 1 and less than or equal to 116 (alternatively written as $1 \leqslant Z \leqslant 116$), since each value of IE / kJ mol^{-1} is associated with only one positive integer in the subset of integers already identified. Hence, in mathematical terms, we say that the relation, or association, just described defines a function, as we have specified both the domain for which the association is valid, and also checked that each number, Z, of the domain has an association with only one decimal number.

In most chemical problems we usually deal with functions that are defined in terms of a formula, in which the permitted values for the variables appearing in the formula (given by symbols) are determined by physical considerations. For example, in the case of temperature on the absolute (Kelvin) scale, negative values have no physical basis in reality.

In the next section, we explore in more detail the role that units play in the relation of formula and function.

2.1.1.1 Understanding the Role of Units – The Mathematically Correct Approach

Consider the ideal gas law, expressed in terms of the simple formula:

$$P = nRT / V \qquad (2.2)$$

where the symbols have the following roles: P is the pressure, n is the amount of gas, V is the volume, T is the temperature and R is the gas constant. Each of the properties listed has associated units, and the units on both sides of eqn (2.2) must be equivalent, or equal. In the S.I. convention, the following choices of units are common, with the equivalent combinations of base S.I. units also given, where appropriate: P in Pa (pascal), atm, or bar, which are names defining appropriate multiples of the base units kg m^{-1} s^{-2}; n in mol; V in m^3; T in K; $R = 8.314$ J K^{-1} mol^{-1}, equivalent to kg m^2 s^{-2} K^{-1} mol^{-1}. Thus, on the left-hand side of the formula the units are kg m^{-1} s^{-2}, and on the right side, we have mol kg m^2 s^{-2} K^{-1} mol^{-1} K / m^3 = kg m^{-1} s^{-2}, as required.

The ideal gas equation is the outcome of a model devised for understanding the properties of a gas, in which there is no interaction between the atoms or molecules occupying the volume, V. In mathematical terms, however, this ideal gas equation remains a formula until we know how to use it as a function, a key aspect of which is developed next.

2.1.1.2 Creating a Function from a Formula

As already noted, the ideal gas formula involves symbols that are associated with a value *and* its associated units. As we know that the units on the left- and right-hand sides of the formula are the same and therefore cancel, we can express each symbol as a value multiplied by appropriate units, leaving us with a relation involving new symbols that stand for numerical values in the ordinary sense of algebra. In particular, if we make the following substitutions:

$$P = p \text{ Pa}; \ n = \bar{n} \text{ mol}; \ R = r \text{ J K}^{-1} \text{ mol}^{-1}; \ V = v \text{ m}^3; \ T = t \text{ K},$$

then the formula:

$$P = \frac{nRT}{V} \tag{2.3}$$

takes the form:

$$p \text{ Pa} = \frac{\bar{n} \text{ mol } r \text{ J K}^{-1} \text{mol}^{-1} t \text{ K}}{v \text{ m}^3} \tag{2.4}$$

which, on cancelling the units, becomes:

$$p = \frac{\bar{n}rt}{v} \tag{2.5}$$

where p, $\bar{n}$, r, t, v are positive numbers, with t also permitted to take the value zero. In this case, we see that p is a function of the three variables, $\bar{n}$, t and v (r is a constant). However, if the amount of gas and either temperature or volume is held constant, then there is only one independent variable and, in these circumstances, we say that p is proportional to t or that p is inversely proportional to v, respectively:

$$p \propto t \text{ or } p \propto 1/v.$$

The constants of proportionality are given by $\bar{n}r / v$ or by $\bar{n}rt$, respectively. In each case, we can collect our constants together and re-label them as a single constant, expressed using a new symbol. Thus, we can express the ideal gas law as:

$$p = bt \text{ or } p = c \,/\, v,$$

where $b = \bar{n}r\,/v$ and $c = \bar{n}rt$.

2.1.1.3 Understanding the Role of Units – The Pragmatic Approach

In the discussion above, we have seen how a formula involving chemical properties may be converted into a function, essentially by removing the units. This procedure works because the units must balance on each side of the equality (=) defining the formula or association. It is very easy to become confused by the distinction between formula and function and the role that units play in defining this distinction. The approach detailed above describes how to treat units in a mathematically correct fashion, but in practice the more pragmatic approach is to simply ignore the units and treat a formula describing some physical relationship as a function (for which the domain is the physically meaningful range of values for the independent variable). Consequently, we find that in most chemistry texts there is an understandable degree of mathematical looseness, which skates over this distinction between formula and function in a chemical context, and frequently results in the units being ignored. For example, it is often stated that the ideal gas law indicates that P is a function of T and V, in which $P \propto T$ and $P \propto 1\,/\,V$. The latter two statements are, of course, true, so long as it is understood that the proportionality constants carry the units of pressure divided by temperature and pressure multiplied by volume, respectively. It is not our intention here to add unnecessary levels of complexity, but it is nevertheless important to be aware of the role that units play and of the distinction between formula and function in the chemical context. We shall return to this problem of units in the next section, where we consider methods used for representing functions.

Worked Problem 2.3

The conventional wisdom in theories of molecular structure presented in organic chemistry is that the strength of a bond between identical atoms increases with increasing bond order and decreasing bond length. Thus, for example, the bond energy, $BE\,/\,\text{J mol}^{-1}$, of a C≡C triple bond is greater than a C=C double bond, which in turn is greater than a C−C single bond. As the bond length is inversely proportional to bond order, we can make a rough approximation that the bond energy, BE / J mol^{-1} is inversely proportional to bond length.

Thus:

$$BE \text{ / J mol}^{-1} \propto \frac{1}{L/\text{m}}.$$

Q. Give the units of the constant of proportionality in terms of the base S.I. units.

A. Writing the formula in terms of base units, we have:

$$\frac{BE}{\text{kg m}^2 \text{ s}^{-2} \text{ mol}^{-1}} \propto \frac{\text{m}}{L}$$

Dividing through by m on both sides gives:

$$\frac{BE}{\text{kg m}^3 \text{ s}^{-2} \text{mol}^{-1}} \propto \frac{1}{L}$$

and so we see that in order for the units to balance on both sides, the constant of proportionality must have units equivalent to the those appearing in the denominator on the left-hand side, *i.e.* kg m^3 s^{-2} mol^{-1}.

A less rigorous, and somewhat more transparent, approach involves writing down the relationship between binding energy and bond length:

$$BE \propto \frac{1}{L}.$$

If we replace each of the symbols BE and L by their appropriate units we have:

$$\text{kg m}^2 \text{ s}^{-2} \text{ mol}^{-1} \propto \frac{1}{\text{m}}$$

and so in order to replace the proportionality symbol ($\propto$) by the equality symbol ($=$) we introduce the constant of proportionality which must have units of:

$$\text{kg m}^2 \text{ s}^{-2} \text{ mol}^{-1} \times \text{m} = \text{kg m}^3 \text{ s}^{-2} \text{ mol}^{-1}.$$

> ## Problem 2.2
>
> The movement of ions with charge ze (where z is a small positive or negative integer, and e is the fundamental unit of charge) through a solution, subject to an external electric field, E, is
>
> determined by the balance between the force arising from the electric field and the viscosity, η, of the solution. The parameter, s, termed the drift speed, gives a measure of the conductivity, and is evaluated using the formula:
>
> $$s = \frac{e\,z\,E}{6\pi\eta a},$$
>
> where a is the effective radius of the ion. Given that the units of e, E, η and a are C, V m^{-1}, m^{-1} kg s^{-1} and m, respectively, give the units of s (remember that C V = J, and J = kg m^2 s^{-2}).

2.2 Representation of Functions

Functions of a single variable, involving a relation between two sets of numbers, may be expressed in terms of a table (expressing an association), formula, prescription or graphical plot. For functions of two independent variables (see below), the preferred representations are formula, prescription or graphical plot; for three or more variables, a formula or prescription are the only realistic representations.

2.2.1 Tabular Representations of Functions of a Single Variable

The function $y = g(x)$, where $g(x) = 2x + 1$, with the domain consisting of the *integers* from −5 to 5, can most easily be expressed in tabular form (see Table 2.1). For each value of x there exists one value of y.

It is clear that there are 11 numbers (elements) in the domain. However, it is not possible to present the function $f(x) = 2x + 1$, with the domain of all real numbers from −5 to 5, in tabular form, as there is an infinite number of elements in the domain. The formula

Table 2.1 The function $g(x) = 2x + 1$, with the domain consisting of the integers from −5 to 5 expressed in tabular form.

x	−5	−4	−3	−2	−1	0	1	2	3	4	5
$g(x)$	−9	−7	−5	−3	−1	1	3	5	7	9	11

$f(x) = 2x + 1$ is the most effective non-graphical way of specifying this function, with the domain as specified above.

2.2.2 Graphical Representations of Functions of a Single Variable

For the function $y = f(x)$, each ordered pair of numbers, (x, y), can be used to define the co-ordinates of a point in a plane, and thus can be represented by a graphical plot, in which the origin, O, with coordinates (0, 0), lies at the intersection of two perpendicular axes. A number on the horizontal x-axis is known as the abscissa, and defines the x-coordinate of a point in the plane; likewise, a number on the (vertical) y-axis is known as the ordinate, and defines the y-coordinate of the point. Thus, an arbitrary point (x, y) in the plane lies at a perpendicular distance $|x|$ from on the y-axis and $|y|$ from the x-axis. If $x > 0$, the point lies to the right of the y-axis and if $x < 0$, it lies to the left. Similarly, if $y > 0$, the point lies above the x-axis, and if $y < 0$, it lies below (see Figure 2.1).

For functions such as $y = g(x)$, where $g(x) = 2x + 1$, the most appropriate type of plot is a point plot if the domain is limited to integers lying within some range. Figure 2.2 displays such a plot for this function, which is defined only at the eleven values of x in its domain (indicated by open circles).

Strictly speaking, it is not appropriate to connect the points with a line, as this would imply that the function is defined at points other

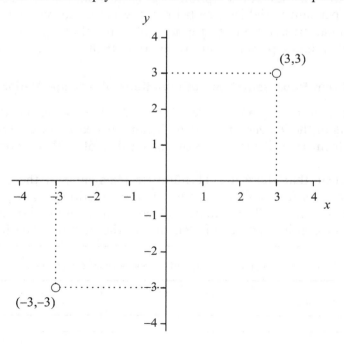

Figure 2.1 The Cartesian coordinate system used to represent the points (3, 3) and (−3, −3) in the plane defined in terms of coordinates referenced to the origin (0, 0).

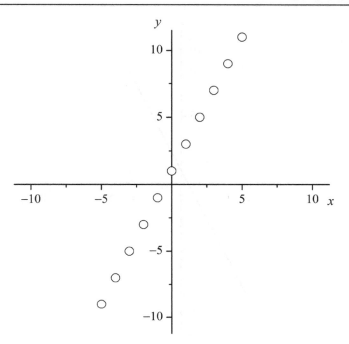

Figure 2.2 A point plot illustrating the values of the function $y = g(x) = 2x + 1$ in its domain [–5, 5].

than at the integers from –5 to +5. However, in some instances, as we shall see below, it may be appropriate to connect the data points with straight line segments in order to guide the eye, but this has no mathematical significance. In contrast, the **line plot** of the function $y = f(x) = 2x + 1$ is created by taking a sufficient number of points in its domain to enable a smooth curve to be drawn (Figure 2.3). In this case, it would be similarly misleading to represent this plot as a series of discrete points, no matter how small the gap between adjacent points. The only correct way of representing this function is by a smoothly varying line plot, but, of course, in practice, we recognise that the logistics of generating a graphical line plot involve arbitrarily selecting discrete points within the domain and then joining the points. This applies equally whether we are drawing the plot by hand or using a computer plotting program.

2.2.2.1 A Chemical Example of a Point Plot

Consider the association between atomic first ionisation energy, *IE*, and atomic number Z (a positive integer). It is convenient to use the electron-volt unit, eV, where $1 eV = 96.485$ kJ mol^{-1}. As there is no formula to express this association, we present the function first in the form of a table and then as a point plot; and in both representations, we take as domain the Z values for the first eighteen elements. Since

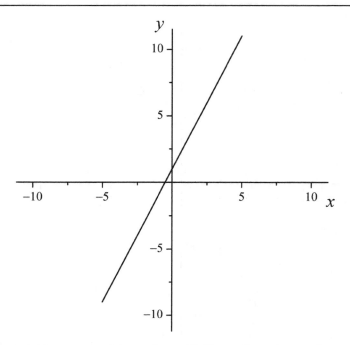

Figure 2.3 Line plot of the function $y = f(x) = 2x + 1$.

units have to be removed in order to define a function, we consider in Table 2.2 the association of Z with IE / eV:

The data in the table are now displayed in graphical form as a point plot (or scatter plot) in Figure 2.4, with points defined by the number pairs (Z, IE / eV). In this procedure, Z is specified as abscissa (x-axis) and IE / eV as ordinate (y-axis).

For every value of Z there is clearly only one value for the ionisation energy, which establishes a function with domain given by the set of the first eighteen positive integers, a subset of the atomic numbers of the 116 elements in the Periodic Table. In this example, the fact that the data points are connected by dashed straight line segments has no mathematical significance; it simply acts as a visual aid to improve the display of the trends in IE / eV values.

There are many situations where we are unable to provide a formula that relates one chemical property with another, although, intuitively, one may be expected. Thus, in the example given above, it

Table 2.2   Atomic number and ionisation for the first eighteen elements.

Z	1	2	3	4	5	6	7	8	9
IE/eV	13.6	24.6	5.4	9.3	8.3	11.3	14.5	13.6	17.4

Z	10	11	12	13	14	15	16	17	18
IE/eV	21.6	5.1	12.8	6.0	8.2	10.5	10.4	13.0	15.8

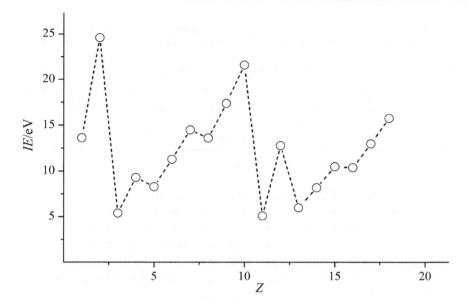

Figure 2.4 A point plot displaying atomic number *vs* ionisation energy in eV.

is not possible to construct a model, based on a formula, that relates Z to IE / eV for atoms containing more than one electron (although a simple relationship does exist for one-electron species, if we ignore relativistic effects). However, regardless of whether a particular problem is as intractable as this one, we can only enhance our understanding of chemistry by using the mathematical tools at our disposal to develop new models to crack particular chemical 'nuts'! As an example of this kind of model development, we now consider some pressure/volume data for a real gas in order to test the ideal gas law, derived from the Boyle model, and to see how we can refine the law to find a better 'fit' to our data.

In Section 2.1.1, we saw that, for an ideal gas, the numerical values of the pressure, p, and volume, v, are related according to $p \propto 1 / v$, or $p = c / v$, where $c = \bar{n}rt$ (a constant). We can now explore how well the ideal gas law works for a real gas by considering experimental data[1] for 1 mol of CO_2 at $T = 313K$. The ideal gas law suggests that pressure is inversely proportional to the volume and so in the first two rows of Table 2.3, we present the variation of p with $1 / v$ for the experimental data (note that the working units for the pressure and volume in this case are atm and dm^3, respectively). In the third row, we show values for $1 / v_B$, obtained using the ideal gas equation, where, in this case, the constant of proportionality $c = 25.6838$ at $T = 313$ K. The data in the fourth and fifth rows derive from a refinement to the model, discussed below.

It should be quite obvious that, although the model provided in the form of the ideal gas law does a reasonable job at lower pressures, it

Table 2.3   A comparison of experimental p vs $1/v$ data for 1 mol of CO_2 at 313K with values for $1/v$ generated: from the ideal gas law; from a fit to the van der Waals equation; and from the van der Waals equation but using the book values[2] for the constants, a and b (see the text for details).

$p=P/atm$	1	10	50	100	200	500
$1/v = 1/(V/dm^3)$	0.0392	0.4083	2.6316	14.300	19.048	22.727
$1/v_B$	0.0389	0.3893	1.9467	3.8934	7.7867	19.467
$1/v_{VDW,\ FIT}$	0.0391	0.4010	2.3294	7.402	19.175	22.730
$1/v_{VDW,\ BOOK}$	0.0391	0.4048	2.5189	11.249	14.184	16.835

rapidly deviates as the pressure increases and the volume decreases. We can see this more clearly in Figure 2.5, in which we compare the real data with that derived from the ideal gas law in a scatter plot of p vs $1/v$. We can see from our plot that the experimental data, shown as solid circles, are modelled reasonably well by a linear (straight line) function, but only for pressures less than 50 atm. The Boyle model is clearly of limited applicability in this case.

2.2.2.2 Improving on the Boyle Model

An example of a model equation for a real gas is provided by the van der Waals equation:

$$P = \frac{RT}{V-b} - \frac{a}{V^2}, \tag{2.6}$$

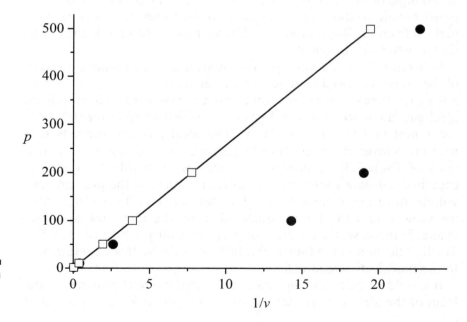

Figure 2.5 Plot of p vs $1/v$, assuming the Boyle model (open box symbols). Experimental data for CO_2 at 313 K are shown as solid circles.

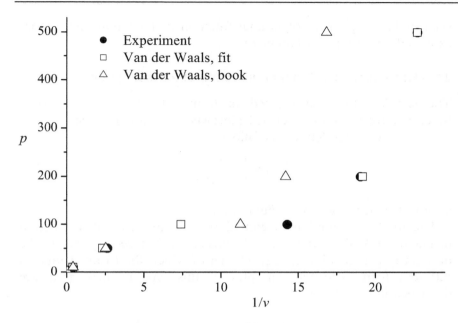

Figure 2.6 Van der Waals (circle symbols) and experimental data (box symbols), used in a pressure *vs* volume plot for CO_2.

in which some account of non-ideality is included through the two parameters a and b. Values of a and b for CO_2 can be obtained by fitting the experimental data to this model expression. For the experimental data set given in Table 2.3, we obtain values for a and b of 2.645 atm dm^6 mol^{-2} and 3.025 $\times$ 10^{-2} dm^3 mol^{-1}, respectively. The book values for a and b are 3.592 atm dm^6 mol^{-2} and 4.267 $\times$ 10^{-2} dm^3 mol^{-1}, respectively.[2] The differences arise from the limited number of data points available to us, but we can see that in spite of this our fitted values for a and b are of the same order of magnitude as the book values. If we now compare a scatter plot of p *vs* $1 / v_{VDW,FIT}$, using our fitted values for a and b (Figure 2.6), we see that, although the fit to our experimental data is really quite good in both the low and high pressure regions, it is quite poor in the region of the critical point where we have a **point of inflection** on our plot (see Chapter 4). Using the book values for the van der Waals constants gives a reasonable fit below 50 atm but increasingly poorer fits to higher pressures; however, the fit in the critical region is much better than that achieved from our fitted values for a and b. While we have not been able to construct a model that fits the experimental data perfectly, it is a considerable improvement on the Boyle model and allows some insight into what factors might be causing the deviation from ideal gas behaviour. Furthermore, our model provides a starting point for further refinements, which might focus, for example, on improving the model for different regions of the domain (such as in the critical region) or even taking a different approach altogether,

The point at which the phase boundary between liquid and gas phases disappears is known as the critical point.

such as looking for a polynomial function in $1/v$ that has a more extended validity (see Chapter 8).

2.2.3 Representing a Function in Terms of a Prescription

The simplest function defined in terms of a prescription (such functions are sometimes termed piecewise functions) is the modulus function, $f(x) = |x|$ defined as follows:

$$f(x) = \begin{cases} x, x \geq 0 \\ -x, x < 0 \end{cases} \tag{2.7}$$

a plot of which is given in Figure 2.7.

The modulus function in eqn (2.7) is an example of a function displaying a 'kink' at the origin. In this case it is necessary to split the domain into two subintervals, in each of which the formula takes a different form. Further examples of this type of behaviour are described in Chapter 3.

Problem 2.3

(a) Give the prescription for the function $f(x) = |x-1|$, and sketch its form.
(b) Sketch the unit pulse function with the prescription

$$g(x) = \begin{cases} 0, x < 1 \\ 1, 1 \leq x < 2 \\ 0, x \geq 2 \end{cases}.$$

Note: The unit pulse function is used for modelling NMR spectra.

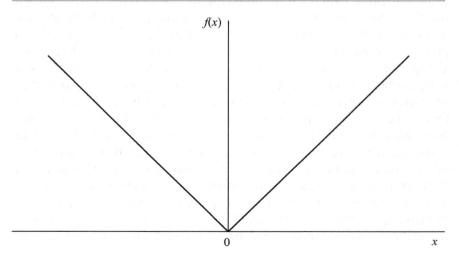

Figure 2.7 The modulus function $y=|x|$.

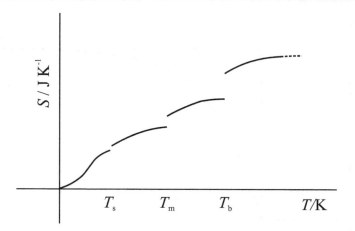

Figure 2.8 A plot of the function describing the change in absolute entropy as a function of temperature. The discontinuities occur at phase changes.

2.2.3.1 Prescription Functions in Chemistry

Functions, specified in the form of a prescription are required when describing properties of chemical systems that undergo phase changes. For example:

- The function describing the change in entropy, as a function of temperature, involves the use of a prescription that contains a formula specific to a particular phase. At each phase transition temperature, the function suffers a finite jump in value because of the sudden change in thermodynamic properties. For example, at the boiling point T_b, the sudden change in entropy is due to the latent heat of evaporation (see Figure 2.8).
- The function describing the change in equilibrium concentration of a given species following a sudden rise in temperature (in a so-called temperature jump experiment), has two parts, corresponding to times before and after the temperature jump (see Figure 2.9).

2.3 Some Special Mathematical Functions

There are many different kinds of function in mathematics, but in this chapter, we shall restrict the discussion to those transcendental functions, such as exponential, logarithm and trigonometric functions, that have widespread use in chemistry.

2.3.1 Exponential Functions

In Chapter 1, we saw that there are 2^n spin states for n equivalent protons, where the physics of such systems requires that $n \geqslant 1$. If we now change the name of the independent variable from n to x, we can

Transcendental functions are mathematical functions which cannot be specified in terms of a simple algebraic expression involving a finite number of elementary operations ($+$, $-$, $\div$, $\times$). By definition, functions which are not transcendental are called algebraic functions.

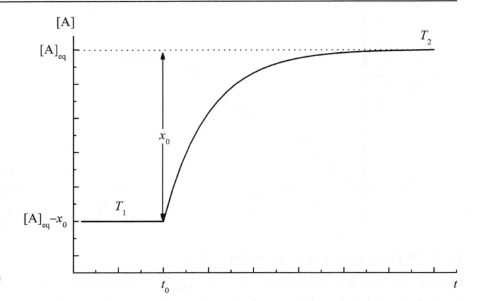

Figure 2.9 The exponential relaxation of the equilibrium concentration to a new equilibrium concentration following a sudden temperature jump from T_1 to T_2.

define the function $y = f(x) = 2^x$ with a domain, for example, initially restricted to the integer values -4 to $+4$. We have displayed this function as a scatter plot in Figure 2.10. If we now extend the domain to any real value for x, we can define the **exponential** function $y = g(x) = 2^x$, with **base** equal to 2, part of which is displayed in Figure 2.10 as the full line plot.

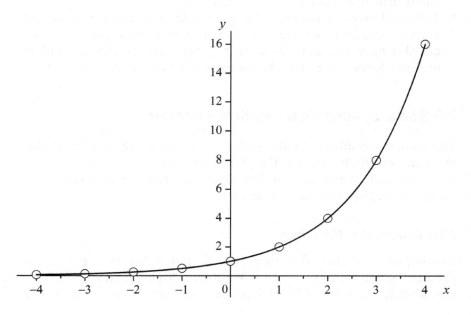

Figure 2.10 Scatter and line plots of the functions $y=f(x)=2^x$, domain $x=-4,-3,-2,-1,0,1, 2, 3, 4$ (open circles) and $y=g(x)=2^x$, domain **R**, (full line).

In chemistry, apart from base 2, which we meet rather infrequently, we also encounter exponential functions with base 10 in, for example relating pH to the activity, a, of hydronium ions in aqueous solutions, using the formula $a = 10^{-\mathrm{pH}}$. However, the base most commonly encountered is provided by the unexpectedly strange, irrational number e, which has the value 2.718 281 828... This base arises when describing growth and decay processes in chemistry, *e.g.* in kinetics, in which changes in concentration with respect to time are the focus of attention, and in quantum chemistry, in which we are interested in the changes in the probability density function for finding an electron at a particular point in space. In a mathematical context, however, e defines the base of the natural logarithm function (see below), and also has a major role in calculus (Chapter 4).

In comparing exponential functions with different bases, the larger the base, the more rapidly the value of the function increases with increasingly large positive values of x, and decreases with increasingly negative values of x. The value of y at $x = 0$ is unity, irrespective of the choice of base. Regardless of the choice of base, exponential functions display a horizontal asymptote at $y = 0$: as x takes on increasingly large negative values, the curve approaches the line $y = 0$ but never crosses it. We explore the limiting behaviour of functions in more detail in Chapter 3.

2.3.1.1 Two Chemical Examples

- In modelling the vibrational "umbrella" mode for ammonia, the potential energy function $V = \frac{1}{2}kx^2 + be^{-cx^2}$ is commonly used, where k, b and c are constants (see Figure 2.11).
- The number of molecular species, n_i, occupying a given energy state, ε_i, is estimated using the Boltzmann distribution function:

$$n_i = n_0 e^{-(\varepsilon_i - \varepsilon_0)/kT}, \tag{2.8}$$

where n_0 is the number of species in the lowest energy state; k is the Boltzmann constant, T the temperature, and the suffix i takes values 0, 1, 2, 3,... Since this function has the domain of positive integers, it can only be visualised graphically using a point plot (see Figure 2.12).

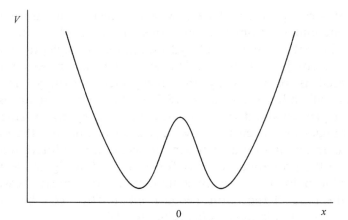

Figure 2.11   A plot of the potential energy function, $V = \frac{1}{2}kx^2 + be^{-cx^2}$, using appropriate values of k, b and c, to describe the umbrella motion in ammonia.

Problem 2.4

If the vibrations of carbon dioxide are assumed to be harmonic, then the energy states are equi-spaced with $\varepsilon_i = \left(i + \frac{1}{2}\right)hv$, where v is one of the four vibration frequencies.

(a) Use the Boltzmann distribution function to show that $\frac{n_i}{n_0} = e^{-ihv/kT}$.

(b) Tabulate the values of $\frac{n_i}{n_0}$ to four significant figures for the Raman active vibration with $v = 0.4032 \times 10^{14}\ s^{-1}$, taking $i = 1,2,3,4,5$, $T = 300\ K$, $k = 1.381 \times 10^{-23}\ J\ K^{-1}$ and $h = 6.626 \times 10^{-34}\ J\ s$.

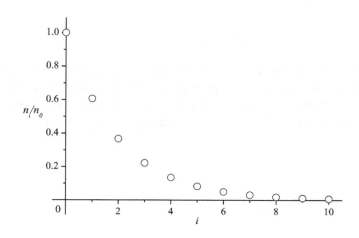

Figure 2.12   A scatter plot of the Boltzmann distribution showing the fractional population of energy levels at a given temperature, T.

2.3.2 Logarithm Functions

Logarithm functions appear widely in a chemical context. For example, in studying:

- The thermodynamic properties of an ensemble of atoms or molecules.
- The model equations for first and second order kinetics.
- The temperature dependence of equilibrium constants.

2.3.2.1 Defining the Logarithm Function

If $y = a^x$ (a is the base), then we define the logarithm to the base a of y to be x, *i.e.*

$$\log_a y = x \tag{2.9}$$

It follows that:

$$a^{\log_a y} = a^x = y \tag{2.10}$$

2.3.2.2 Properties of Logarithms

Given two numbers y_1, y_2, such that $y_1 = a^{x_1}$ and $y_2 = a^{x_2}$, we have from the definition:

$$\log_a(y_1 y_2) = \log_a(a^{x_1 + x_2}) = x_1 + x_2 \tag{2.11}$$

However, again from the definition, we have $\log_a y_1 = x_1$ and $\log_a y_2 = x_2$, and hence:

$$\log_a(y_1 y_2) = \log_a y_1 + \log_a y_2. \tag{2.12}$$

By a simple extension of this argument, we find that:

$$\log_a(y^n) = n \log_a y. \tag{2.13}$$

Note that this applies equally if the index is negative; thus:

$$\log_a(y^{-n}) = -n \log_a y. \tag{2.14}$$

Similarly, by using the laws of indices and the defining relations for logarithms above, we have:

$$\log_a\left(\frac{y_1}{y_2}\right) = \log_a\left(\frac{a^{x_1}}{a^{x_2}}\right) = \log_a(a^{x_1 - x_2}) = x_1 - x_2 = \log_a y_1 - \log_a y_2. \tag{2.15}$$

Finally, to convert the logarithm from base a into base b, we can use the initial equality:

$$y = a^{x_1} = b^{x_2}, \tag{2.16}$$

to give:

$$\log_b y = \log_b(a^{x_1}) = x_1 \log_b a, \text{ and } x_1 = \log_a y, \tag{2.17}$$

and hence:

$$\log_b y = \log_a y \, \log_b a. \tag{2.18}$$

Logarithm functions with the bases e and 10 are usually designated by ln and log, respectively.

Worked Problem 2.4

Q. Using appropriate properties of logarithms listed above, and, without the aid of a calculator, evaluate log1 – log100.

A. $\log 1 - \log 100 = \log \dfrac{1}{100} = \log 100^{-1} = -1 \times \log 100 = -2.$

Problem 2.5

(a) Express the following in terms of log 2:

   (i) log 4; (ii) log 8; (iii) log 6 – log 3; (iv) ln 8; (v) $\ln \dfrac{1}{2}$.

Hint: For part (iv), you will need to convert from base e logs into base 10 logs using eqn (2.18).

(b) Simplify the following expressions:

   (i) log 2 + log 3;   (ii) ln 3 – ln 6.

Problem 2.6

(a) Given that pH $= -\log a_H$, where a_H is the activity of hydronium ions, derive an expression for pH in terms of $\ln a_H$.

(b) In electrochemistry, the standard electromotive force, $E^{\ominus}$, of a cell is related to the equilibrium constant, K, for the cell reaction according to the formula:

$$E^{\ominus} = -\frac{RT}{nF}\ln K,$$

where n is the number of electrons involved and F is the Faraday constant. Find an equivalent expression in terms of $\log K$.

(c) The strength of a weak monobasic acid HA, with dissociative equilibrium constant, K, is measured in terms of a value of pK, where p$K = -\log K$. Find the value of K to 4 significant figures, for ethanoic acid, given that p$K = 4.756$.

2.3.3 Trigonometric Functions

Consider the right-angled triangle shown in Figure 2.13:

The basic **trigonometric functions sine** and **cosine**, given the names sin and cos, respectively, are defined using the ratios of the side-lengths of a right-angled triangle as:

$$\sin\theta = \frac{BC}{AB} \qquad (2.19)$$

and:

$$\cos\theta = \frac{BC}{AB} \qquad (2.20)$$

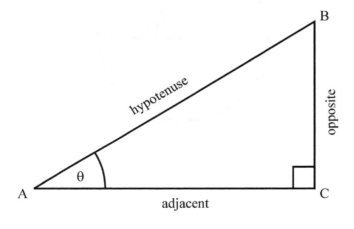

Figure 2.13 A right-angled triangle, with angle BAC specified as θ.

The sides of a right-angled triangle are referred to as the adjacent or base (AC), opposite or perpendicular (BC) and hypotenuse (AB), opposite the right-angle. The tangent of the angle θ is given by the quotient of $\sin \theta$ and $\cos \theta$:

$$\tan \theta = \frac{\sin \theta}{\cos \theta} = \frac{BC}{AC}. \tag{2.21}$$

2.3.3.1 The Question of Angle

Figure 2.13 shows a circle of radius r and an arc (a portion of the circumference) of length s, subtended by the angle θ. There are two basic measures of the angle θ. The first, and probably more familiar, is the degree. The angle, θ, has the value of one degree if the arc-length s is equal to one 360th of the circumference of the circle; and so a complete revolution corresponds to 360 degrees, with half a revolution corresponding to 180 degrees, and a quarter to 90 degrees (a right angle) (see Figure 2.14). The second measure of angle is the radian; one radian is the angle made when the arc length is equal to the radius of the circle; in other words, it is defined in terms of the ratio of arc length to radius, *i.e.* $\theta = s/r$. As the circumference of the circle is equal to $2\pi r$, it follows that there must be 2π radians in one complete revolution, π radians in half a revolution, and $\pi/2$ in a quarter revolution. Since π radians is equivalent to 180 degrees, we can see that one radian must equal $180/\pi = 57.296$ degrees (to 5 significant figures).

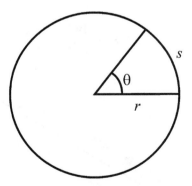

Figure 2.14 A circle of radius r and an arc of length s, subtended by the angle θ.

2.3.3.2 Angle Measure in S.I. Units

Since radians are defined in terms of the ratio of two lengths, the values associated with an angle carry no units and are said to be dimensionless. Similarly, the degree measure of angle is also dimensionless. We can reinforce this by remembering that the sine or cosine of an angle, whether measured in degrees or radians, is defined as the ratio of the lengths of two sides of a triangle. However, in order to indicate which form of angle measure is in use, it is common practice to attach the S.I. symbol 'rad' (as a quasi unit), or to place a small circle as a superscript to indicate degrees (°). For example, we have $\theta = \pi / 2$ rad or $\theta = 90°$ or, equivalently, $\theta / \text{rad} = \pi/2$ or $\theta/° = 90$.

2.3.3.3 Sign Conventions for Angles and Trigonometric Functions

The geometric definition given above for the trigonometric functions in terms of the ratio of the sides of a right-angled triangle, restricts the angles to values in the range $0°$ to $90°$ or, alternatively, to:

$$0 \leq \theta \leq \frac{\pi}{2} \text{ rad.}$$

The definition of angle may be broadened, however, by considering the location of a point (x, y) on the circumference of a circle, with centre at the origin of a Cartesian xy-coordinate system (sometimes referred to as a rectangular coordinate system) (see Figure 2.15). The

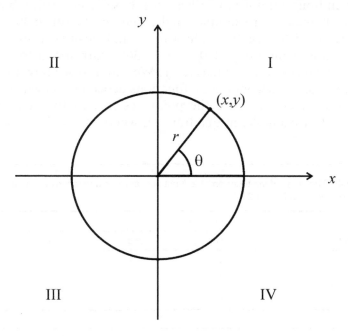

Figure 2.15 The angle θ represented in terms of a circle placed on a Cartesian coordinate system.

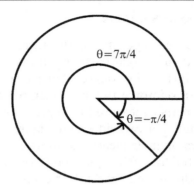

Figure 2.16 Positive values for angle are generated by rotation in an anti-clockwise sense, whereas negative values imply clockwise rotation.

line joining the point on the circle to the origin is of length r, and equal to the radius of the circle.

A zero value for the angle corresponds to the point lying on the positive x-axis. The angle increases in a positive sense as the point circulates in an anti-clockwise direction; circulation in a clockwise sense is indicated by a negative value of θ. Thus, for example, $\theta = -\pi/4$ is equivalent to $\theta = 7\pi/4$ (see Figure 2.16).

We can now redefine the trigonometric functions in terms of the radius r, and the co-ordinates x and y, of a point on the circle:

$$\sin\theta = \frac{y}{r}, \ \cos\theta = \frac{x}{r}, \ \text{and} \ \tan\theta = \frac{y}{x} \tag{2.22}$$

These definitions are not in conflict with those given earlier, but now allow for all angles; for example, angles lying in the range $90° < \theta < 180°$ correspond to negative values for x and positive values for y, whereas those in the range $270° < \theta < 360°$ correspond to positive values for x and negative values for y. We can also see how the signs of the values of the trigonometric functions depend upon which of the four quadrants of the circle the point lies in (see Figure 2.15 and Table 2.4). For example, in quadrant II, where $90° < \theta < 180°$, and

Table 2.4 The signs of the trigonometric functions, sin, cos and tan in each of the four quadrants shown in Figure 2.15.

Quadrant	Function		
	sin	cos	tan
I	+	+	+
II	+	−	−
III	−	−	+
IV	−	+	−

where x is negative and y positive (and remembering that r is always positive):

$$\sin \theta > 0, \ \cos \theta < 0 \text{ and } \tan \theta < 0.$$

2.3.3.4 Special Values for Trigonometric Functions

There are only a few special cases where trigonometric functions have exact values, all of which are obtained easily without reference to tables or resorting to the use of a calculator. For example, $\sin(\pi/4)$ is calculated from the definition of the sine function and use of **Pythagoras' Theorem**. An angle of $\theta = \pi / 4$ (or $\theta = 45°$), requires the magnitudes of x and y to be equal, which implies the length of the hypotenuse (given by r) to be a factor of $\sqrt{2}$ larger than either x or y. Thus, if $\sin \theta = y / r$, then $\sin(\pi/4) = 1/\sqrt{2}$. Table 2.5 lists some of the special values for the sine, cosine and tangent functions.

Pythagoras' Theorem states that for a right-angled triangle, the square of the length of the hypotenuse is equal to the sum of the squares of the lengths of the remaining two sides. Using Figure 2.13 for reference, we can write this more succinctly as $AB^2 = AC^2 + BC^2$.

Problem 2.7

(a) Given that the PH_2 radical has a 'V' shape, with an $H\hat{P}H$ angle of 123°, and a P–H bond length of 140 pm in its ground state, calculate the H–H distance.

(b) Given that the bond angle and P–H bond length change to 107° and 102 pm, respectively, in the first electronic excited state, calculate the change in the H–H distance.

Hint: you may find it helpful to draw an isosceles triangle, and drop a perpendicular from the P atom to the line joining the two hydrogen atoms.

An **isosceles** triangle is a triangle with at least two sides of equal length and with two equal angles. The name derives from the Greek *iso* (same) and *skelos* (leg). A triangle having all sides of equal length is called an **equilateral** triangle, but because it has two sides of equal length is also a special case of an isosceles triangle. A triangle with no equal length sides is called **a scalene triangle**.

Table 2.5 Some special values for the sin, cos and tan functions.

	$\theta = 0$	$\pi/6$	$\pi/4$	$\pi/3$	$\pi/2$	π	$3\pi/2$	2π
$\sin \theta$	0	$\dfrac{1}{2}$	$\dfrac{1}{\sqrt{2}}$	$\dfrac{\sqrt{3}}{2}$	1	0	-1	0
$\cos \theta$	1	$\dfrac{\sqrt{3}}{2}$	$\dfrac{1}{\sqrt{2}}$	$\dfrac{1}{2}$	0	-1	0	1
$\tan \theta$	0	$\dfrac{1}{\sqrt{3}}$	1	$\sqrt{3}$	∞	0	$-\infty$	0

2.3.3.5 Reciprocal Trigonometric Functions

Three further trigonometric functions, cosecant, secant and cotangent, are provided by the reciprocals of the basic functions:

$$\text{cossec}\,\theta = \frac{1}{\sin\theta}, \;\; \sec\theta = \frac{1}{\cos\theta}, \;\; \cot\theta = \frac{1}{\tan\theta} = \frac{\cos\theta}{\sin\theta} \quad (2.23)$$

2.3.3.6 Domains and Periodic Nature of Trigonometric Functions

Thus far we have considered angles ranging from 0 to 2π (0 to 360°), but we can further extend this range by allowing additional complete rotations about the origin. Each additional rotation, anti-clockwise or clockwise, adds or subtracts 2π to or from the angle, with the value of the sine and cosine trigonometric functions simply repeating with each full rotation. The tangent function repeats every half rotation. Thus, for the angles $\theta \pm 2n\pi$, where $n = 0, 1, 2, 3, \ldots.$

$$\sin(\theta \pm 2n\pi) = \sin\theta \;\text{ and }\; \cos(\theta \pm 2n\pi) = \cos\theta \quad (2.24)$$

and for the angles $\theta \pm n\pi$:

$$\tan(\theta \pm n\pi) = \tan\theta. \quad (2.25)$$

Plots of the three trigonometric functions are shown in Figure 2.17.

Functions having a property $f(x \pm a) = f(x)$ are known as **periodic functions** with a **period** a, and are said to be **many-to-one functions**. In the examples given above, the period for the sine and cosine functions is 2π, whereas that for the tangent function is π.

We can see from Table 2.5 and Figure 2.17 that the sine and cosine functions both have as domain the set of real numbers. The domains of the tangent and reciprocal trigonometric functions are different, however, because we must exclude values of θ for which the denominator of the defining formula is zero. Thus, for example, since $\cos\theta = 0$ for $\theta = (2n - 1)\pi/2$, where n is any integer (including zero), the domains for the secant (sec) and tangent (tan) functions consist of the set of real numbers, with the exclusion of $\theta = (2n - 1)\pi/2$ with n defined as above. For the tan and sec functions, the lines at $\theta = (2n - 1)\pi/2$ are known as **vertical asymptotes** because the curves of the respective functions approach these lines without ever crossing them (see Figure 2.17). In some situations it is necessary to limit the domains so that the functions are so called 1:1 functions (as opposed to many-to-one). The **principal branches** for the sine, cosine

Many-to-one functions are those for which more than one value of x is associated with one value of $f(x)$: thus, for $f(x)=x^2$, the numbers ± 2 are both associated with the number 4, and so this function is 2:1.

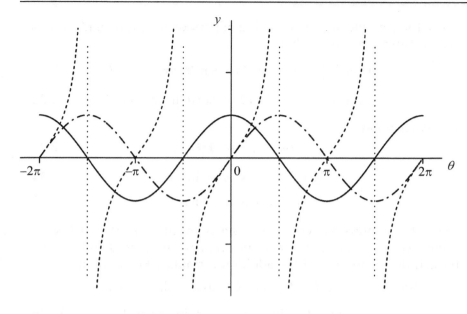

Figure 2.17 Plots of the trigonometric functions sin θ ($\cdot$—$\cdot$—), cos θ (—), and tan θ (- - - -) for $-2\pi \leqslant \theta \leqslant 2\pi$. The principal branch of each function is shown by the thick lines. The dotted vertical lines at odd multiples of $\pi/2$ indicate the points of discontinuity in the tangent function at these values of θ.

and tangent functions, chosen by convention to define them as 1:1 functions, are

$$\sin \theta : \quad -\frac{\pi}{2} \leq \theta \leq \frac{\pi}{2} \tag{2.26}$$

$$\cos \theta : \quad 0 \leq \theta \leq \pi \tag{2.27}$$

$$\tan \theta : \quad -\frac{\pi}{2} \leq \theta \leq \frac{\pi}{2} \tag{2.28}$$

Problem 2.8

Give the domains of the cotangent (cot) and cosecant (cosec) functions.

2.3.3.7 Important Identities Involving Trigonometric Functions

The addition formulae

Expressions for the sine and cosine of the sum or difference of two angles are given by the following formulae:

$$\sin(A \pm B) = \sin A \cos B \pm \sin B \cos A \tag{2.29}$$

$$\cos(A \pm B) = \cos A \cos B \mp \sin A \sin B. \tag{2.30}$$

Thus, for example, in the discussion of periodicity above, and with the use of Table 2.5, we see that:

$$\sin(\theta+2\pi) = \sin\theta\cos 2\pi + \sin 2\pi\cos\theta = \sin\theta; \quad (2.31)$$

$$\cos(\theta+2\pi) = \cos\theta\cos 2\pi - \sin\theta\sin 2\pi = \cos\theta. \quad (2.32)$$

Useful identities

$$\cos^2 A + \sin^2 A = 1 \quad (2.33)$$

$$\cos 2A = \cos^2 A - \sin^2 A \quad (2.34)$$

$$\sin 2A = 2\cos A\sin A \quad (2.35)$$

where the expressions $\cos^2 A$ and $\sin^2 A$ mean $(\cos A)^2$ and $(\sin A)^2$, respectively. All the other identities that we may need follow from these three identities and the addition formulae. For example:

$$\cos 3A = \cos(2A+A) = \cos 2A\cos A - \sin 2A\sin A$$

$$= \cos^3 A - \sin^2 A\cos A - 2\sin^2 A\cos A \quad (2.36)$$

$$= \cos^3 A - 3\sin^2 A\cos A$$

but since $\sin^2 A = 1 - \cos^2 A$, we can rewrite this as:

$$\cos 3A = 4\cos^3 A - 3\cos A. \quad (2.37)$$

2.3.3.8 Further Important Properties of Trigonometric Functions

Since negative angles arise when using trigonometric functions, it is important to establish how, for example, $\sin(-\theta)$ is related to $\sin\theta$. The periodicity of the sine function yields the equality:

$$\sin(-\theta) = \sin(2\pi-\theta) \quad (2.38)$$

and so using the sine addition rule, we obtain:

$$\sin(-\theta) = \sin 2\pi\cos\theta - \sin\theta\cos 2\pi = -\sin\theta. \quad (2.39)$$

Problem 2.9

Repeat the above example for $\cos(-\theta)$ and $\tan(-\theta)$, remembering the definition of $\tan\theta$.

2.3.4 Exponential Functions with Base e Revisited

The hyperbolic sine and cosine functions **sinh** and **cosh** are defined in terms of the sum and difference of the exponential functions e^x and e^{-x} respectively:

$$\sinh x = \frac{1}{2}(e^x - e^{-x}) \tag{2.40}$$

$$\cosh x = \frac{1}{2}(e^x + e^{-x}) \tag{2.41}$$

and have the graphical forms depicted in Figure 2.18.

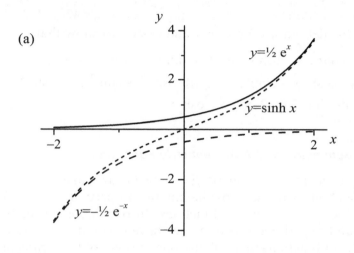

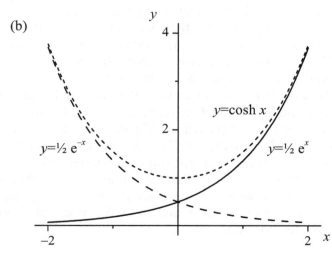

Figure 2.18 Plots of the hyperbolic functions (a) $y = \sinh x$ and (b) $y = \cosh x$ compared with the exponential functions $y = \frac{1}{2}e^x$, $y = \frac{1}{2}e^{-x}$ and $y = -\frac{1}{2}e^{-x}$.

The other hyperbolic functions, tanh, cosech, sech, and coth, defined in terms sinh and cosh, are the hyperbolic analogues of the functions tan, cosec, sec, and cotan, and are defined as follows:

$$\tanh x = \frac{\sinh x}{\cosh x}, \ \operatorname{cosech} x = \frac{1}{\sinh x}, \ \operatorname{sech} x = \frac{1}{\cosh x},$$

$$\coth x = \frac{\cosh x}{\sinh x}.$$

(2.42)

The coth and tanh functions play an important role in the modelling of the magnetic behaviour of transition metal complexes.

Problem 2.10

Use the definitions for sinh x and cosh x to show that:

(a) (i) $\sinh x + \cosh x = e^x$, (ii) $\sinh x - \cosh x = -e^{-x}$.

(b) (i) $\cosh^2 x - \sinh^2 x = 1$, (ii) $\cosh^2 x + \sinh^2 x = \cosh 2x$,
   (iii) $\sinh 2x = 2 \sinh x \cosh x$.

2.3.4.1 Symmetric and Antisymmetric Functions

Functions having the property $f(-x) = f(x)$ are called symmetric, or even, functions, whereas those having the property $f(-x) = -f(x)$ are called antisymmetric or odd functions. In our discussion of trigonometric and hyperbolic functions, we have encountered a number of examples of functions that fall into one or other of these categories, as well as some that fall into neither. Symmetric and antisymmetric functions are so called because they are symmetric or antisymmetric with respect to reflection in the y-axis. If we look closely at Figure 2.17 we see that, since $\cos\theta = \cos(-\theta)$, and $\sin\theta = -\sin(-\theta)$, the cos and sin functions are symmetric and antisymmetric, respectively. Likewise, we can classify the cosh and sinh functions as symmetric and antisymmetric, respectively (see Figure 2.18). The exponential functions displayed in Figure 2.18 above are neither symmetric nor antisymmetric. In Chapter 6 we shall meet these ideas again, when we consider the integration of functions having well-defined symmetry: a feature that has important applications in quantum mechanics, where we consider the physical significance of whether certain integrals involving wave functions of atoms and molecules are zero or non-zero.

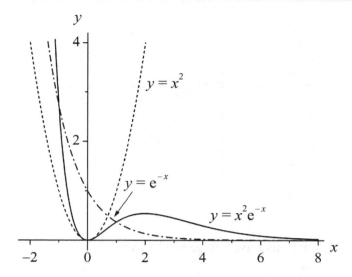

Figure 2.19   Plots of the functions $y=x^2$, $y=e^{-x}$, and the product $y=x^2e^{-x}$.

2.3.4.2 The Product Function x^2e^{-x}

The function $y=x^2e^{-x}$ is a product of two functions, x^2 and e^{-x}; the former *increases* rapidly with increasing x, but the latter *decreases* even more rapidly with increasing x. The result is that the value of the product function, which is initially dominated by the quadratic term, will quite rapidly be overcome by the exponential term as x increases. In fact this is true regardless of the degree of the power term: it does not matter whether we consider the function $y = x^2e^{-x}$ or $y = x^{20}e^{-x}$ or $y = x^{200}e^{-x}$. Eventually, and for surprisingly small values of x, the exponential term will always dominate. In fact even in the last example, the function starts to become overwhelmed by the exponential term around $x = 200$. However, for increasingly negative values of x, both terms in the product function are positive and increasing, thus ensuring that the product function increases more rapidly than either of its component terms. All of these features are apparent in Figure 2.19.

2.3.4.3 Product of a Polynomial or Trigonometric Function with an Exponential Function

The most common types of expression of this kind found in chemistry typically have one of the following forms:

- $P_n(x)e^{-x}$, where $P_n(x)$ is a polynomial function of degree n,
- $\sin(nx)e^{-x}$.

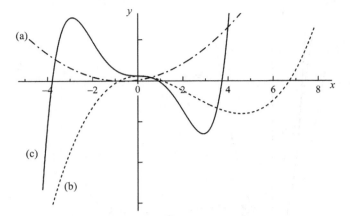

Figure 2.20 Plots of the polynomial functions (a) $y=3x^2+4x+1$ (degree 2) (b) $y=x^3-7x^2+x+6$ (degree 3) and (c) $y=\frac{1}{2}x^5-7x^3-x+6$ (degree 5): the latter two display finite regions of oscillation.

Polynomial functions have the general form:

$$P_n(x) = c_0 + c_1 x + c_2 x^2 + c_3 x^3 + \ldots + c_n x^n \tag{2.43}$$

where $c_0, c_1, \ldots, c_n$, are constants and n is a positive integer, the largest value of which defines the degree of the polynomial. Polynomial functions of degree 3 or higher may display finite regions of oscillation (see Figure 2.20); in contrast, trigonometric functions sin and cos oscillate indefinitely (see Figure 2.17).

When we form a product of either a polynomial or a trigonometric function with the exponential function $y = e^{-x}$, the rapid decline in value of the exponential function as x increases from zero results in a rapid damping of the oscillation (see Figure 2.21):

In the case of polynomials of higher degree, a *finite* number of oscillations occur before they become overwhelmed by the exponential function, whereas for the product of sine or cosine with an exponential function, the number of oscillations is *infinite*, with their amplitude decreasing with increasing positive x. For negative values of x the opposite occurs, with the amplitude of the oscillations increasing with increasingly negative values of x.

2.3.4.4 A Chemical Example: The 3s Atomic Radial Wave Function for the Hydrogen Atom

The radial part of the 3s atomic orbital function for the hydrogen atom provides a good chemical example of a product of a polynomial with an exponential function, and takes the form:

$$R_{3s} = P_2\left(\frac{r}{a_0}\right) e^{-\frac{r}{3a_0}} = N\left\{27 - \frac{18r}{a_0} + 2\frac{r^2}{a_0^2}\right\} e^{-\frac{r}{3a_0}} \tag{2.44}$$

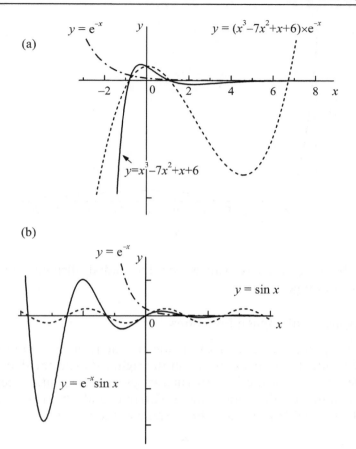

$y = e^{-x}$

$y = (x^3 - 7x^2 + x + 6) \times e^{-x}$

$y = x^3 - 7x^2 + x + 6$

(a)

(b)

$y = e^{-x}$

$y = \sin x$

$y = e^{-x}\sin x$

Figure 2.21 The product of (a) a polynomial or (b) a trigonometric function with the exponential function $y = e^{-x}$, results in a rapid damping of the oscillation as x increases.

where N is a constant, having the form $\dfrac{1}{81a_0^{3/2}\sqrt{3\pi}}$. This may look rather complicated but it is in fact relatively straightforward in its form, comprising a second degree polynomial function and an exponential function, both of which are expressed as a function of the independent variable, r (the distance of the electron from the nucleus). In fact, wherever r appears in both parts of the product, it is divided by a_0, the Bohr radius, and we say in the case of the polynomial function that it is second degree in $\dfrac{r}{a_0}$ (the independent variable). A plot of $R_{3s}a_0^{3/2}$ *versus* $\dfrac{r}{a_0}$ is given in Figure 2.22. If we compare the plot with the function displayed in Figure 2.19, it should be clear that they both have essentially the same form. The main difference in the case of the radial wave function is that we consider only values of $r \geqslant 0$,

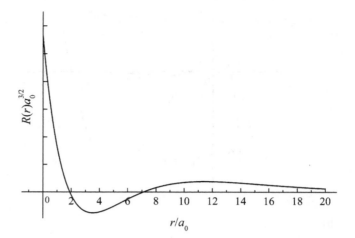

Figure 2.22 A plot of the radial function for a 3s hydrogen atomic orbital.

simply because negative values for the radial distance have no physical meaning.

2.3.5 Explicit and Implicit Functions

Up to this point, we have met functions of the form $y=f(x)$ in which the independent variable appears on the right side and the dependent variable appears on the left. In such cases, the association between a given value of the independent variable and the value of the dependent variable is explicit. For example, the function:

$$y=e^x$$

is an example of one in which y is an **explicit function** of x. However, we can always express such functions in the form $f(x, y) = 0$ in which y is an **implicit function** of x. For example, the function $y = e^x$ may be presented in an implicit form as:

$$\ln y - x = 0.$$

In this example, the implicit form of the function may be rearranged into a form in which either variable is an explicit function of the other. However, sometimes we meet functions which are impossible to arrange into an explicit form. The function:

$$y + e^y = x^5$$

is an example of an implicit function for which there is one unique value of y associated with each value of x but which cannot be expressed in the form of an explicit relationship between y and x. It is nevertheless possible in this case to compute each value of y

associated with a particular value of x using numerical methods. An example from chemistry is the van der Waals eqn (2.6) in which both P and T can be expressed as an explicit function of the other;

$$P = \frac{RT}{V_m - b} - \frac{a}{V_m^2}; \quad T = \frac{(V_m - b)}{R}\left(P + \frac{a}{V_m^2}\right)$$

However, it is most convenient to consider the molar volume V_m as an *implicit function* of P and T [see eqn (2.53)].

2.4 Equations

Consider the plots of the quadratic polynomial functions $y = x^2 - 4x + 3$, $y = x^2 - 4x + 4$, and $y = x^2 - 4x + 6$ in Figure 2.23. Curve (a) cuts the x-axis ($y = 0$) at $x = 3$ and $x = 1$, values which correspond to the two solutions (or **roots**) of the **quadratic equation** $x^2 - 4x + 3 = 0$. In this example, we can more easily obtain the two roots by factorising the **polynomial equation**, rather than by plotting the function. Thus $x^2 - 4x + 3$ can be expressed as the product of two linear factors:

$$x^2 - 4x + 3 = (x - 3)(x - 1),$$

and we can see that this will equal zero when either of the two linear factors equals zero;

$$\textit{i.e. when } x - 3 = 0 \Rightarrow x = 3$$

$$\text{or when } x - 1 = 0 \Rightarrow x = 1.$$

In cases where factorisation proves difficult, it is always possible to use the formula for the roots of a quadratic equation, $ax^2 + bx + c = 0$:

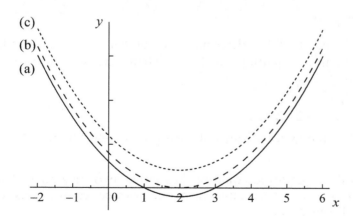

Figure 2.23 Plots of quadratic polynomial functions (a) $y = x^2 - 4x + 3$, (b) $y = x^2 - 4x + 4$ and (c) $y = x^2 - 4x + 6$.

$$x = \frac{-b \pm \sqrt{b^2 - 4ac}}{2a}. \tag{2.45}$$

In this example, the coefficients a, b and c have values equal to 1, –4 and 3 and substituting these into our formula gives:

$$x = \frac{4 \pm \sqrt{16 - (4 \times 3)}}{2} = 2 \pm \frac{\sqrt{4}}{2} = 2 \pm 1.$$

The quantity $b^2 - 4ac$ is known as the **discriminant**, and its value can be positive, zero or negative. In cases where it is positive, the equation has two real and different roots; if it is zero then the equation will have two, identical roots, and, if it is negative, then there are no real roots, as the formula involves the square root of a negative number, for which there is no real result. A way around this latter difficulty is described in Chapter 9, where **complex numbers** are introduced.

The value of the discriminant for the equation $x^2 - 4x + 3 = 0$ is positive, and we see that there are clearly two different roots, as indicated in plot (a) (Figure 2.23), which shows the curve cutting the x-axis at $x = 1$ and $x = 3$. The curve of the function $y = x^2 - 4x + 4$, shown in plot (b), touches the x-axis at $x = 2$. In this case, the discriminant is zero, and we have two equal roots, given by $x = \frac{4}{2} \pm \sqrt{0} = 2 \pm 0$. Note that although the curve only touches the x-axis in one place, the equation $x^2 - 4x + 4 = 0$ still has two roots – they just happen to be identical. Finally, in the case of curve (c), there are no values of x corresponding to $y = 0$, indicating that there are no real roots of the quadratic equation $x^2 - 4x + 6 = 0$, as the discriminant is equal to –8.

Problem 2.11

(a) Use eqn (2.45) to determine the number of real roots of the quadratic equations, $f(x) = 0$, where $f(x)$ is given by:
(i) $x^2 + x - 6$;
(ii) $x^2 - 1$;
(iii) $x^2 - 2\sqrt{2}x + 2$;

(b) Give the factored form of each polynomial function $f(x)$.

In general, a quadratic equation has either two or zero real roots. However, a **cubic equation** may have one or three real roots, as seen in Worked Problem 2.5.

Worked Problem 2.5

Q. Use a graphical method to find the number of roots of the polynomial equations:

(a) $x^3 - 7x + 6 = 0$; (b) $x^3 - 4x^2 - 2x - 3 = 0$.

A. (a) 3; (b) 1. See Figure 2.24

2.4.1 An Algebraic Method for Finding Roots of Polynomial Equations

For a given polynomial function $y = f(x)$, one (or more) roots of the polynomial equation $f(x) = 0$ can often be found by an algebraic method. Suppose the polynomial $f(x)$ is of degree n. If $x = \lambda$ is a root of the polynomial equation, then $f(\lambda) = 0$, and $(x - \lambda)$ is a factor of the polynomial:

$$f(x) = (x - \lambda)(c_1 x^{n-1} + c_2 x^{n-2} + \ldots + c_n) \qquad (2.46)$$

The truth of the previous statement follows by substituting $x = \lambda$ into the above equation, where we see that, irrespective of the value of the second expression in parentheses, which is a polynomial of degree $n - 1$, the first term in parentheses is zero, thus implying that $f(\lambda) = 0$. If there is a root with integer value, then it can sometimes be found by trial and error, using $\lambda = \pm 1, \pm 2, \ldots$ and the polynomial of degree $n - 1$ can then be treated in the same way. If no further roots can be found algebraically at any stage in the iterative procedure, then the current polynomial can be plotted to exhibit the existence, or otherwise, of remaining roots. The key requirement is that, at each step, the

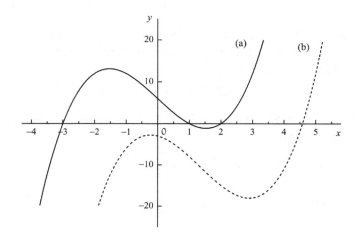

Figure 2.24 Plots of the polynomial functions (a) $y = x^3 - 7x + 6$ and (b) $y = x^3 - 4x^2 - 2x - 3$.

coefficients c_i are found, in order to facilitate the recovery of another root. Once the polynomial of degree two is reached, it is easiest to use the formula given in eqn (2.45) to test for the existence of a further two or zero roots.

Worked Problem 2.6

Q. (a) Use the algebraic method to find the roots of the polynomial equation $f(x) = 0$, where
$f(x) = 2x^3 + 11x^2 + 17x + 6$.
  (b) Give the factored form of $f(x)$.
  (c) Sketch a graph of the function $y = f(x)$.

A. (a) Simple trial and error shows that $x = -2$ is a root of $f(x)$, since $f(-2) = 0$. The polynomial equation may now be written in the form:

$$(x + 2)(c_1 x^2 + c_2 x + c_3) = 0.$$

On multiplying out the brackets, and collecting terms, we have:

$$c_1 x^3 + (c_2 + 2c_1)x^2 + (c_3 + 2c_2)x + 2c_3 = 0$$

Comparing coefficients of the powers of x with the original polynomial equation, we find:

$$c_1 = 2, \tag{2.47}$$

$$c_2 + 2c_1 = 11, \tag{2.48}$$

$$c_3 + 2c_2 = 17, \tag{2.49}$$

$$2c_3 = 6 \tag{2.50}$$

Eqns (2.47) and (2.50) give the values $c_1 = 2$ and $c_3 = 3$, respectively. It then follows, by substituting the value of c_1 in (2.48) that $c_2 = 7$, and we then have:

$$(x + 2)(2x^2 + 7x + 3) = 0.$$

(b) The solutions of $(2x^2 + 7x + 3) = 0$ are then found using eqn (2.45):

$$x = -\frac{7}{4} \pm \frac{1}{4}\sqrt{49 - 24} = -\frac{7}{4} \pm \frac{5}{4} = -3 \text{ or } -\frac{1}{2}$$

Thus, $f(x) = (x + 2)(x + \tfrac{1}{2})(x + 3)$.

(c) From (b) we know that the curve crosses the x-axis at $x = -2, -3, -\frac{1}{2}$; in addition, for $x > -\frac{1}{2}$, all three brackets are positive and increase in value as x increases. Likewise, for $x < -3$, all brackets have increasing negative values, and therefore $f(x)$ is negative for these values. For $-3 < x < -2$, $(x + 3)$ is positive, and $(x + 2)$, $\left(x + \frac{1}{2}\right)$ are both negative, and hence $f(x) > 0$. A similar argument shows that $f(x) < 0$ for $-2 < x < -\frac{1}{2}$, and it is then an easy matter to sketch the form of the cubic polynomial function (Figure 2.25):

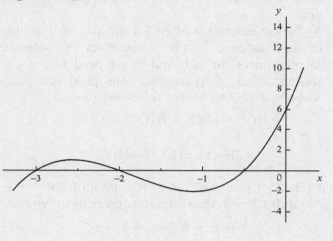

Figure 2.25   Plot of the function $f(x) = 2x^3 + 11x^2 + 17x + 6$.

2.4.2 Solving Polynomial Equations in a Chemical Context

In practice, the solution of polynomial equations is problematic if no simple roots are found by trial and error. In such circumstances the graphical method may be used or, in the case of a quadratic or cubic equation, there exist algebraic formulae for determining the roots. Alternatively, computer algebra software (such as Maple or Mathematica, for example) can be used to solve such equations explicitly. In Worked Problem 2.7, we show how the calculation of the pH of 10^{-6} mol dm^{-3} HCl (aq) requires the solution of a quadratic equation.

Worked Problem 2.7

Q. Calculate the pH of 10^{-6} mol dm^{-3} HCl (aq), taking into account the hydronium (H_3O^+) ions from:

    (a) HCl alone;

    (b) HCl and the dissociation of water (equilibrium constant, $K_w = 10^{-14}$).

A. (a) The simple formula pH $= -\log([H_3O^+]\ /\ \text{mol dm}^{-3})$, leads to a value for the pH of 6, since $[H_3O^+] = 10^{-6}$ mol dm^{-3}.

    (b) As the concentration of HCl is so small, it is appropriate to take account of the dissociation of water in our calculation of the pH, and so we need to consider the concentration of hydronium ions produced from two sources, described by the following processes:

$$HCl + H_2O \rightarrow H_3O^+ + Cl^- \ ;$$

$$2H_2O \overset{K_w}{\rightleftharpoons} H_3O^+ + OH^-$$

Thus, if $[H_3O^+]\ /\ \text{mol dm}^{-3} = h$, $[Cl^-]\ /\ \text{mol dm}^{-3} = c$, and $[OH^-]\ /\ \text{mol dm}^{-3} = b$, then charge conservation requires:

$$h = c + b \Rightarrow b = h - c.$$

Charge conservation requires that there is the same number of cations as anions in the solution. Thus the sum of the concentrations of the OH$^-$ and Cl$^-$ ions must be the same as the hydronium ion concentration and so $h = c + b$.

where c $= 10^{-6}$. The equilibrium constant for the dissociation of water is given by $K_w = hb$, which we can now rewrite as:

$$K_w = hb = h(h - c) = h^2 - ch \Rightarrow h^2 - ch - K_w = 0 \quad (2.51)$$

Eqn (2.51) is a quadratic equation in h, and the two roots may be found using eqn (2.45). Thus:

$$h = \frac{c}{2} \pm \frac{\sqrt{c^2 + 4K_w}}{2}$$

and, on substituting for c and K_w, we find $h = 1.099 \times 10^{-6}$ or $h = -9.902 \times 10^{-9}$. The first solution yields pH $= 5.996$; the second solution, although mathematically required, does not correspond to an acceptable physical result, as the logarithm of a negative number is not defined as a real number and thus has no physical significance.

The logarithm of a negative number is a so-called complex number which we discuss in some detail in Chapter 8.

Problem 2.12

The radial function of the 3s atomic orbital for the hydrogen atom has the form given in eqn (2.44).

(a) Calculate the value of R_{3s} at $r = 0$ and as r tends to infinity. Note that the exponential term will always dominate the term in parentheses (see Section 2.3.4), and so its limiting behaviour alone will determine the behaviour of the function as r tends to infinity (see also Chapter 3 for a more detailed discussion of limits).

(b) Calculate the values of r / a_0, and hence of r, for which $R_{3s} = 0$, by solving the quadratic polynomial equation

$$\left\{ 27 - 18\left(\frac{r}{a_0}\right) + 2\left(\frac{r}{a_0}\right)^2 \right\} = 0.$$

(c) Sketch the form of R_{3s} for $0 \le \dfrac{r}{a_0} \le 12$, and then compare your result with that displayed in Figure 2.22.

2.4.2.1 Polynomial Equations of Higher Degree in Chemistry

Polynomial equations of degree three (cubic equations) arise in a number of areas of classical physical chemistry; Higher degree equation arise in the modelling of:

- Electronic structures, through the determination of molecular orbitals, constructed as linear combinations of atomic orbitals (LCAO); thus, for example, the determination of the simplest σ-type molecular orbitals for HCN, in its linear configuration (as in the ground state), involves the use of the seven σ atomic orbitals $1s_H$, $1s_C$, $1s_N$, $2s_C$, $2s_N$, $2p\sigma_C$ and $2p\sigma_N$, and leads to the solution of a polynomial equation of degree seven for the molecular orbital energies.

- Characteristic frequencies of molecular vibrations: in the case of HCN, for example, there are four vibrational frequencies that may be calculated from a polynomial equation of degree four, by making appropriate assumptions about the stiffness of bond stretching and bond angle deformation.

> ### Problem 2.13
>
> Give the degree of the polynomial equation that arises in calculating the molecular orbitals for the following species in their ground states (σ or π bonding, as indicated):
>
> (a) Carbon dioxide (σ only);
> (b) Benzene (π bonds only).

2.4.2.2 Examples of Cubic Polynomial Equations in Physical Chemistry

The van der Waals Equation Revisited

Consider the relationship between pressure, temperature and volume provided by the van der Waals equation used to model the physical properties of a real, rather than an ideal, gas:

$$P = \frac{RT}{V_m - b} - \frac{a}{V_m^2}. \tag{2.52}$$

Here, a and b are parameters for a specific gas, and V_m is the molar volume. If we now multiply both sides of this equation by $V_m^2(V_m - b)$ and rearrange the terms, the following cubic equation results:

$$V_m^3 - V_m^2\left(b + \frac{RT}{P}\right) + V_m\frac{a}{P} - \frac{ab}{P} = 0. \tag{2.53}$$

We can use this third degree polynomial to find the molar volume of a gas at a given temperature, T, and pressure, P. For example,[3] we can estimate the molar volume of CO_2 at 500 K and 100 atm using the literature values[2] for $a = 3.592$ atm dm^6 mol^{-1} and $b = 0.04267$ dm^6 mol^{-1} and taking $R = 0.082058$ atm dm^6 K^{-1} mol^{-1}. The solution to eqn (2.53) yields only *one* real root (the other two roots are complex), and we obtain a value $V_m = 0.3663$ dm^3 (found using Maple, the computer algebra software). The plot of the function (Figure 2.26) confirms this finding.

The 4s Radial Wave Function for the Hydrogen Atom

A node is a point where the wavefunction passes through zero

In order to locate the nodes in the radial part of the hydrogen 4s atomic orbital:

$$R_{4s} = N\left\{24 - 18\left(\frac{r}{a_0}\right) + 3\left(\frac{r}{a_0}\right)^2 - \frac{1}{8}\left(\frac{r}{a_0}\right)^3\right\}. \tag{2.54}$$

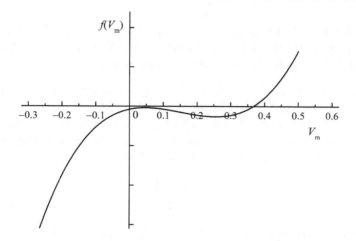

Figure 2.26 Plot of the van der Waals polynomial function for CO_2 using $P=100$ atm and $T=500$ K.

We need to solve a cubic polynomial equation for the three values of r/a_0, and hence r, as a multiple of a_0. Unlike the simple expressions for the solutions of a quadratic equation given in eqn (2.45), and the cubic equation in Worked Problem 2.6, a more involved algebraic procedure is required to solve the cubic equation given in eqn (2.54). However, we know that a hydrogen 4s atomic orbital has 3 radial nodes ($n - l - 1$), and since there are three roots to the third order polynomial equation $R_{nl}(r) = 0$, we conclude that all three roots are real. In this case, therefore, the graphical method will give good estimates for the location of the roots, which can then be improved by trial and error; alternatively, computer algebra software can be used to determine the roots to any sensible number of decimal places.

Problem 2.14[4]

When 1.00 mol of sodium ethanoate (NaAc) is dissolved in 1.00 dm^3 of water, Na^+ ions are liberated and some of the Ac^- species combine with H_3O^+ to form ethanoic acid (HAc). The following equilibria are set up:

$$AcOH + H_2O \overset{K}{\rightleftharpoons} H_3O^+ + AcO^- \text{ and } 2H_2O \overset{K_w}{\rightleftharpoons} H_3O^+ + OH^-$$

where:

$$K = \frac{a_{H_3O^+} a_{AcO^-}}{a_{AcOH}} \qquad (2.55)$$

and:

$$K_w = a_{H_3O^+} a_{OH^-} \qquad (2.56)$$

If the activities a_{AcOH}, $a_{H_3O^+}$, a_{OH^-} and a_{AcO^-} are designated by the four unknowns e, h, b and c, then four equations need specifying in order to solve for the unknowns.

(a) Use eqns (2.55) and (2.56) to give the defining relations for K and K_w in terms of e, h, b and c.

(b) Confirm that the conservation requirements for AcO^- and charge (concentrations of cations and anions are equal) yield eqns (2.57) and (2.58).

$$c + e = 1 \tag{2.57}$$
$$c + b = 1 + h \tag{2.58}$$

(c) Rearrange eqn (2.57) to find an expression for e.

(d) Obtain an expression for c in terms of K and h, by substituting for e in eqn (2.55).

(e) Substitute the expression for c into eqn (2.58), and show that $b = 1 + h - \dfrac{K}{K+h}$.

(f) Substitute for b in eqn (2.56), and show that $K_w = h + h^2 - \dfrac{Kh}{K+h}$.

(g) Multiply the first two terms on the right side of the result in (f) by $\dfrac{K+h}{K+h}$, and rearrange the expression in (f) to show that the following cubic equation is obtained for determining h:

$$h^3 + h^2(1 + K) - K_w h - K_w K = 0 \tag{2.59}$$

(h) Substitute the values $K = 1.8 \times 10^{-5}$ and $K_w = 10^{-14}$, into eqn (2.59) to obtain a cubic equation with numerical coefficients.

(i) Given that eqn (2.59) has $h = 4.243 \times 10^{-9}$ as its only physically meaningful root (the other two roots are negative and therefore meaningless), find the values for e, b and c, and hence of a_{AcOH}, a_{OH^-} and a_{AcO^-} to 3 significant figures.

The exact definition of the equilibrium constant given by IUPAC requires it to be defined in terms of fugacity coefficients or activity coefficients in which case it carries no units. This convention is widely used in popular physical chemistry texts, but it is also common to find the equilibrium constant specified in terms of molar concentrations, pressure or molality in which cases the equilibrium constant will carry appropriate units.

Summary of Key Points

This chapter revolves around the important concepts of function, equation and formula. The key points discussed include:

1. Function as an association between one number and another; the domain as a set which specifies the numbers for which the association applies.
2. The independent and dependent variables in an association and their identification with particular, yet arbitrary, symbols in a formula.
3. The role of units in working with functions in a chemical context; creating a function from a formula.
4. Representing functions in tabular, formula, prescription or graphical forms; how the choice of domain affects the appearance of a plot.
5. The definition of the domain for a chemical function as opposed to an abstract function;
6. Testing a chemical formula by comparison of real experimental data with that generated from a model.
7. Special mathematical functions: exponential, logarithm (base 10 and base e), trigonometric, reciprocal trigonometric and hyperbolic trigonometric.
8. Working with the properties of logarithms and trigonometric identities.
9. Measurement of angles: degrees and radians.
10. Symmetric, antisymmetric and periodic functions; product functions; the product of a polynomial function or a trigonometric function with an exponential function.
11. Equations; solving quadratic and higher order equations; finding the factors and roots of simple polynomial equations using either algebraic or graphical procedures.

References

1. Data from: W. R. Moore, in *Physical Chemistry*, Longman, London, 5th edn., 1976, p.23 (Longman, 1976).
2. Data from: *Handbook of Chemistry and Physics*, ed. R. C. Weast, CRC Press, Florida, 58th edn., 1977–1978, Table D178.

3. Example from: P. W. Atkins and J. de Paula, in *Physical Chemistry*, Oxford University Press, Oxford, 7th edn., 2002, p.20.

4. Adapted from: M. J. Sienko, in *Equilibrium (Part 2)*, W.A. Benjamin, New York, 1964, p.511.

3

Limits

The concept of the limit is a fairly broad one, commonly used for probing the behaviour of mathematical functions as the independent variable approaches a particular value. This is particularly useful in exploring errant or unexpected behaviour in a function as well as in examining the behaviour of a function as the independent variable takes on increasingly large or small positive or negative values. More importantly, limits are central to our understanding of differential calculus, as is seen in the work of Fermat who, in the early 17th century, used the concept of the limit for finding the slope of the tangent at a point on a curve (a topic discussed in Chapter 4). Likewise, in Chapter 6, we shall see how the concept of the limit provides a foundation for integral calculus.

Aims:

By the end of the chapter you should be able to:

- Understand the principles involved in defining the limit.
- Understand the notions of continuity and discontinuity.
- Use limits to examine the point behaviour of functions which might display unexpected characteristics.
- Investigate asymptotic behaviour of functions as the independent variable takes on increasingly large positive or negative values.
- Use limits to improve your understanding of how physical processes may change as experimental conditions change from one extreme to another.

3.1 Mathematical and Chemical Examples

3.1.1 Point Discontinuities

The function shown in Figure 3.1 shows a break at $x = 3$, where the value of y is $0/0$, and is therefore indeterminate. In this situation, the function is said to exhibit a discontinuity at $x = 3$, which means that it

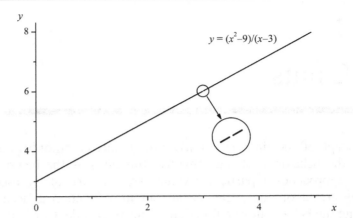

Figure 3.1 A plot of the function $y=(x^2-9)/(x-3)$ over the sub-interval $0 \leqslant x \leqslant 5$.

is impossible to sketch the plot of the function by hand without taking the pencil off the paper.

A chemical example is shown schematically in Figure 2.8, where discontinuities are seen in the entropy function at the melting and boiling points, T_m and T_b respectively, as well as at a temperature T_s where a change in crystal structure occurs in the solid state. Although the entropy function is undefined at these three transition temperatures, the discontinuities are finite in nature, as the corresponding changes in S are finite in size. We can see from this example that S is continuous only over sub-intervals of the domain; furthermore at each of the transition temperatures, T_m, T_b and T_s, the value of S is ambiguous. This situation arises because two values of S result depending on whether we approach a transition temperature from higher or lower values of T.

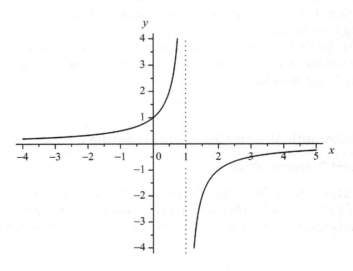

Figure 3.2 A plot of the function $y=1/(1-x)$.

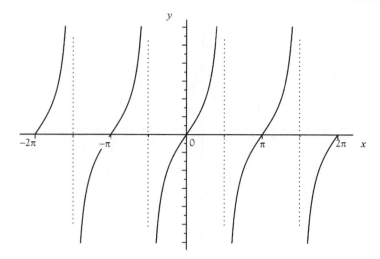

Figure 3.3 A plot of the function $y=\tan x$.

Sometimes we meet functions displaying an **infinite discontinuity**. For example, the function $y = f(x) = 1 / (1 - x)$, shown in Figure 3.2, displays such a discontinuity at $x = 1$ because, as we approach $x=1$ from higher and lower values of x, the value of $f(x)$ tends towards infinitely large values in negative and positive senses, respectively. In this example, the line $x = 1$ is known as a vertical asymptote (see Sections 2.3.1 and 2.3.3 for further discussion of asymptotic behaviour).

The tangent function, $\tan x = \sin x / \cos x$, as shown in Figure 3.3, is interesting because it exhibits infinite discontinuities whenever x passes through an odd multiple of $\frac{\pi}{2}$.

3.1.2 Limiting Behaviour for Increasingly Large Positive or Negative Values of the Independent Variable

We now turn to examining the **limiting behaviour** of functions as the independent variable takes on increasingly large positive or negative values. As an illustration, consider the function shown in Figure 3.2. We see from the form of $f(x)$ that the value of y approaches zero as x becomes increasingly large in both positive and negative senses: the line $y=0$ is an **asymptote**. In the former case, the values of y are increasingly small negative numbers and in the latter, they are increasingly small positive numbers. The **limiting values** of $f(x)$ are therefore zero in both cases.

Periodic functions such as $\sin x$ or $\cos x$ have no asymptotes (no single limiting value), because their values oscillate between two limits as the independent variable increases in a positive or negative sense:

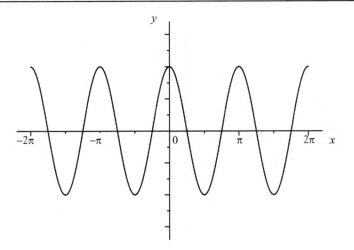

Figure 3.4   A plot of the function $y = \cos(2x)$.

for example, the value of the function $f(x) = \cos(2x)$, oscillates between $+1$ and -1 as $x \to \infty$ (see Figure 3.4).

Problem 3.1

Find the limiting values for (a) $x^2 e^{-x}$ and (b) $\cos(2x)e^{-x}$ as $x \to \infty$.

3.1.3 Limiting Behaviour for Increasingly Small Values of the Independent Variable

Frequently, the context of a particular problem requires us to consider the limiting behaviour of a function as the value of the independent variable approaches zero. For example, consider the physical measurement of heat capacity at absolute zero. Since it is impossible to achieve absolute zero in the laboratory, a natural way to approach the problem would be to obtain measurements of the property at increasingly lower temperatures. If, as the temperature is reduced, the corresponding measurements approach some value m, then it may be assumed that the measurement of the property (in this case, heat capacity) at absolute zero is also m, so long as the specific heat function is continuous in the region of study. We say in this case that the limiting value of the heat capacity, as the temperature approaches absolute zero, is m. As we shall see in Section 3.2, the notation we use to describe this behaviour is:

$$\lim_{T \to 0} C_V(T) = m \tag{3.1}$$

where, in this case, $m=0$, because the limiting value of the heat capacity as $T \rightarrow 0$ K is zero. It is also important to note that it is only possible to approach absolute zero from positive values of T; thus, in this situation, the 'right' limit, usually written as $\lim_{T \rightarrow 0^+} C_V(T) = m$, is the only one of physical significance.

Problem 3.2

Find the limiting values for (a) $x^2 e^{-x}$ and (b) $\cos(2x)e^{-x}$ as $x \rightarrow 0$.

3.2 Defining the Limiting Process

For a function of a single variable x, symbolised, as usual, by $y = f(x)$, we are interested in the value of $f(x)$ as x approaches a particular value, a, but never takes the value a. Points where the function is not defined, as seen, for example at $x = 1$ in Figure 3.2, are excluded from the domain of the function; at other points, the function is continuous.

   Limits play an important role in probing the behaviour of a function at any point in its domain, and the notation we use to describe this process is:

$$\lim_{x \rightarrow a} f(x) = m \qquad (3.2)$$

Note: in this symbolism, the suffix to the symbol lim indicates that, although x approaches a, it *never* actually takes the value a. For the limit to exist, the same (finite) result must be obtained whether we approach a from smaller or larger values of x. Furthermore, if $m = f(a)$, then the function is said to be continuous at $x = a$.

3.2.1 Finding the Limit Intuitively

Consider the plot of the function:

$$y = f(x), \text{ where } f(x) = \frac{x^2 - 9}{x - 3} \qquad (3.3)$$

shown in Figure 3.1. It is evident that $f(x)$ is continuous (unbroken) for all values of x except $x = 3$. Since the denominator and numerator of the function are both zero at $x = 3$, we see that the function is

Table 3.1   Values of $f(x)=(x^2-9)/(x-3)$ in the vicinity of $x=3$.

x	x^2-9	$x-3$	$\dfrac{x^2-9}{x-3}$
4	7	1	7
3.5	3.25	0.5	6.5
3.1	0.61	0.1	6.1
3.01	0.0601	0.01	6.01
3	0	0	*Indeterminate*
2.99	−0.0599	−0.01	5.99
2.9	−0.59	−0.1	5.9
2.5	−2.75	−0.5	5.5
2	−5	−1	5

indeterminate at this value of x; however, as seen in Table 3.1, the *ratio* of the numerator and denominator seems to be approaching the value $y = 6$ as $x \rightarrow 3$ from smaller or larger values.

Taking even smaller increments either side of 3, say $x = 3 \pm 0.0001$, we find that $f(3.0001) = 6.0001$ and $f(2.9999) = 5.9999$. These results suggest that for smaller and smaller increments in x, either side of $x = 3$, the values of the function become closer and closer to 6. Thus we say that, in the limit $x \rightarrow 3$, m takes the value 6:

$$\lim_{x \to 3} \frac{x^2-9}{x-3} = 6 \tag{3.4}$$

3.2.2 An Algebraic Method for Evaluating Limits

In practice, it is often easiest when evaluating limits to write $x = a \pm \delta$, and consider what happens as $\delta \rightarrow 0$, *but never takes the value zero*. This procedure allows us to let x become as close as we like to the value a, *without* it taking the value $x = a$.

Worked Problem 3.1

Q: Evaluate $\lim_{x \to 3} f(x)$, where $f(x) = \dfrac{x^2-9}{x-3}$.

A: By substituting $x = 3 + \delta$ in the expression for $f(x)$, and expanding the square term in the numerator, we obtain:

$$\lim_{x\to 3}\frac{x^2-9}{x-3}=\lim_{\delta\to 0}\frac{(3+\delta)^2-9}{3+\delta-3}$$

$$=\lim_{\delta\to 0}\frac{9+6\delta+\delta^2-9}{\delta}=\lim_{\delta\to 0}\frac{6\delta+\delta^2}{\delta}$$

$$=\frac{6-\delta}{1}=6,$$

where, in the last step, δ can be cancelled in every term of the numerator and denominator as its value is never zero. Thus we obtain the expected result that $f(x)$ approaches the limiting value of 6 as x tends to the value 3, irrespective of the sign of δ. In this situation, m in the definition of the limit has the value 6.

Problem 3.3

For each of the following functions, $f(x)$, identify any points of discontinuity (those values of x where the function is of indeterminate value) and use the method described in Worked Problem 3.1, where appropriate, to find the limiting values of the functions at your suggested points of discontinuity.

(a) $f(x)=\dfrac{2x}{x-4}$

(b) $f(x)=\dfrac{x^2-4}{x-2}$

(c) $f(x)=\dfrac{x-1}{x^2-1}$

(d) $f(x)=3x^2-\dfrac{2}{x}-1$

3.2.3 Evaluating Limits for Functions Whose Values Become Indeterminate

Whenever the value of a function becomes indeterminate for particular limiting values in the independent variable (for example, division by zero or expressions such as ∞ / ∞ or $\infty - \infty$), we need to adopt alternative strategies in determining the limiting behaviour. Such situations arise quite commonly in chemistry, especially when we are interested in evaluating some quantity as the independent variable takes on increasingly large or small values. Good examples occur in dealing with mathematical expressions arising in:

- Manipulating the solutions of rate equations in kinetics.
- Determining high or low temperature limits of thermodynamic properties.

Worked Problem 3.2

Q: Find $\lim\limits_{x\to\infty} \dfrac{2x^2+4}{x^2-x+1}$.

A: Both the numerator and denominator tend to infinity as $x\to\infty$, but their ratio remains finite. There are two ways of handling this situation:

First, we note that as x becomes very large, $2x^2 + 4$ is increasingly well approximated by $2x^2$, and $x^2 - x + 1$ by x^2 as, in both expressions, the highest power of x dominates as x becomes indefinitely large. Thus, as x increases without limit, we find:

$$\lim_{x\to\infty} \frac{2x^2+4}{x^2-x+1} = \lim_{x\to\infty} \frac{2x^2}{x^2} = \lim_{x\to\infty} 2 = 2.$$

Second, we could divide the numerator and denominator by the highest power of x, before taking the limit:

$$\lim_{x\to\infty} \frac{2x^2+4}{x^2-x+1} = \lim_{x\to\infty} \frac{2+4/x^2}{1-1/x+1/x^2} = 2,$$

and, again we see that as x increases without limit, the ratio of numerator to denominator tends to 2.

Problem 3.4

Evaluate the following limits:

(a) $\lim\limits_{x\to\infty} \dfrac{5}{x+1}$;

(b) $\lim\limits_{x\to\infty} \dfrac{3x}{x-4}$;

(c) $\lim\limits_{x\to\infty} \dfrac{x^2}{x+1}$;

(d) $\lim\limits_{x\to\infty} \dfrac{x+1}{x+2}$.

The limiting behaviour of functions for increasingly small values of the independent variable can be found in a similar way by applying exactly the same principles, except that, now, the lowest power of x provides the largest term in both numerator and denominator.

Worked Problem 3.3

Q: Find $\lim\limits_{x\to0} \dfrac{x^2+x}{x^3-1}$.

A: This time, for increasingly small values of x, the numerator and denominator are dominated by x and -1, respectively. Consequently, the ratio of the numerator to denominator tends to $\dfrac{x}{-1}$, which leads to a limiting value of zero: $\lim\limits_{x\to0} \dfrac{x}{-1} = 0$.

Problem 3.5

Evaluate the limit $\lim\limits_{x\to0}(\ln x - \ln 2x)$.

Hint: Remember that $\ln a - \ln b = \ln\dfrac{a}{b}$ (see Chapter 2).

Problem 3.6

The Einstein model for the molar heat capacity of a solid at constant volume, C_V, yields the formula:

$$C_V = 3R(ax)^2\left\{\frac{e^{\frac{ax}{2}}}{e^{ax}-1}\right\}^2 .$$

where $a = \dfrac{hv}{k}$ and $x = \dfrac{1}{T}$. Find the limiting value of C_V as $T \to 0$ K, remembering that $x = \dfrac{1}{T}$.

Note: We shall revisit this problem in Chapter 8 in which we explore the limiting behaviour for high values of T (Problem 8.10).

Problem 3.7

The radial function for the 3s atomic orbital of the hydrogen atom has the form:

$$R_{3s} = N \left(\frac{r}{a_0} \right)^2 e^{-r/a_0},$$

where N is a constant. Find the values of R_{3s} as:

(a) $r \to 0$,
(b) $r \to \infty$.

Hint: See your answers to Problems 3.1(a) and 3.2(a).

3.2.4 The Limiting Form of Functions of More Than One Variable

Sometimes, we are interested in how the form of a function might change for limiting values in one or more variables. For example, consider the catalytic conversion of sucrose into fructose and glucose by the enzyme invertase (β-fructo-furanidase). The rate of formation of product P for this reaction varies in a rather complicated way with the sucrose concentration [S]. At low [S], the reaction is first order in [S], and at high [S], it is zero order. The behaviour observed in Figure 3.5 is established by investigating the form of the function describing the rate of reaction for the two limiting cases where [S]

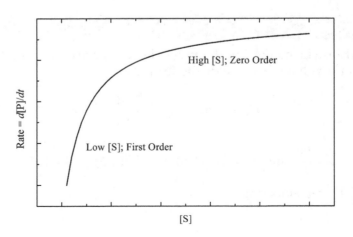

Figure 3.5 The variation in rate of enzymolysis for low and high sucrose concentration, [S], where the reaction is first and zero order, respectively.

approaches either very large or very small values, rather than the absolute value of the function as in the examples discussed above. This is a consequence in this case of the rate equation being a function of more than one variable.

Worked Problem 3.4

Q. The rate of formation of the product P in the catalytic conversion of sucrose into fructose and glucose by the enzyme invertase (β-fructo-furanidase) is given by:

$$\frac{d[P]}{dt} = \frac{k_2[E]_0[S]}{K_M + [S]},$$

where k_2 is a rate constant, K_M is known as the Michaelis constant, $[E]_0$ the initial enzyme concentration and $[S]$ the sucrose concentration. Find the order of reaction with respect to $[S]$ when (a) $[S] \gg K_M$ and (b) $[S] \ll K_M$.

A. (a) For $[S] \gg K_M$, $K_M + [S] \approx [S]$ and so

$$\frac{d[P]}{dt} = \frac{k_2[E]_0[S]}{K_M + [S]} \approx \frac{k_2[E]_0[S]}{[S]} = k_2[E]_0 \quad - \text{ zeroth order in } [S].$$

(b) For $[S] \ll K_M$, $K_M + [S] \approx K_M$ and

$$\frac{d[P]}{dt} = \frac{k_2[E]_0[S]}{K_M + [S]} \approx \frac{k_2[E]_0[S]}{K_M} \quad - \text{ first order in } [S].$$

Problem 3.8

A rate law derived from a steady-state analysis of a reaction mechanism proposed for the reaction of H_2 with NO is given by:

$$\frac{d[N_2]}{dt} = \frac{k_1 k_2 [H_2][NO]^2}{k_{-1} + k_2[H_2]}.$$

Find the limiting form of the rate law when (a) $k_{-1} \gg k_2[H_2]$ and (b) $k_{-1} \ll k_2[H_2]$.

Summary of Key Points

This chapter introduces the concept of the limit, with a view, not only to probing limiting behaviour of functions but also as a foundation to the development of differential and integral calculus in the following chapters. The key points discussed include:

1. The principles involved and notation used in defining a limit.
2. Point discontinuities; infinite discontinuities and asymptotic behaviour.
3. Finding a limit intuitively and algebraically.
4. Investigating the limiting value of functions for increasingly large and small values of the independent variable.
5. Finding the limiting forms of functions of more than one variable.

4
Differentiation

A great deal of chemistry is concerned with processes in which properties change as a function of some variable. Good examples are found in the field of chemical kinetics, which is concerned with measuring and interpreting changes in concentrations of reactants or products with time, and in quantum mechanics, which describes how wavefunctions associated with electronic, vibrational and rotational degrees of freedom of an atom or molecule change as a function of distance, length, angle or time.

Aims:

Calculus is of fundamental importance in chemistry because it underpins so many key chemical concepts. In this chapter, we discuss the foundations and applications of differential calculus; by the end of the chapter you should be able to:

- Describe processes involving change in one independent variable.
- Define the average rate of change of the dependent variable.
- Use the concepts of limits to define the instantaneous rate of change.
- Differentiate most of the standard mathematical functions by rule.
- Differentiate a sum, product or quotient of functions.
- Apply the chain rule to non-standard functions.
- Understand the significance of higher order derivatives and identify maxima, minima and points of inflection.
- Understand the concept of the differential operator.
- Understand the basis of the eigenvalue problem and identify eigenfunctions, eigenvalues and operators.
- Differentiate functions of more than one variable.

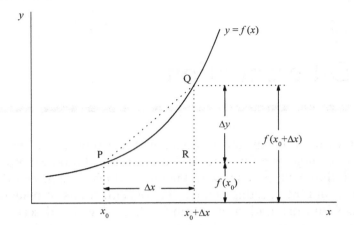

Figure 4.1 Defining the average rate of change of $f(x)$ as x is incremented from x_0 to $x_0+\Delta x$.

4.1 The Average Rate of Change

Consider the plot of the function $y = f(x)$, in which x is the independent variable, shown in Figure 4.1.

The **average rate of change** of $f(x)$ over the increment Δx in x is given by:

$$\frac{QR}{PR} = \frac{f(x_0 + \Delta x) - f(x_0)}{\Delta x} = \frac{\Delta y}{\Delta x}, \tag{4.1}$$

where $f(x_0)$ and $f(x_0 + \Delta x)$ are the values of $f(x)$ at the points x_0 and $x_0 + \Delta x$, and Δy is the change in y, that results in the change Δx in x.

This average rate of change corresponds to the slope of the **chord** PQ; that is, the slope of the straight line (sometimes termed the **secant**) joining P and Q. In chemical kinetics, we can draw a direct analogy by equating the concentration of a species A at time t, often designated by [A], to the dependent variable (designated as y in Figure 4.1), and the time after initiation of the reaction, t, to the independent variable (designated by x in Figure 4.1). Consequently, if we measure the concentration of a reaction product at two intervals of time, say 1 minute apart, we might conclude that over that interval, the concentration of the product had changed by 1.00 mol dm^{-3}. In this case, we could state that the average rate of reaction in this interval is 1.00 mol dm^{-3} per min. The problem here is that we know nothing about how the reaction rate changes in detail during that interval of 1 minute, and it is this detail that so crucial to our understanding of the kinetics of the reaction. Consequently, what we need, in general, is to be able to quantify the rate of change of the dependent variable at a *particular* value of the independent variable, rather than simply the average rate of change over some increment in the independent variable. This equates in our chemical analogy, to being able to measure the instantaneous reaction rate at a given instant in time (and consequently for a given concentration of reactant or product), rather than the average rate of reaction over some extended period of time. However, before we can determine these instantaneous chemical rates, we must first establish some mathematical principles.

4.2 The Instantaneous Rate of Change

4.2.1 Differentiation from First Principles

If we now reconsider the general situation shown in Figure 4.1, we can determine the instantaneous rate of change by examining the limiting behaviour of the ratio, QR / PR, the change in y divided by the change in x, as Δx tends to zero:

$$\lim_{\Delta x \to 0} \left\{ \frac{QR}{PR} \right\} = \lim_{\Delta x \to 0} \left\{ \frac{\Delta y}{\Delta x} \right\} = \lim_{\Delta x \to 0} \left\{ \frac{f(x_0 + \Delta x) - f(x_0)}{\Delta x} \right\} \quad (4.2)$$

The limiting value defined in eqn (4.2) exists if:

- The function does not undergo any abrupt changes at x_0 (it is continuous at the point x_0).
- It is independent of the direction in which the point x_0 is approached.

If the limit in eqn (4.2) exists, it is called the derivative of the function $y = f(x)$ at the point x_0. The value of the derivative varies with the choice of x_0, and we define it in general terms as:

$$\left(\frac{dy}{dx} \right)_{x=x_0} = \lim_{\Delta x \to 0} \left\{ \frac{f(x_0 + \Delta x) - f(x_0)}{\Delta x} \right\}, \quad (4.3)$$

where $\left(\dfrac{dy}{dx} \right)_{x=x_0}$ is the name given to the value of the derivative at the point x_0. The derivative of the function $y = f(x)$ at $x = x_0$ in Figure 4.1 corresponds geometrically to the slope of the tangent to the curve $y = f(x)$ at the point P (known as the gradient).

The basic formula in eqn (4.3) for the derivative is often given in the form:

$$\frac{dy}{dx} = \lim_{\Delta x \to 0} \left\{ \frac{f(x + \Delta x) - f(x)}{\Delta x} \right\}, \quad (4.4)$$

for an arbitrary value of x.

We should also note that:

- $\frac{dy}{dx}$ is the name of the derivative function, commonly also represented as $f'(x)$.
- The domain of the derivative function is not necessarily the same as that of $y = f(x)$ (see Table 4.1).

The requirement that, for the limit in eqn (4.2) to exist, the function does not undergo any abrupt changes is sometimes overlooked, yet it is an important one. An example of a function falling into this category is the modulus function, $y = |x|$, defined by:

$$y = f(x) = |x| = \begin{cases} x \text{ if } x \geqslant 0 \\ -x \text{ if } x < 0 \end{cases}.$$

This function is continuous for all values of x [Figure 4.2 (a)], but there is no unique slope at the point $x = 0$ as the derivative is undefined at this point [Figure 4.2(b)].

Chemical examples showing this type of behaviour include processes associated with sudden changes in concentration, phase, crystal structure, temperature, *etc*. For example, Figure 2.9 shows how the equilibrium concentration of a chemical species changes suddenly

(a)

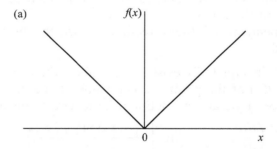

(b)

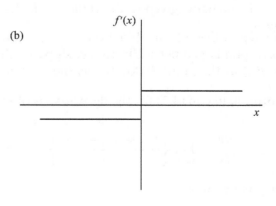

Figure 4.2 (a) The modulus function $y=f(x)=|x|$; (b) the derivative of the modulus function.

when a temperature jump is applied at time t_0. Although there are no discontinuities in this function, its derivative is undefined at time t_0.

Worked Problem 4.1

Q. Differentiate $y = f(x) = x^2$, using the definition of the derivative given in eqn (4.4).

A. $\dfrac{dy}{dx} = \lim\limits_{\Delta x \to 0} \left\{ \dfrac{f(x+\Delta x) - f(x)}{\Delta x} \right\} = \lim\limits_{\Delta x \to 0} \left\{ \dfrac{(x+\Delta x)^2 - x^2}{\Delta x} \right\}$

$= \lim\limits_{\Delta x \to 0} \left\{ \dfrac{x^2 + 2x\Delta x + (\Delta x)^2 - x^2}{\Delta x} \right\} = \lim\limits_{\Delta x \to 0} \left\{ \dfrac{2x\Delta x + (\Delta x)^2}{\Delta x} \right\}$

Since Δx tends to zero, but never takes the value zero, cancellation of Δx from all terms in the in the numerator and denominator yields:

$$\frac{dy}{dx} = \lim\limits_{\Delta x \to 0} \{2x + \Delta x\} = 2x$$

Problem 4.1

Differentiate the function $y = f(x)$, where $f(x) = 3$ using the definition of the derivative given in eqn (4.3).

Hint: The function $y = f(x) = 3$ requires that $y = 3$ for all values of x – thus if $f(x) = 3$, then $f(x + \Delta x)$ must also equal 3.

Problem 4.2

Use eqn (4.4) to find the derivative of the function $y = f(x)$, where:

(a) $f(x) = 3x^2$
(b) $f(x) = 1/x^2$

Hint: In your answer to (b), you will need to remember how to subtract fractions, $i.e.$:

$$\frac{1}{a} - \frac{1}{b} = \frac{b-a}{ab}.$$

4.2.2 Differentiation by Rule

4.2.2.1 Some Standard Derivatives

The derivatives of all functions can be found using the limit method described in Section 4.2.1. Some of the more common functions, and their derivatives, are listed in Table 4.1. Unless otherwise indicated, the respective domains (Dom) are 'all values of x'.

Table 4.1   Derivatives of some common functions and their respective domains.

$f(x)$	$f'(x)$	$Dom(f(x))$	$Dom(f'(x))$	Notes
c	0			1
x^n	$nx^{n-1}(n\neq0)$	$x\neq0$ for $n<0$	$x\neq0$ for $n<1$	2
$\sin ax$	$a\cos ax$			3
$\cos ax$	$-a\sin ax$			4
$\tan ax$	$a\sec^2 ax$	$x\neq(2n+1)\pi/2$	$x\neq(2n+1)\pi/2$	5
$\sec ax$	$a\sec ax\tan ax$	$x\neq(2n+1)\pi/2$	$x\neq(2n+1)\pi/2$	6
$\ln ax$	$1/x$	$x>0$	$x\neq0$	7
e^{ax}	ae^{ax}			

Notes:(1) The constant function, c; (2) $n=0$ corresponds to the constant function; (3) $a\neq0$; for a 1:1 function, Dom $(f(x)) =[-\pi/2, \pi/2]$; (4) $a\neq0$; for a 1:1 function, Dom $(f(x))=[0, \pi]$; (5-7) $a\neq0$.

However, as we have seen above, and in Table 4.1, we do meet functions for which the derivative $f'(x)$ does not exist at selected values of x. The functions $y = f(x) = \ln x$ at $x = 0$ and $y = f(x) = \tan x$ at $x = (2n + 1)\pi / 2$, both listed in Table 4.1, fall into this category. Naturally, since the derivative does not exist in these cases at selective values of x, the domain of the derivatives of these functions will not be the same as the original functions. The restrictions on the respective domains are best seen in sample plots of these functions shown in Figure 4.3.

4.2.2.2 An Introduction to the Concept of the Operator

The notation $\dfrac{dy}{dx}$ (or sometimes dy / dx) for the derivative is just one of a number of different notations in widespread use, all of which are equivalent:

$$\frac{dy}{dx}, dy/dx, f'(x), f^{(1)}(x), \hat{D}f(x)$$

The more commonly used notations are $\dfrac{dy}{dx}$ and $f'(x)$, but expressing the derivative in the form $\hat{D}f(x)$ provides a useful reminder that the

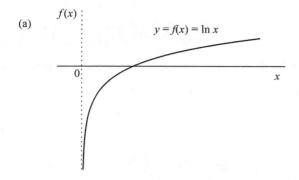

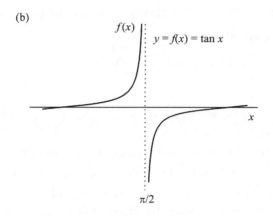

Figure 4.3 The functions (a) $y=f(x)=\ln x$ and (b) $y=f(x)=\tan x$ are both examples of functions for which the derivative does not exist at certain values in the independent variable (see Table 4.1).

derivative function is obtained from the function $y = f(x)$ by the *operation* "differentiate with respect to x". Thus, we express this instruction in symbols as:

$$\hat{D}f(x) \equiv \frac{d}{dx}f(x) = \frac{d}{dx}y = \frac{dy}{dx} \tag{4.5}$$

It is worth emphasising that the symbol $\frac{dy}{dx}$ does not mean dy divided by dx in this context, but represents the *limiting* value of the quotient $\frac{\Delta y}{\Delta x}$ as $\Delta x \to 0$.

In general, an **operator**, $\hat{A}$, is represented by a symbol with a caret ('hat') denoting an instruction to undertake an appropriate action on the object to its right [here $f(x)$]. In eqn (4.5), we consider $\frac{dy}{dx}$ to be the differentiation operator $\frac{d}{dx}$ acting on the function $f(x)$, which we have labelled y, to give a new function, say $g(x)$:

$$\hat{A}f(x) = g(x). \tag{4.6}$$

Worked Problem 4.2

Q. For the function $f(x) = x^2$, find $\hat{A}(f(x))$ where $\hat{A} = \dfrac{d}{dx}$.

A. For $f(x) = x^2$, $\dfrac{d}{dx}(f(x)) = 2x$.

Problem 4.3

For each of the following functions, $f(x)$, use the information in Table 4.1 to find $\hat{A}(f(x))$ where $\hat{A} = \dfrac{d}{dx}$

(a) $x^{3/4}$, (b) e^{-3x}, (c) $1/x$ and (d) $a \cos ax$.

Problem 4.4

Use the information in Table 4.1 to demonstrate that, when the operator $\hat{A} = \dfrac{d}{dx} + 2$ acts on $f(x) = e^{-2x}$, the function is annihilated [*i.e.* the null function, $g(x) = 0$ results].

We will come to appreciate the full significance of the concept of the operator in Section 4.3.1, when we consider the *eigenvalue problem*.

4.2.3 Basic Rules for Differentiation

Although all functions can be differentiated from first principles using eqn (4.4), this can be a rather long-winded process in practice. In this chapter, we deal with the differentiation of more complicated functions with the aid of a set of rules, all of which may be derived from the defining relation (4.4). In many cases, however, we simply need to learn what the derivative of a particular function is, or how to go about differentiating a certain class of function. For example, we learn that the derivative of $y = f(x) = \sin x$ is $\cos x$, but that the derivative of $y = f(x) = \cos x$ is $-\sin x$. Similarly we can differentiate any function of the type $y = f(x) = x^n$ by remembering the rule that we reduce the index of x by 1, and multiply the result by n; that is

$$\frac{d}{dx}x^n = nx^{n-1} (n \neq 0). \tag{4.7}$$

For functions involving a combination of other elementary functions, we follow another set of rules: if u and v represent functions $f(x)$ and $g(x)$, respectively, then the rules for differentiating a sum, product, or quotient can be expressed as:

$$\frac{d}{dx}(u+v) = \frac{du}{dx} + \frac{dv}{dx} \tag{4.8}$$

$$\frac{d}{dx}(uv) = v\frac{du}{dx} + u\frac{dv}{dx} \tag{4.9}$$

$$\frac{d}{dx}\left(\frac{u}{v}\right) = \frac{v\dfrac{du}{dx} - u\dfrac{dv}{dx}}{v^2} \tag{4.10}$$

Problem 4.5

Differentiate the following, using the appropriate rules:

(a)$(x-1)(x^2+4)$; (b)$\dfrac{x}{(x+1)}$; (c) $\sin^2 x$; (d)$x \ln x$; (e)$e^x \sin x$.

4.2.4 Chain Rule

Quite frequently we are faced with the problem of differentiating functions of functions, such as $y = \ln(x^2 + x + 1)$. The derivative of this function is not immediately obvious, and so we use a strategy known as the **chain rule** to reduce the problem to a more manageable form. We can proceed as follows:

- Introduce a new variable $u = x^2 + x + 1$ to transform the function $y = \ln(x^2 + x + 1)$ into the simpler form $y = \ln u$.
- Determine the derivative of y with respect to u:

$$\frac{dy}{du} = \frac{1}{u}.$$

- Determine the derivative of u with respect to x:

$$\frac{du}{dx} = 2x + 1.$$

- Combine the two derivatives using:

$$\frac{dy}{dx} = \frac{dy}{du}\cdot\frac{du}{dx} = \frac{1}{u}\cdot(2x+1).$$

- Eliminate the variable u:

$$\frac{dy}{dx} = \frac{2x+1}{x^2+x+1}.$$

Problem 4.6

Apply the chain rule to find the derivative of $y = e^{x \sin x}$, using the substitution $u = x \sin x$.

Problem 4.7

Use the chain rule to find the derivative of the following functions:

(a) $y = \ln(2 + x^2)$, (b) $y = 2\sin(x^2 - 1)$.

4.3 Higher Order Derivatives

In general, when we differentiate a function $y = f(x)$ another function of x is obtained:

$$\frac{dy}{dx} = f'(x)$$

If this derivative function is specified, say, by the relation $h = f'(x) = g(x)$, then, so long as $g(x)$ is not zero, h may be differentiated again to yield the *second derivative* of $f(x)$:

$$\frac{dh}{dx} = \frac{d}{dx}h = \frac{d}{dx}\frac{dy}{dx} = \frac{d^2y}{dx^2} \equiv f''(x) \text{ or } f^{(2)}(x) \qquad (4.11)$$

When we write $h = f'(x) = g(x)$, we are simply labelling the function $g(x)$ that results from differentiation of $f(x)$ arbitrarily with the letter h in the same way that we labelled $f(x)$ with y.

This process may usually be repeated to determine higher order derivatives, if they exist. Thus, for example, if $f(x) = x^3 - x + 1$, then:

$$\frac{dy}{dx} = 3x^2 - 1; \quad \frac{d^2y}{dx^2} = 6x; \quad \frac{d^3y}{dx^3} = 6, \text{ and } \frac{d^ny}{dx^n} = 0 \text{ for } n > 3.$$

Worked Problem 4.3

Q. Given $y = f(x) = (1+x)^4$ find $\dfrac{d^2y}{dx^2}$. Deduce for what value

of n, $\dfrac{d^n y}{dx^n} = 0$.

A. Let $u = (1 + x)$ to transform the function into a simpler form, $y = u^4$, and use the chain rule to find $\dfrac{dy}{dx}$:

- $\dfrac{dy}{dx} = \dfrac{dy}{du} \cdot \dfrac{du}{dx} = 4u^3 \cdot 1 = 4u^3 = 4(1+x)^3$

- Let $h = \dfrac{dy}{dx} = 4(1+x)^3$ and use the chain rule again:

  $\dfrac{d^2y}{dx^2} = \dfrac{dh}{dx} = \dfrac{dh}{du} \cdot \dfrac{du}{dx} = 12u^2 \cdot 1 = 12(1+x)^2.$

In this example, we can see that each act of differentiation decreases the index of $(1 + x)$ by one and so it follows that the fifth derivative will be zero.

We can gain some useful insight into what exactly the first and second derivatives of a function tell us by looking at the form of the three functions $f(x)$, $f'(x)$ and $f''(x)$, as shown in Figure 4.4.

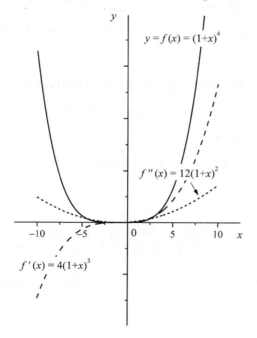

Figure 4.4 Plots of the function $f(x) = (1+x)^4$ and its first two derivatives.

The original function $y = f(x) = (1 + x)^4$ must be positive for all values of x and has a minimum value of zero at $x = -1$. The first derivative $f'(x) = 4(1 + x)^3$ gives us the rate of change (slope of the tangent) of the function $f(x)$ for any value of x. For $x < -1$, the value of $f'(x) = 4(1 + x)^3$ is negative, which means the slope of the original function is also negative (which we can see for ourselves by inspection of the plot). For $x > -1$, the first derivative is positive and so the slope of the original function is also positive. The fact that the value of $f'(x) = 4(1 + x)^3$ is zero at $x = -1$ indicates that the slope of the function is zero at this point. Such a point is identified as a **stationary point**, which, in this case, corresponds to a minimum (as we can see from the plot). We shall see later in Section 4.4 how to prove whether a stationary point is a maximum or minimum (or point of inflection) without needing to plot the function. Similarly, the form of the second derivative, $f''(x) = 12(1 + x)^2$, gives us the slope, or rate of change, of the first derivative and by extension the slope of the slope of the original function $f(x)$. The form of the second derivative provides us with the means to characterise the nature of any stationary points in the original function, whereas that of the first derivative tells us if and where the stationary points exist (see Section 4.4).

Problem 4.8

Find the second and third derivatives of:

(a) $y = 1/x$ and (b) $y = N \sin ax$ (N, a are constants).

4.3.1 Operators Revisited: An Introduction to the Eigenvalue Problem

In Section 4.2.2 we defined the act of differentiation as an operation in which the operator $\hat{D} = \dfrac{d}{dx}$ acts on some function $f(x)$. Similarly, we can express the act of differentiating twice in terms of the operator $\hat{D}^2 = \dfrac{d^2}{dx^2}$.

Worked Problem 4.4

Q. For the function $f(x) = \cos kx$, find $\hat{A}(f(x))$, where $\hat{A} = \dfrac{d^2}{dx^2}$.

A. For $f(x) = \cos\ kx$, $\dfrac{d}{dx}(f(x)) = -k\sin kx$ and

$\dfrac{d^2}{dx^2}(f(x)) = -k^2\cos kx$ and so $\hat{A}\cos kx = -k^2\ \cos\ kx$.

4.3.1.1 The Eigenvalue Problem

A problem common to many areas in physical chemistry is the following: given an operator, $\hat{A}$, find a function $\phi(x)$, and a constant a, such that $\hat{A}$ acting on $\phi(x)$ yields a constant multiplied by $\phi(x)$. In other words, the result of operating on the function $\phi(x)$ by $\hat{A}$ is simply to return $\phi(x)$, multiplied by a constant factor, a. This type of problem is known as an **eigenvalue problem**, and the key features may be described schematically as follows:

The **eigenfunction** of the operator $\hat{A}$

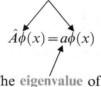

$$\hat{A}\phi(x) = a\phi(x)$$

The **eigenvalue** of the operator $\hat{A}$

The solution to Worked Problem 4.4 is an example of an eigenvalue problem.

The key eigenvalue equation in chemistry is the Schrödinger equation, $\hat{H}\psi = E\psi$. The solution of this equation for a particular system (such as an electron bound by the field of a nucleus) yields so-called wavefunctions, ψ, that completely describe the system of interest and from which any property of the system can be extracted.

Worked Problem 4.4 *revisited*

For $f(x) = \cos kx$ and $\hat{A} = \dfrac{d^2}{dx^2}$:

$$\hat{A}\cos kx = -k^2\cos kx$$

In this example, we see that by differentiating the function $f(x) = \cos kx$ twice, we regenerate our original function multiplied by a constant which, in this case, is $-k^2$. Hence, $\cos kx$ is an eigenfunction of $\hat{A}$, and its eigenvalue is $-k^2$.

Problem 4.9

Perform the following operations:

(a) For $f(x) = x^3$ find $\hat{A}(f(x))$ where $\hat{A} = \dfrac{d^2}{dx^2}$.

(b) For $f(x) = \sin kx$, find $\hat{A}(f(x))$ where $\hat{A} = \dfrac{d^2}{dx^2}$.

(c) For $f(x) = \sin kx + \cos kx$, find $\hat{A}(f(x))$ where $\hat{A} = \dfrac{d^2}{dx^2}$.

(d) For $f(x) = e^{ax}$, find $\hat{A}(f(x))$ where $\hat{A} = \dfrac{d}{dx}$.

Which of (a)–(d) would be classified as eigenvalue problems? What is the eigenfunction and what is the eigenvalue in each case?

Problem 4.10

Show that $y = f(x) = e^{mx}$ is an eigenfunction of the operator $\hat{A} = \dfrac{d^2}{dx^2} - 2\dfrac{d}{dx} - 3$, and give its eigenvalue. For what values of m does $\hat{A}$ annihilate $f(x)$?

Annihilation of a function implies that the null function is produced after application of an operator.

Problem 4.11

The lowest energy solution of the Schrödinger equation for a particle (mass, m) moving in a constant potential (V) and in a one-dimensional box of length (L) takes the form:

$$\psi = \sqrt{\frac{2}{L}}\sin\frac{\pi x}{L}.$$

If we take V as the zero of energy, then ψ satisfies the Schrödinger equation:

$$-\frac{h^2}{8\pi^2 m}\frac{d^2\psi}{dx^2} = E\,\psi,$$

Find an expression for the total energy E in terms of L and the constants π, m and h.

Hint: You may have noticed that the expression above is an example of an eigenvalue problem where the eigenfunction is $\psi = \sqrt{\dfrac{2}{L}}\sin\dfrac{\pi x}{L}$ and the eigenvalue is E. In this case, the total energy E is determined by operating on the function ψ using the operator $-\dfrac{h^2}{8\pi^2 m}\dfrac{d^2}{dx^2}$.

In quantum mechanics, the operator $-\dfrac{h^2}{8\pi^2 m}\dfrac{d^2}{dx^2}$ is called the *Hamiltonian* and is given the symbol $\hat{H}$.

4.4 Maxima, Minima and Points of Inflection

We often encounter situations in the physical sciences where we need to establish at which value(s) of an independent variable a maximum or minimum value in the function occurs. For example:

- The probability of finding the electron in the ground state of the hydrogen atom between radii r and $r + dr$ is given by $D(r)dr$, where $D(r)$ is the radial probability density function shown in Figure 4.5.

  The most probable distance of the electron from the nucleus is found by locating the maximum $D(r)$ (see Problem 4.12 below). It should come as no surprise to discover that this maximum occurs at the value $r = a_0$, the Bohr radius.

- When we attempt to fit a theoretical curve to a set of experimental data points, we typically apply a least squares fitting technique which seeks to minimise the deviation of the fit from the experimental data. In this case, differential calculus is used to find the minimum in the function that describes the deviation between fit and experiment (see Chapter 13) .

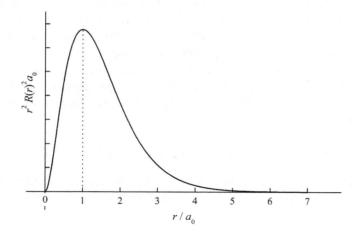

Figure 4.5 The radial probability density function for the 1s atomic orbital of the hydrogen atom.

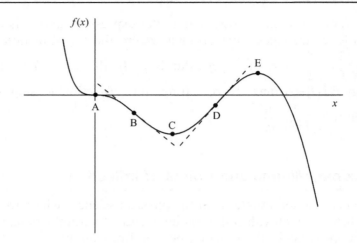

Figure 4.6   A plot of the function $y = f(x)$. Points A, C and E are all stationary points, for which $f'(x) = 0$, while points C and E are also turning points (minimum and maximum, respectively). Points A, B and D are all points of inflection, but B and D are neither stationary points nor turning points. Note that at the points of inflection, the tangents ($- - - -$) cut the curve.

4.4.1 Finding and Defining Stationary Points

Consider the function $y = f(x)$ in Figure 4.6. As we saw in our discussion of Worked Problem 4.3, values of x for which $f'(x) = 0$ are called **stationary points**. A stationary point may be:

- A maximum (point E, a turning point) or a minimum (point C, also a turning point). The value of dy/dx changes sign on passing through these points.
- A point of inflection: the tangent cuts the curve at this point (points A, B and D).

4.4.1.1 Turning Points (Maxima and Minima)

E and C are called **turning points** because, in passing through E and C, the value of dy/dx changes sign. The existence and nature of stationary points, which are also turning points, may be identified through the first and second derivatives of the function. If we consider point C, we see that as we pass through this point the gradient becomes less negative as we approach C, passes through zero at point C, and then becomes positive. Clearly the *rate of change* of the gradient is positive at point C (because the gradient changes from negative to positive), which suggests that the function has a minimum at this point:

$$\text{A } \textbf{minimum} \text{ exists if } f'(x) = 0 \text{ and } f^{(2)}(x) > 0.$$

Similarly, on passing through point E, the gradient becomes less positive, passes through zero at E and then becomes negative. In this case, the rate of change in the gradient is negative and we can identify point E as a maximum:

$$\text{A maximum exists if } f'(x) = 0 \text{ and } f^{(2)}(x) < 0.$$

In general $y = f(x)$ will display a number of turning points within the domain of the function.

Turning points corresponding to maxima and minima may be classified as either:

- A global maximum or minimum which has a value greater or smaller than all other points within the domain of the function.
- A local maximum or minimum which has a value greater or smaller than all neighbouring points.

4.4.1.2 Points of Inflection

At a point of inflection (A, B or D), which may or may not be a stationary point:

- The tangent cuts the curve.
- The slope of the tangent does not change sign.

Note that A is both a point of inflection and a stationary point, but while B and D are both points of inflection, they are not stationary points because $f'(x) \neq 0$.

Points of inflection occur when the gradient is a maximum or minimum. This requires that $f^{(2)}(x) = 0$ but this in itself is not sufficient to characterise a point of inflection. We achieve this through the first non-zero higher derivative.

If $f'(x) = 0$, $f^{(2)}(x) = 0$ but $f^{(3)}(x) \neq 0$, then we have a point of inflection which is also a stationary point (such as point A). However, if $f'(x) \neq 0$, $f^{(2)}(x) = 0$ and $f^{(3)}(x) \neq 0$, then we have a point of inflection which is *not* a stationary point (B or D). The rules for identifying the location and nature of stationary points, turning points and points of inflection are summarised in Table 4.2.

Interestingly, in the last row of Table 4.2 we see that a turning point may exist for which $f^{(2)}(x) = 0$. In such cases, $f^{(3)}(x) = 0$ and the nature of the turning point is determined by the sign of the fourth derivative. An example of a function for which this latter condition applies is $y = f(x) = (x-1)^4$. If there is any doubt over the nature of a stationary point, especially if the second derivative vanishes, it is always helpful to sketch the function!

Table 4.2 The location and nature of turning points, stationary points and points of inflection are given by the first, second and, where appropriate, third and fourth derivatives.

	$f'(x)$	$f^{(2)}(x)$	$f^{(3)}(x)$	$f^{(4)}(x)$
Minimum	0	>0	-	-
Maximum	0	<0	-	-
Inflection point (stationary)	0	0	$\neq 0$	-
Inflection point (not stationary)	$\neq 0$	0	$\neq 0$	-
Turning points where $f^{(2)}(x) = 0$	0	0	0	$\neq 0$

Worked Problem 4.5

Q. Consider the function $y = f(x)$, where $f(x) = x^2 - x^3/9$.

(a) Plot the function for selected values of x in the interval $-3.5 \leqslant x \leqslant 10$.

(b) Identify possible values of x corresponding to turning points and points of inflection.

(c) Derive expressions for the first and second derivatives of the function.

(d) Identify the nature of the turning points (*e.g.* maximum, minimum, global or local).

(e) Verify that there is a point of inflection where $f'(x) \neq 0$, $f^{(2)}(x) = 0$ and $f^{(3)}(x) \neq 0$.

A. (a)

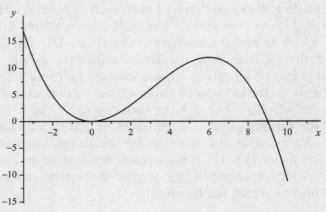

Figure 4.7 A plot of the function $y = f(x) = x^2 - x^3/9$ for $-3.5 \leqslant x \leqslant 10$.

(b) By inspection, we can identify turning points at $x = 0$ and in the vicinity of $x = 6$; there is no turning point corresponding to a point of inflection.

(c) $f'(x) = 2x - x^2/3$; $f^{(2)}(x) = 2 - 2x/3$.

(d) $f'(x) = x\left(2 - \dfrac{x}{3}\right) = 0$ at $x = 0$ $\left[\text{local minimum}; f^{(2)}(x) > 0\right]$

and at $x = 6$ $\left[\text{local maximum}; f^{(2)}(x) < 0\right]$.

(e) $f^{(2)}(x) = 2 - 2x/3 = 0$ when $x = 3$

$f'(x) = 2x - x^2/3 \neq 0$ when $x = 3$

$f^{(3)}(x) = -2/3 \neq 0$ when $x = 3$

corresponding to a point of inflection.

Problem 4.12

The radial probability density function for the electron in the ground state of the hydrogen atom takes the form:

$$D(r) = Nr^2 e^{-2r/a_0},$$

where N is a constant.

(a) Use the product rule to show that:

$$\frac{dD(r)}{dr} = 2Ne^{-2r/a_0}\left(r - \frac{r^2}{a_0}\right).$$

(b) Identify the non-zero value of r at which D displays a turning point, and give the value of D at this point.

(c) Demonstrate, by examining the sign of the second derivative of D, that the turning point corresponds to a maximum.

(d) Show that points of inflection, which are not stationary points, occur at $r = \left(1 \pm \dfrac{\sqrt{2}}{2}\right)a_0$.

4.5 The Differentiation of Functions of Two or More Variables

In chemistry we frequently meet functions of two or more variables. For example:

- The pressure (P) of an ideal gas depends upon the two independent variables, temperature (T) and volume (V):

$$P = \frac{nRT}{V}$$

- The electron probability density function, $\rho(x, y, z)$ for a molecule depends upon three spatial coordinates (x, y, z) to specify its value at a chosen position.
- The entropy, S, for a system containing three species A, B and AB at a given temperature and pressure depends upon five variables: N_A, N_B, N_{AB}, T and P, where N_X is the number of moles of A, B or AB.

(a)

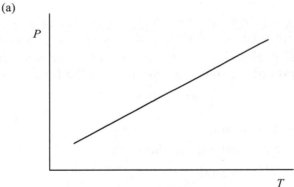

(b)

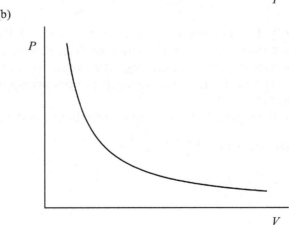

Figure 4.8 The ideal gas equation $P = \dfrac{nRT}{V}$ shows that (a) P varies linearly with T at constant volume but (b) in a non-linear way with V at constant temperature.

When we explore the nature and form of these and other multi-variable functions, we need to know how to locate specific features, such as maximum or minimum values. Clearly, functions of two variables, such as in the ideal gas equation above, require plots in three dimensions to display all their features (such plots appear as surfaces). Derivatives of such functions with respect to one of these (independent) variables are easily found by treating all the other variables as constants and finding the *partial derivative* with respect to the single variable of interest. For example, we can see from the ideal gas equation:

$$P = \frac{nRT}{V}$$

that P varies linearly with T but in a non-linear way with V, as shown in Figure 4.8.

If we were to differentiate this expression with respect to T and V, we would be able to evaluate precisely the rate at which P varied with respect to T (constant), or, with respect to V (variable) (see Problem 4.13).

Worked Problem 4.6

Q. For the function $z = xy + y^2$ find:

    (a) The partial derivative of z with respect to x.
    (b) The partial derivative of z with respect to y.

A. (a) The partial derivative of z with respect x is found by treating y as a constant. In order to make it clear that several variables are present, we sometimes use the following notation:

$$\left(\frac{\partial z}{\partial x}\right)_y$$

which means differentiation with respect to x, keeping y constant. We often drop the y suffix and the brackets because there is usually no problem in recognising which variables are held constant. In this case, the y^2 term is a constant and will vanish and so the derivative is given by:

$$\frac{\partial z}{\partial x} = y.$$

(b) Similarly, the partial derivative with respect to y, keeping x constant, is:

$$\frac{\partial z}{\partial y} = x + 2y.$$

Problem 4.13

For n mol of an ideal gas $P = \dfrac{nRT}{V}$, where P is a function of the two variables T and V (R and n are constants). Write down $\partial P / \partial T$ and $\partial P / \partial V$.

Summary of Key Points

Differential calculus is inextricably linked to the notion of rates of change. This is especially important to our understanding of chemical kinetics and other areas of chemistry, such as thermodynamics, quantum mechanics and spectroscopy. This chapter concerns the application of differential calculus to problems involving rates of change of one property with respect to another. The key points discussed include:

1. A comparison of average and instantaneous rates of change.
2. The use of limits to define the instantaneous rate of change as the derivative of a function.
3. Differentiation from first principles and by rule.
4. Differentiation of a sum, product or quotient of functions
5. Discussion of operators, the eigenvalue problem and associated eigenfunctions.
6. The use of the chain rule for differentiating functions of functions.
7. Higher order derivatives to locate and identify maxima, minima and points of inflection.
8. Differentiation of functions of more than one variable.

5
Differentials

In many areas of chemistry (*e.g.* error analysis; thermodynamics, *etc.*) we are concerned with the consequences of small (and, sometimes, not so small) changes in a number of variables and their overall effect upon a property depending on these variables. For example, in thermodynamics, the temperature dependence of the equilibrium constant, K, is usually expressed in the form:

$$K = e^{-\Delta G^\ominus / RT},$$

where the change in Gibbs energy, $\Delta G^\ominus = \Delta H^\ominus - T\Delta S^\ominus$, itself depends upon temperature, both explicitly through the presence of T, and implicitly, as the changes in $\Delta H^\ominus$ and $\Delta S^\ominus$ are, in general, both temperature-dependent. However, if we assume that $\Delta H^\ominus$ and $\Delta S^\ominus$ are, to a good approximation, independent of temperature, then for small changes in temperature, we obtain the explicit formula relating K and T:

$$K = e^{-(\Delta H^\ominus - T\Delta S^\ominus)/RT} = e^{-(\Delta H^\ominus/T - \Delta S^\ominus)/R} \qquad (5.1)$$

Quite frequently, we are interested in the effect of *small changes* in the temperature on the equilibrium constant. We could, of course, use eqn (5.1) to calculate K at two different temperatures for any reaction which satisfies the requirements given above and determine the change in K by subtraction. However, in practice, a much more convenient route makes use of the properties of differentials. This Chapter is concerned with exploring what effect small changes in one or more independent variables have on the dependent variable in expressions such as eqn (5.1). We shall see that this is particularly useful in determining how errors propagate through expressions relating one property to another. However, before discussing further the importance of differentials in a chemical context, we need to discuss some of the background to the method of differentials.

Aims:

By the end of this chapter you should be able to:

- Understand the definition of change defined by the differential and the concept of infinitesimal change.
- Understand the difference between the differential dy representing an approximate change in the dependent variable resulting from a small change in the independent variable, and the actual change in the dependent variable, Δy.
- Calculate the differentials and the errors in approximating the differential to the actual change in a dependent variable.
- Define the differential of a function of more than one variable.
- Use differentials to calculate relative and percentage errors in one property deriving from those in other properties.

5.1 The Effects of Incremental Change

We recall from Chapter 4 (Figure 4.1) that if Δy is the change in y that accompanies an *incremental* change Δx in x, then:

$$\Delta y = f(x + \Delta x) - f(x) \tag{5.2}$$

For example, if we consider the function $y = f(x) = x^3$, the incremental change in y that accompanies a change in Δx in x is given as:

$$\Delta y = (x + \Delta x)^3 - x^3$$

which, on expanding, yields:

$$\Delta y = 3x^2 \Delta x + 3x(\Delta x)^2 + (\Delta x)^3.$$

For sufficiently small values of Δx, the power terms in Δx decrease very rapidly in magnitude. Thus, for example, if $\Delta x = 10^{-2}$, then $\Delta x^2 = 10^{-4}$ and $\Delta x^3 = 10^{-6}$. This may be expressed algebraically as:

$$(\Delta x)^3 \ll (\Delta x)^2 \ll \Delta x,$$

and, if we neglect Δx raised to power 2 or higher, we can approximate the expression for Δy by:

$$\Delta y \approx 3x^2 \Delta x$$

The appearance of $3x^2$ in this expression is no accident. If we rewrite the expression for Δy as:

$$\Delta y = \left(\frac{f(x + \Delta x) - f(x)}{\Delta x} \right) \times \Delta x, \qquad (5.3)$$

then it is clear that, for a very small Δx, the term in parentheses is an approximation for the derivative of $f(x)$, which, for the present choice of function, is $3x^2$. We can therefore rewrite the general result in the form $\Delta y \approx f'(x)\Delta x$.

5.1.1 The Concept of Infinitesimal Change

An *infinitesimal* change in x, known as the *differential* dx, gives rise to a corresponding change in y that is well represented by the differential dy:

$$dy = f'(x)dx. \qquad (5.4)$$

The concept of an infinitesimal change is not soundly based mathematically: we interpret such changes as being very very small (non-zero) increments in the specified variable.

We can see from the defining eqn (5.4), and from Figure 5.1, that $f'(x)$ is the slope of the tangent to the curve $y = f(x)$ at the point P. We can also see that dy represents the change in the dependent variable y that results from a change, Δx, in x, as we move along the tangent to the curve at point P. It is important to stress that, although dy is *not* the same as Δy, for small enough changes in x it is reasonable to assume that the two are equivalent. Consequently, the difference between Δy and dy is simply the error in approximating Δy to dy. However, the same is not true of the differential dx, because at all times, $\Delta x = dx$.

5.1.1.1 The Origins of the Infinitesimal

The concept of the infinitesimal first arose in 1630 in Fermat's 'Method of Finding Maxima & Minima'. This work marks the beginning of differential calculus. The ideas introduced by Fermat lead to speculation about how we can evaluate *'just' before* or *'just' after*. In the 17th century, the infinitesimal was known as the 'disappearing' and tangents as 'touchings'. Leibniz thought infinitesimals *"well-founded fictions"*, whereas the philosopher Berkeley attacked differentials as *"neither finite quantities, not quantities infinitely small, not yet nothing. May we not call them the ghosts of departed quantities?"*. Newton described them as *"quantities which*

(a)

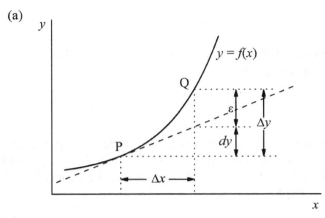

(b)

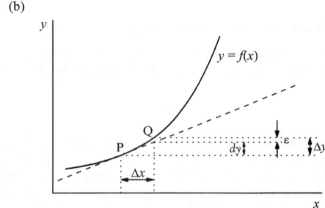

Figure 5.1 (a) The differential *dy*, for a change Δx in *x*, for the function $y = f(x)$. The actual change in *y* is given by $\Delta y = dy + \varepsilon$, where ε is the difference between Δy and *dy*. (b) As $\Delta x \rightarrow 0$, the error ε gets proportionately smaller and Δy becomes increasingly well approximated by *dy*.

diminish not to the point where they have disappeared, nor to the point before, but to the point where they are disappearing". Today, Borowski and Borwein in their Dictionary of Mathematics[1] regard an infinitesimal as *"a paradoxical conception ... largely abandoned in favour of the EPSILON-DELTA treatment of limits, ... but made their reappearance in the formulation of hyper-real numbers"*. Whether or not you believe in their existence, they are clearly capable of producing extraordinary results!

5.1.2 Differentials in Action

The use of the differential is important in the physical sciences because fundamental theorems are sometimes expressed in differential form. In chemistry, the laws of thermodynamics are nearly always expressed in terms of differentials. For example, it is common to work with the following formula as a means of expressing how the molar specific heat capacity at constant pressure, C_P, of a substance varies, with temperature, T:

$$C_P = g(T), \text{ where } g(T) = \alpha + \beta T + \gamma T^2. \tag{5.5}$$

The parameters in an expression such as eqn (5.5) allow the expression to be tailored to fit experiment to some reasonable accuracy.

The optimum values of the parameters α, β and γ are found by fitting measured values of C_P over a range of temperatures to eqn (5.5). Thus, if we know the value of C_P at one temperature we can evaluate it at another temperature, and thereby determine the effect of that incremental (or decremental) change in temperature, ΔT, upon C_P, given by ΔC_P. Alternatively, we can use the properties of differentials given in eqn (5.4) to evaluate the differential of C_P, dC_P, in terms of the differential dT as:

$$dC_P = g'(T)dT = (\beta + 2\gamma T) \times dT \tag{5.6}$$

For small enough changes in T, it is reasonable to make the approximation that the differential dC_P is equivalent to the actual change ΔC_P, and we can use the expression above as a simple one-step route to evaluating the effect of small changes in T upon C_P.

Worked Problem 5.1

Q. (a) Find dy and Δy for the function $y = f(x)$, where $f(x) = x^3$, given that $x = 4$, and $\Delta x = -0.1$.
  (b) Give the actual and approximate values of y at the point $x = 3.9$.
  (c) Calculate the percentage error in your approximate value from (b).

A. (a) $f'(x) = 3x^2 \Rightarrow f'(4) = 48$. It follows that $dy = f'(4)\Delta x = 48 \times -0.1 = -4.8$. The actual change in y is given by $\Delta y = f(3.9) - f(4) = -4.681$;
  (b) The actual and approximate values of y at $x = 3.9$ are 59.319 and 59.2, respectively.
  (c) The percentage error is given by $\dfrac{59.319 - 59.2}{59.319} \times 100 = 0.201\%$.

Sometimes, Δy will be smaller than dy, as in Worked Problem 5.1, but sometimes it can be larger: examples include functions whose slope decreases with increasing values of the independent variable, such as $y = f(x) = \ln x$ and $y = \sqrt[n]{x}$ where $n > 1$.

Problem 5.1

For the function $y = x^{1/3}$, find the values of the differential, dy, and the actual change, Δy, when the value of x is increased:

(a) from 27 to 30, and
(b) from 27 to 27.1.

Give the percentage error in each case in approximating Δy by dy.

Problem 5.2[2]

The variation of the molar heat capacity at constant pressure for CH_4 (g) is described by eqn (5.5), with $\alpha = 14.143$ J $K^{-1}mol^{-1}$, $\beta = 75.495 \times 10^{-3}$ J K^{-2} mol^{-1}, $\gamma = -179.64 \times 10^{-7}$ J $K^{-3}mol^{-1}$.

(a) Use eqn (5.5) to calculate the value of C_P at $T = 500$ K and at $T = 650$ K.
(b) Use eqn (5.6) to evaluate dC_P for an incremental change in T, dT, of 150 K at $T = 500$ K. Hence, estimate the value of C_P at $T = 650$ K.
(c) Compare the value for C_P obtained in (b) with the value calculated directly from eqn (5.5).

5.2 The Differential of a Function of Two or More Variables

We have seen in eqn (5.4) that the differentials dy and dx are related through the derivative $dy = f'(x)dx$ which we can rewrite as:

$$dy = \frac{dy}{dx} dx. \tag{5.7}$$

We can now extend this principle to define differentials for functions of two or more variables. If $z = f(x, t)$ is a general function of two independent variables x and t, then there are two contributions to the differential dz: one from the change in x and the other from the change in t:

$$dz = \frac{\partial z}{\partial x} dx + \frac{\partial z}{\partial t} dt \qquad (5.8)$$

This result extends readily to functions of n independent variables $x_1, x_2, x_3, \ldots, x_n$. Thus, if $z = f(x_1, x_2\ x_3, \ldots x_n)$, the differential of z is built up from contributions associated with each independent variable, as a straightforward generalisation of the result for two independent variables:

$$dz = \frac{\partial z}{\partial x_1} dx_1 + \frac{\partial z}{\partial x_2} dx_2 + \cdots + \frac{\partial z}{\partial x_n} dx_n = \sum_{i=1}^{n} \frac{\partial z}{\partial x_i} dx_i. \qquad (5.9)$$

Examples of functions of two or more variables expressed in differential form are common in thermodynamics: for example, the equation:

$$dG = dH - TdS$$

relates the consequence of very small changes in the enthalpy, H, and entropy, S, on the Gibbs energy, G (here G is the dependent variable, and H and S are the independent variables). As we shall see below, the use of differentials helps us to study such effects, if the changes are small. However, for large changes in the defining variables, we have to evaluate the overall change in the property with the aid of *integral calculus*, which we meet in Chapters 6 and 7.

Worked Problem 5.2

Q. Given the function $z = x^2 y + y^2 x - 2x + 3$, express dz in terms of dx and dy.

A. $dz = \frac{\partial z}{\partial x} dx + \frac{\partial z}{\partial y} dy = (2xy + y^2 - 2) dx + (x^2 + 2xy) dy.$

Problem 5.3

If $z = xy / w$, express dz in terms of the differentials of the three independent variables.

Problem 5.4

(a) For a non-reacting system, the internal energy, $U = f(V, T)$ is a function of both V and T. By analogy with eqn (5.8), write down an expression for the differential dU in terms of the differentials dV and dT.

(b) In thermodynamics, the expression derived in part (a) is commonly written as:[3]

$$dU = \pi_T dV + C_V dT,$$

(c) where π_T and C_V are the internal pressure and specific heat capacity at constant volume.

   (i) Use your answer to part (a) to find expressions for π_T and C_V.

   (ii) Assuming that $\Delta U \approx dU$, calculate the change in U that results when a sample of ammonia is heated from 300 K to 302 K and compressed through 100 cm^3, given that $C_V = 27.32$ J K^{-1} and $\pi_T = 840$ J m^{-3} at 300 K. Comment on the relative magnitudes of the two contributions to dU.

5.3 The Propagation of Errors

In many chemical situations we deduce a value for a property of interest by placing experimentally measured values in the right-hand side of an appropriate formula. For example, if we use the ideal gas equation:

$$P = n\frac{RT}{V} \tag{5.10}$$

to calculate the pressure, P, from a knowledge of volume, temperature, amount of substance, and the gas constant, R, we might wish to know how the errors in the measured property values (n, T and V) propagate through to errors in the calculation of the pressure, P. If, for simplicity, we assume that n and R are fixed (given) constants, how can we estimate the error, dP, in P that results from errors, dT and dV, in the measurement of T and V, respectively? The answer lies in adapting eqn (5.8) to obtain dP in terms of dV and dT:

$$dP = \frac{\partial p}{\partial T}dT + \frac{\partial p}{\partial V}dV \tag{5.11}$$

If dV and dT are the estimated errors in the measured values of V and T, then we need to know the two partial derivatives, so that we can estimate the error dP in P. However, in this and other instances, the differentials themselves do not provide realistic measure of the errors. For example, an absolute error of 10 cm in a measured length is insignificant if we are talking about the shortest distance from Berlin to Moscow, but highly significant if a furniture van driver has enough clearance to pass under a low bridge in a country lane. For this reason, the relative error, or the closely related percentage error, give much more useful measures of error than absolute errors. Thus, in the context of the ideal gas example, the two kinds of error are defined as follows:

- The relative error in P is given by $\dfrac{dP}{P}$.
- The percentage error in P is given by $\dfrac{dP}{P} \times 100$.

Propagation of errors will be explored in more detail in Chapter 13. In particular, we will see that the approach described above, in which the relationship involves more than one independent variable, actually provides an upper limit to the absolute and relative errors. This method is really best used where the errors in the different independent variables are correlated: i.e. all contribute in the same direction in a positive or negative sense.

Worked Problem 5.3

Q. For a right-angled triangle with adjacent sides a, b and hypotenuse c, we have the relation:

$$c = \left(a^2 + b^2\right)^{\frac{1}{2}}.$$

Find the relative and percentage errors in c when $a = 3$ cm, $b = 4$ cm, $da = 0.1$ cm and $db = 0.1$ cm.

A. Using the chain rule, with the substitution $u = a^2 + b^2$, we initially define the partial derivatives of u with respect to a and b, respectively:

$$\frac{\partial u}{\partial a} = 2a; \ \frac{\partial u}{\partial b} = 2b.$$

Differentiating c with respect to the *single* variable, u gives:

$$\frac{dc}{du} = \frac{1}{2}u^{-1/2}.$$

Finally, we use the chain rule to obtain the partial derivatives of c with respect to a and b:

$$\frac{\partial c}{\partial a} = \frac{\partial u}{\partial a} \times \frac{dc}{du} = 2a \times \frac{1}{2}u^{-1/2} = 2a \times \frac{1}{2}\left(a^2+b^2\right)^{-1/2} = a\left(a^2+b^2\right)^{-1/2}.$$

$$\frac{\partial c}{\partial b} = \frac{\partial u}{\partial b} \times \frac{dc}{du} = 2b \times \frac{1}{2}u^{-1/2} = 2b \times \frac{1}{2}\left(a^2+b^2\right)^{-1/2} = b\left(a^2+b^2\right)^{-1/2}.$$

The differential dc is then given by:

$$dc = \frac{\partial c}{\partial a}da + \frac{\partial c}{\partial b}db = a\left(a^2+b^2\right)^{-1/2}da + b\left(a^2+b^2\right)^{-1/2}db$$

and so:

$$dc = 3(9+16)^{-1/2} \times 0.1 + 4(9+16)^{-1/2} \times 0.1 = 0.06 + 0.08 = 0.14\,\text{cm}.$$

Thus the relative error $\dfrac{dc}{c} = \dfrac{0.14}{5} = 0.028$ and the percentage error $\dfrac{dc}{c} \times 100 = 2.8\%$.

> Note that we have taken the product of the partial derivatives, $\frac{\partial u}{\partial a}$ and $\frac{\partial u}{\partial b}$ with the derivative $\frac{dc}{du}$. This is perfectly legitimate because $\frac{dc}{du} \equiv \frac{\partial c}{\partial u}$ in the context of the original expression involving two independent variables.

Problem 5.5

The volume, V, of an orthorhombic unit cell with edges of length a, b and c, and all internal angles between vertices of 90°, is given by $V = abc$.

(a) Find the approximate change in volume, dV, when a, b and c, change by da, db and dc, respectively.
(b) Give an expression for the percentage error in V in terms of the percentage errors in a, b and c.

Problem 5.6

Calcium carbonate crystallises in several different forms. In aragonite,[4] there are four formula units in an orthorhombic primitive unit cell with dimensions $a = 4.94 \times 10^{-10}$ m, $b = 7.94 \times 10^{-10}$, and $c = 5.72 \times 10^{-10}$ m.

(a) Calculate the mass, M, of a unit cell in kg using molar atomic masses as follows: Ca, 40.08 g mol^{-1}; C, 12.01 g mol^{-1}; O, 16.00 g mol^{-1}; ($N_A = 6.022 \times 10^{23}$ mol^{-1}).

(b) Calculate the volume, V, of the unit cell, using the values of a, b and c above, and hence determine the density, ρ, of aragonite, using the formula $\rho = M / V$.

(c) Since the values of the unit cell parameters have been given to two decimal places, the error in their values is $\pm 0.005 \times 10^{-10}$ m. Ignoring the effects of the analogous errors associated with the masses of the atoms, give the relative and percentage errors in the volume of the unit cell.

(d) Find the greatest and smallest estimated unit cell volumes, and give the corresponding greatest and smallest estimates of the density (again ignoring errors associated with the relative atomic masses). Using the value of the density calculated in part (b), find the percentage errors and compare your answers to part (c).

Summary of Key Points

Differentials provide a means to quantify the effect of small changes in one or more variables upon a property that depends on those variables. The key points discussed include:

1. An illustration of the use of differentials in the mathematical and chemical context; in particular, many of the fundamental laws of thermodynamics are expressed in terms of differentials.

2. A review of the concept of infinitesimal change and its relevance in chemistry, in view of the links to the concept of reversibility in thermodynamics.

3. The distinction between approximate and exact changes in the dependent variable, resulting from changes in one or more independent variables.

4. The use of differentials in assessing how errors in one or more properties of a system propagate through to errors in a related property.

5. How differentials associated with each variable in a function of two or more variables contribute to the differential associated with the dependent variable.

References

1. E. J. Borowski and J. M. Borwein, *Collins Dictionary of Mathematics* , Harper Collins, New York, 1989, p. 294.
2. The data for Problem 5.2 were taken from: R. A. Alberty and R. J. Silbey, *Physical Chemistry*, Wiley, New York, 1992, p.52.
3. See, for example: P. W. Atkins and J. de Paula, *Physical Chemistry,* 8th edn, Oxford University Press, Oxford, 2006, p.60.
4. See: H. D. Megaw, *Crystal Structures: A Working Approach*, Saunders, Philadelphia, 1973, p. 247.

6
Integration

In the earlier chapters on arithmetic, algebra and functions, we saw examples of actions for which there was another action available to reverse the first action; such a reversing action is called an **inverse**. Some examples of mathematical actions and their inverses are listed in Table 6.1:

Table 6.1 A selection of mathematical actions and their inverses.

Start	→	Action	→	Result	→	Inverse action	→	Result
2		Add 3		5		Subtract 3		2
x^2		Subtract $2x$		x^2-2x		Add $2x$		x^2
$(x-1)$		Multiply by x^3		$(x-1)x^3$		Divide by x^3 †		$(x-1)$
x		Logarithm		$\ln(x)$		Exponential $\exp(\ln(x))$		x
		x^3-x^2+1				Differentiate		
		$3x^2-2x$		Integrate		x^3-x^2+C		

†division by x^3 requires that $x \neq 0$

The final example listed above proposes that the inverse to the operation of differentiation is known as **integration**. The field of mathematics which deals with integration is known as integral calculus and, in common with differential calculus, plays a vital role in underpinning many key areas of chemistry.

A differentiation/integration cycle involving a chosen initial function will lead to the appearance of an unspecified constant, C (as we shall see later on).

Aims:

In this chapter we define and discuss integration from two perspectives: one in which integration acts as the inverse, or reverse, of differentiation and the other in which integration provides a means to finding the area under a curve. By the end of the chapter, you should be able to:

- Understand the concept of integration as the reverse of differentiation.
- Find the indefinite integral of a number of simple functions from first principles.
- Integrate standard functions by rule.

- Understand why the results of integration are not unique, unless constraints are placed on the integrated function.
- Apply the integration by parts and substitution methods to integrate more complicated functions.
- Understand the concept of the definite integral and be able evaluate a wide range of definite integrals using the methods discussed above.

6.1 Reversing the Effects of Differentiation

Integration is used frequently in kinetics, thermodynamics, quantum mechanics and other areas of chemistry, in which we work with models that deal with changing quantities. Thus, if we know the rate of change of a property, y (the dependent variable), with respect to x (the independent variable), in the form of dy/dx, then integral calculus provides us with the tools for obtaining the form of y as a function of x. We see that integration reverses the effects of differentiation.

Consider, for example, a car undergoing a journey with an initial speed, u, and moving with a constant acceleration, a. The distance, s, travelled after time t is given by:

$$s = ut + \frac{1}{2}at^2. \tag{6.1}$$

Differentiating eqn. 6.1 gives the rate of change of distance with time, or speed, v, at time, t:

$$\frac{ds}{dt} = u + at = v. \tag{6.2}$$

However, the reverse process, in going from speed to distance involves integration of the rate equation (eqn 6.2). In chemistry, the concept of rate is central to an understanding of chemical kinetics, in which we have to deal with analogous rate equations which typically involve the rate of change of concentration, rather than the rate of change of distance. For example, in a first order chemical reaction, in which the rate of loss of the reactant is proportional to the concentration of the reactant, the rate equation takes the form:

$$-\frac{d[A]}{dt} = k[A], \tag{6.3}$$

where k, the constant of proportionality, is defined as the rate constant. The concentration of the reactant at a given time is found by integrating the rate eqn (6.3), and the relationship between the differentiated and integrated forms of the rate equation is given schematically by:

$$-\frac{d[A]}{dt} = k[A]$$

differentiate ↑   ↓ *integrate*

$$[A] = [A]_0 e^{-kt}$$

where $[A]_0$ is the initial concentration of reactant A. We will discuss the integration methods required to obtain the solution of this type of problem in some detail when we discuss differential equations in Chapter 7.

6.2 The Definite Integral

6.2.1 Finding the Area Under a Curve - The Origin of Integral Calculus

The concept of integration emerges when we attempt to determine the area bounded by a plot of a function $f(x)$ [where $f(x) > 0$] and the x axis, within an interval $x = a$ to $x = b$ (written alternatively as $[a, b]$). Clearly, if the plot gives a straight line, such as for the functions $y = 4$ or $y = 2x + 3$, as shown in Figure 6.1, then measuring the area is straightforward, as the two areas are rectangular and trapezoidal in shape, respectively. However, for areas bounded by a curve and three straight lines, the problem is more difficult. The three situations are shown in Figure 6.1.

The area of a trapezium is given by half the sum of the parallel sides, multiplied by the distance between them.

The solution to the general problem of determining the area under a curve arises directly from differential calculus, the concept of limits, and the infinitesimal. Seventeenth century mathematicians began to think of the area, not as a whole, but as made up of a series of rectangles, of width Δx, placed side by side, and which together cover the interval $[a, b]$ (see Figure 6.2).

With this construction, there are two ways of estimating the area under the curve: first, the interval $[a, b]$ is divided into n sub-intervals of width $\Delta x = (b - a) / n$. The area of each rectangle is obtained by multiplying its width, Δx, by its height on the left-hand vertical side, as shown in Figure 6.3(a).

In this case, the total area is given by:

$$A_L(n) = f(a)\Delta x + f(a + \Delta x)\Delta x + f(a + 2\Delta x)\Delta x + \cdots$$

$$\cdots + f(a + [n-1]\Delta x)\Delta x = \sum_{k=0}^{n-1} f(a + k\Delta x)\Delta x \tag{6.4}$$

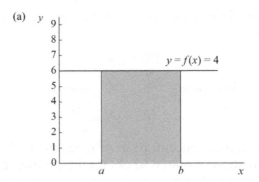

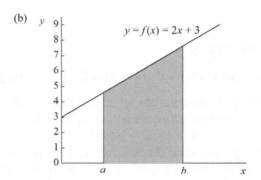

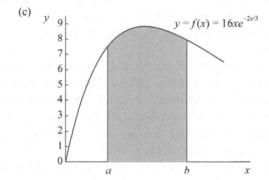

Figure 6.1 Plots of the three functions (a) $y = 4$, (b) $y = 2x + 3$ and (c) $y = 16xe^{-2x/3}$. Evaluating the area bound by the straight line functions and the x-axis in the interval $x = a$ to $x=b$ in (a) and (b) is straightforward but, in (c), where the plot is a curve, we need to make use of the definite integral.

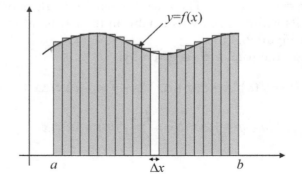

Figure 6.2 Approximating the area under a curve by a contiguous sequence of rectangles of width Δx.

Alternatively, we might have used the height of the right-hand vertical side in computing the total area [Figure 6.3(b)], in which case the total area is given by:

$$A_R(n) = f(a+\Delta x)\Delta x + f(a+2\Delta x)\Delta x + f(a+3\Delta x)\Delta x + \cdots$$

$$\cdots + f(a+[n-1]\Delta x)\Delta x = \sum_{k=1}^{n} f(a+k\Delta x)\Delta x \qquad (6.5)$$

The two estimates we obtain for the area will be different, but if we decrease the sub-interval width, thereby increasing the number, n, of sub-intervals, then in the limit $n \to \infty$, $A_L(n)$ and $A_R(n)$ converge to the same limiting value, A, which is the area under the curve:

$$A = \lim_{n\to\infty} A_L(n) = \lim_{n\to\infty} A_R(n). \qquad (6.6)$$

From the definition of $A_R(n)$, and with an analogous expression involving the limit of $A_L(n)$, we can now write:

$$A = \lim_{n\to\infty} \sum_{k=1}^{n} f(x_k)\Delta x, \qquad (6.7)$$

where $f(x_k) = f(a + k\Delta x)$.

In order to symbolise this sum, Leibniz introduced an elongated S which gives the familiar integral sign $\int$. Thus we can rewrite our equation as:

$$A = \int_a^b f(x)dx = \lim_{n\to\infty} \sum_{r=1}^{n} f(x_r)\Delta x_r \qquad (6.8)$$

where x takes all values between the lower and upper limits a and b, respectively. This integral is known as the **definite integral** because we

If $A_L(n)$ and $A_R(n)$ do not converge to the same value, A, then the integral is said to diverge, *i.e.* it is not defined.

Note that the area, A, under the curve is the sum of an infinite number of rectangles of infinitesimally narrow width, an addition involving not a finite number of finite values, but an infinity of infinitesimal values, yielding a finite value!

(a)

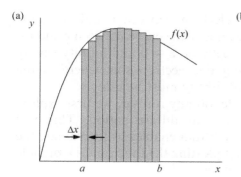

(b)

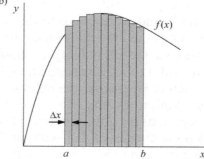

Figure 6.3 Choice of rectangles for estimating the area under the curve: (a) using the left-hand side and (b) using right-hand side.

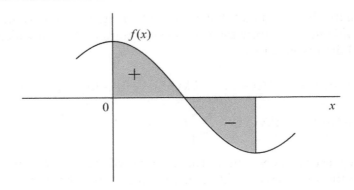

Figure 6.4 In this plot of the function $f(x)=\cos x$, the definite integral has a positive value over the interval $\left[0, \frac{\pi}{2}\right]$; a negative value over the interval $\left[\frac{\pi}{2}, \pi\right]$, and a net zero value over $[0, \pi]$.

have restricted x to the interval $[a, b]$ and, as seen in Figure 6.3, we can use the concept of area under the curve of $y = f(x)$ to give a visualisation of the value of the integral.

6.2.1.1 Negative 'Areas'

Attractive though the concept of area is when $f(x) \geqslant 0$, for x restricted to $[a, b]$, we do need to be careful if $f(x)$ also takes negative values in $[a, b]$. It turns out that, for those regions where the curve lies below the x axis, the contribution from $f(x)$ to the definite integral is negative. If it transpires that $A = 0$, this is perfectly acceptable, as the definite integral has equal positive and negative contributions (see Figure 6.4); likewise, if the curve lies below the x axis, the definite integral will have a negative value.

6.2.2 A Chemical Example: Where is the Electron in the Hydrogen Atom?

Consider the radial probability density function, $D(r)$ for the ground state of the hydrogen atom. This function describes the probability per unit length of finding an electron at a radial distance between r and $r + dr$ (see Figure 6.5).

The radial probability density function is sometimes called the radial distribution function.

The probability of finding the electron between r and $r + dr$ is $D(r)dr$, and corresponds to the area under the curve between r and $r + dr$. Thus the area under the curve between $r = 0$ and infinity simply gives us the probability of finding the electron somewhere in the interval 0 to ∞, which we know intuitively must be unity.

Before we discuss the definite integral any further, we first explore integration as the inverse operation to differentiation. This will prepare us for a most important result that enables us to evaluate the definite integral of $f(x)$, without first plotting the function as a prelude to computing the area under the curve.

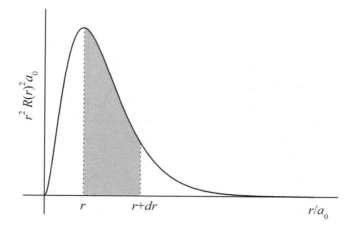

Figure 6.5 A plot of the radial probability density, $D(r) = Nr^2e^{-2r/a_0}$, for the 1s orbital of the hydrogen atom, where a_0 is the Bohr radius (units, m) and N has units m^{-3}.

6.3 The Indefinite Integral

The indefinite integral of a function $y = f(x)$ is usually written as:

$$\int f(x)dx = F(x) + C, \tag{6.9}$$

where:

- $f(x)$ is known as the integrand
- C is an arbitrary constant called the constant of integration
- $F(x) + C$ is known as the indefinite integral.

The new function, $y = F(x) + C$, which we obtain after integration, must be such that its derivative is equal to $f(x)$ to ensure that the definition conforms with the requirement that integration is the reverse (or inverse) of differentiation. Thus, we must have:

$$\frac{d}{dx}(F(x) + C) = F'(x) = f(x). \tag{6.10}$$

The relation between the indefinite integral of $f(x)$ and $f(x)$ itself is shown schematically in Figure 6.6 for the functions $f(x) = 18x^2$ and $F(x) = 6x^3$.

So, to summarise: the indefinite integral is determined by finding a suitable function, $F(x)$, which, on differentiation, yields the function we are trying to integrate, and to which we then add a constant. In common with the strategies described in Chapter 4 for finding the derivative of a given function, an analogous set of strategies can be constructed for finding the indefinite integral of a function. For simple

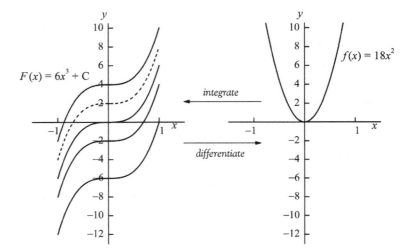

Figure 6.6 Integration of the function $f(x) = 18x^2$ (right-hand side) yields a family of functions given by the indefinite integral $F(x) = 6x^3 + C$ (left-hand side) where C can take any value. Differentiation of $F(x)$ yields the original function, $f(x)$.

functions, a set of standard indefinite integrals can be constructed without too much difficulty, some of which are listed in Table 6.2:

Worked Problem 6.1

Q. (a) Show that:
$$\frac{d}{dx}\ln(1-2x) = -\frac{2}{1-2x}.$$

(b) Deduce that:
$$\int \frac{1}{1-2x}\,dx = -\frac{1}{2}\ln(1-2x) + C.$$

A. (a) Since the first step involves establishing that the derivative of $y = \ln(1-2x)$ is $-\dfrac{2}{1-2x}$, it is simplest to use the chain rule (see Section 4.2.4). If $u = 1 - 2x$, then:
$$\frac{dy}{dx} = \frac{d}{dx}\ln(1-2x) = \frac{dy}{du}\frac{du}{dx} = \frac{1}{u}\cdot -2 = -\frac{2}{1-2x}$$

(b) Reversing the procedure by integration yields the following result, where B is the constant of integration:
$$-\int \frac{2}{1-2x}\,dx = \ln(1-2x) + B$$

$$\Downarrow$$

$$\int \frac{1}{1-2x}\,dx = -\frac{1}{2}\ln(1-2x) + C \quad (\text{where } C = -B/2).$$

Table 6.2   A selection of functions, $f(x)$, and their indefinite integrals, $F(x)+C$.

$f(x)$	$\xrightarrow{\text{integrate}}$	$F(x)+C$
$x^a (a \neq -1)$		$\dfrac{x^{a+1}}{a+1} + C$
$\dfrac{1}{x}$		$\ln x + C$
$\dfrac{1}{x+a}$		$\ln (x+a) + C$
$\cos(ax)$		$\dfrac{1}{a}\sin(ax) + C$
$\sin(ax)$		$-\dfrac{1}{a}\cos(ax) + C$
e^{ax}		$\dfrac{1}{a}e^{ax} + C$
$\sec^2(x)$		$\tan x + C$
$f(x)$	$\xleftarrow{\text{differentiate}}$	$F(x)+C$

Problem 6.1

(a) Evaluate $\dfrac{d}{dx}e^{2x}$ and hence deduce that $\displaystyle\int e^{2x}\,dx = \dfrac{1}{2}e^{2x} + C$.

(b) Show that $\dfrac{d}{dx}\left(\dfrac{1}{1+e^x}\right) = -\dfrac{e^x}{(1+e^x)^2}$ and hence find
$\displaystyle\int \dfrac{e^x}{(1+e^x)^2}\,dx$.

6.4 General Strategies for Solving More Complicated Integrals

Integrals involving complicated forms for $f(x)$ require strategies for reducing the integral to one or more integrals of simpler (standard) form, thus making it possible to find $F(x)$. If all else fails, or we do not have an explicit form for $f(x)$, then numerical integration must be carried out using methods described elsewhere.[1]

Some of the strategies involved in simplifying the form of an integral are quite straightforward; for example:

A chemical example of a function which does not have an explicit form can be found in thermodynamics, where the entropy is determined by integrating C_P/T, which may be known only at selected temperatures.

• If $f(x)$ is in the form of a linear combination of simpler functions, *e.g.*

$$\int (3x^2 + 2x + 1)\,dx \qquad (6.11)$$

then we may be able to rewrite such an integral as a sum of standard integrals that are immediately recognisable:

$$\int (3x^2 + 2x + 1)dx = \int 3x^2 dx + \int 2x dx + \int 1 dx \qquad (6.12)$$

- Integrals can be simplified by placing constant terms outside the integral, *e.g.*

$$\int (3x^2 + 2x + 1)dx = 3\int x^2 dx + 2\int x dx + \int 1 dx \qquad (6.13)$$

Problem 6.2

Integrate the function $y = f(x) = 9x^2 + 2e^{2x} + \dfrac{1}{x}$.

In practice, we may find ourselves faced with more complicated functions, the solutions to which require us to use methods involving adaptation of some of the rules for differentiation. The choice of method more often than not involves some guesswork, but coming up with the correct guesses is all part of the fun! In addition, it may be necessary to use a combination of several methods. In the following two sections, we discuss integration by parts and the substitution method.

6.4.1 Integration by Parts

The integration by parts method is appropriate for integrands of product form. The method starts from the familiar product rule, used in differential calculus [eqn (4.9)]:

$$\frac{d}{dx}(uv) = v\frac{du}{dx} + u\frac{dv}{dx}$$

Integration over x yields:

$$\int \frac{d}{dx}(uv)dx = \int v\frac{du}{dx}dx + \int u\frac{dv}{dx}dx \qquad (6.14)$$

and, on using the properties of differentials, the left side $\int \dfrac{d}{dx}(uv)dx$ becomes $\int d(uv) = uv$. It follows that re-arrangement of the above expression yields:

$$\int u\frac{dv}{dx}dx = uv - \int v\frac{du}{dx}dx. \qquad (6.15)$$

Eqn (6.15) shows that the integral on the left-hand side, which is the one sought, is replaced by two terms, one of which is another integral which we hope is more tractable than the initial integral. The success of the method relies on making the right choices for u and $\frac{dv}{dx}$. The term identified as u is differentiated to form part of the integrand on the right-hand side of eqn (6.15); the other part of the integrand is formed by integrating the term identified as $\frac{dv}{dx}$.

Worked Problem 6.2

Q. Given the integrand $f(x) = x\cos x$, find the indefinite integral.

A. The integrand is the product of x and $\cos x$, and in this case we identify x with u and $\frac{dv}{dx}$ with $\cos x$ in eqn (6.15):

$$u = x \text{ and } \frac{dv}{dx} = \cos x.$$

Thus, $\frac{du}{dx} = 1$ and $v = \sin x$, and so eqn (6.15) becomes:

$$\int x \cos x \, dx = x \sin x - \int \sin x \, dx.$$

The final step simply requires us to evaluate $\int \sin x \, dx$ which we know by reference to Table 6.2 to be $-\cos x + C$. Thus:

$$\int x \cos x \, dx = x \sin x + \cos x + D$$

where $D = -C$. If, on the other hand, we had identified u and $\frac{dv}{dx}$ the other way round, we end up with a more complicated integral to evaluate:

$$\int x \cos x \, dx = x^2 \cos x + \int \frac{x^2}{2} \sin x \, dx$$

Clearly, some practice is required in identifying u and $\frac{dv}{dx}$ for use in eqn (6.15), when it seems that integration by parts is appropriate.

Problem 6.3

Use the method of integration by parts to evaluate $\int xe^{-x}dx$ assuming:

(a) $u = x$ and $\dfrac{dv}{dx} = e^{-x}$;

(b) $u = e^{-x}$ and $\dfrac{dv}{dx} = x$.

Comment on which choice you think is the most appropriate for this integral.

6.4.2 Integration Using the Substitution Method

The second integration technique, known as the substitution method, derives from the inversion of the chain rule for differentiation described in Chapter 4. The objective here, once again, is to transform the integrand into a simpler or, preferably, a standard form. However, similar to the integration by parts method, there is usually a choice of substitutions and, although in some cases, different substitutions yield different answers, these answers must only differ by a constant (remember that, for an indefinite integral, the answer is determined by inclusion of a constant). The substitution method is best illustrated using a worked problem:

Worked Problem 6.3

Q. Evaluate $\int xe^{ax^2}dx$.

A. Here $f(x) = xe^{ax^2}$. Let us try the substitution $u = ax^2$ in order to transform the integral over x to an integral over u. From the properties of differentials we know that:

$$du = \frac{du}{dx}dx = 2axdx$$

This result enables us to express dx in terms of du, according to $dx = \dfrac{1}{2ax}du$, thus transforming the integral into:

$$\int xe^u \frac{du}{2ax} = \frac{1}{2a}\int e^u du = \frac{1}{2a}e^u + C$$

We now express the result in terms of the original variable, x, by substituting back for u:

$$\int xe^{ax^2}\,dx = \frac{1}{2a}e^{ax^2} + C.$$

At this point, it is good practice to check the result by differentiating the function $F(x) = \frac{1}{2a}e^{ax^2}$, to ensure that the original integrand $f(x)$ is regenerated [see eqn (6.10)]:

$$F'(x) = \frac{d}{dx}\left(\frac{1}{2a}e^{ax^2}\right) = \frac{2ax}{2a}e^{ax^2} = xe^{ax^2},$$

as required.

Problem 6.4

Repeat Worked Problem 6.3, using the substitution $u = x^2$.

Worked Problem 6.4

Q. Evaluate $\displaystyle\int \frac{x}{(1-x)^{1/2}}\,dx$.

A. A possible substitution is given by $u = (1 - x)^{1/2}$, from which it follows that:

$$u^2 = 1 - x \Rightarrow x = 1 - u^2$$

Differentiating the last equation with respect to u gives:

$$\frac{dx}{du} = -2u \Rightarrow dx = -2u\,du$$

Hence,

$$\int \frac{x}{(1-x)^{1/2}}\,dx = \int \frac{1-u^2}{u}\cdot -2u\,du$$

$$= -2\int (1-u^2)\,du = -2u + \frac{2}{3}u^3$$

$$\Rightarrow \int \frac{x}{(1-x)^{1/2}}\,dx = -\frac{2}{3}(1-x)^{1/2}(2+x) + C.$$

Problem 6.5

Evaluate the indefinite integral $\int \frac{x}{(1-x)^{1/2}} dx$, using the substitution $u = 1 - x^2$.

Problem 6.6

Find:

(a) $\int x(x^2 + 4)^{1/2}\, dx$, using the substitution $u = x^2 + 4$.

(b) Show that $\int \frac{1}{x \ln x} dx = \ln(\ln x) + C$, using the substitution $u = \ln x$,

6.4.2.1 Use of Trigonometrical Substitutions

The integrand in Problem 6.5 is of a form which suggests that a trigonometrical substitution might be appropriate. Bearing in mind the key identity $\cos^2 u + \sin^2 u = 1$, the appearance of a factor such as $(1-x^2)^{1/2}$ in the integrand suggests the substitutions $x = \cos u$ or $x = \sin u$. Thus, for the substitution $x = \cos u$, the factor $(1 - x^2)^{1/2}$ becomes $(1 - \cos^2 u)^{1/2} = \sin u$.

Problem 6.7

(a) Repeat Problem 6.5, using the trigonometrical substitution $x = \cos u$.
Hint: You will need to remember that $\sin^2 u = 1 - \cos^2 u$ and consequently that $\sin u = (1 - \cos^2 u)^{1/2}$ for the final step of your integration. You should have obtained the same result as your answer to Problem 6.5.

(b) Show that $\int \frac{\cos x}{\sin x} dx = \ln(\sin x) + C$ using the substitution $u = \sin x$.

6.4.2.2 General Comment

The choice of method for evaluating indefinite integrals relies on experience to a large extent. Sometimes, integration by parts and the substitution methods are equally applicable, but in many cases they are not. For example, the integration by parts method is much more suited to finding the integral of the function $f(x) = x \cos x$ described in Worked Problem 6.2, than the substitution method (which would prove frustrating and fruitless in this case). It may also be necessary to use several applications of one or both methods before the answer is accessible. However, whichever method is used, the answer may always be checked by verifying that $F'(x) = f(x)$.

6.5 The Connection Between the Definite and Indefinite Integral

As we saw in Section 6.2.1, the concept of integration emerged from attempts to determine the area bounded by a plot of a function $f(x)$, and the x-axis, within some interval $[a, b]$. This area is given by the definite integral, the definition of which derives from numerical methods involving limits (see Section 6.2.1). Such numerical methods can be tedious to apply in practice (although instructive) but, fortunately, there is a direct link between the indefinite integral, $F(x) + C$, of a function, $f(x)$, and the definite integral, in which x is constrained to the interval $[a, b]$. The relationship between the two forms of integration is provided by the fundamental theorem of calculus:

$$\int_a^b f(x)dx = (F(b) + C) - (F(a) + C)$$

$$= F(b) - F(a)$$

(6.16)

where $F(a)$ is the value of $F(x)$ at $x = a$ and $F(b)$ is the value of $F(x)$ at $x = b$. In other words, the definite integral over the interval $[a, b]$ is obtained by subtracting the indefinite integral at the point $x = a$ from that at $x = b$. Furthermore, we see that the constant of integration, which appears in the indefinite integral, does not appear in the final result [see eqn (6.16)].

Worked Problem 6.5

Q. Evaluate $\int_0^1 \dfrac{x}{1+x}\,dx$.

A. The first step requires us to find the indefinite integral $\int \dfrac{x}{1+x}\,dx$. Using the substitution $u = 1 + x$, the integral becomes:

$$\int \frac{u-1}{u}\,du = \int \left(1 - \frac{1}{u}\right)du = u - \ln u + C = (1+x) - \ln(1+x) + C.$$

Thus identifying $F(x)$ with $(1 + x) - \ln(1 + x)$, the definite integral can be evaluated from:

$$\int_0^1 \frac{x}{1+x}\,dx = F(1) - F(0) = 2 - \ln 2 - 1 - 0 = 1 - \ln 2$$

Problem 6.8

(a) Evaluate (i) $\int_1^2 \dfrac{1}{x^3}\,dx$ and (ii) $\int_0^2 x\left(x^2+4\right)^{1/2}dx$ [see Problem 6.6(a)].

(b) Show that $\int_0^2 \dfrac{x}{(x^2+4)}\,dx = \dfrac{1}{2}\ln 2$, using an appropriate substitution.

Problem 6.9

For the expansion of a perfect gas at constant temperature, the reversible work is given by the expression:

$$W = \int_{V_a}^{V_b} P\,dV$$

where $P = nRT / V$ and V_a and V_b are the initial and final volumes, respectively. Derive an expression for the work done by evaluating the integral between the limits V_a and V_b.

Problem 6.10

Let K be the equilibrium constant for the formation of CO_2 and H_2 from CO and H_2O at a given temperature T. From thermodynamics, we know that:

$$\frac{d}{dT} \ln K = \Delta H^{\ominus}/RT^2 \qquad (6.17)$$

(a) Assuming that $\Delta H^{\ominus}$ is independent of temperature, integrate eqn (6.17) to find how $\ln K$ varies with T.
(b) Given $\Delta H^{\ominus} = 42.3$ kJ mol^{-1}, find the change in $\ln K$ as the temperature is raised from 500 K to 600 K.

Summary of Key Points

This chapter provides an introduction to integral calculus, together with examples set in a chemical context. As we shall see in the following chapter, we also need integral calculus to solve the differential equations which appear in chemical kinetics, quantum mechanics, spectroscopy and other areas of chemistry. The key points discussed in this chapter include:

1. The definition of integration as the inverse of differentiation, yielding the indefinite integral.
2. The definition of integration as a means of evaluating the area bounded by a plot of a function over a given interval, and the x-axis − yielding the definite integral.
3. The use of integration by parts method for integrating products of functions.
4. The use of the substitution method for reducing more complicated functions to a simpler or standard form.
5. The use of trigonometric substitutions in the substitution method.

References

1. See for example: M.J. Englefield, *Mathematical Methods for Engineering and Science Students*, Arnold, London 1987, ch. 15.

7
Differential Equations

In Chapter 2 we explored some of the methods used for finding the roots of algebraic equations in the form $y = f(x)$. In all of the examples given, we were seeking to determine the value of an unknown (typically the value of the independent variable, x) that resulted in a particular value for y, the dependent variable. In general, the methods discussed can be used to solve algebraic equations where the dependent variable takes a value other than zero, because the equation can always be rearranged into a form in which $y = 0$. For example, if we seek the solution to the equation:

$$4 = x^2 - 5$$

then we can rearrange it to:

$$0 = x^2 - 9$$

by subtracting 4 from both sides. The problem now boils down to one in which we search for the two roots of the equation which, in this case, are $x = \pm 3$.

In this chapter, we are concerned with equations containing derivatives of functions. Such equations are termed **differential equations**, and arise in the derivation of model equations describing processes involving rates of change, as in, for example:

- Chemical kinetics (concentrations changing with time).
- Quantum chemical descriptions of bonding (probability density changing with position).
- Vibrational spectroscopy (atomic positional coordinates changing with time).

In these three, as well as in other examples, we are trying to determine how the chosen property (such as concentration, probability density or atomic position) varies with respect to time, position, or some other variable. This is a problem which requires the solution of one or more differential equations in a procedure that is made possible by using the tools of differentiation and integration discussed in Chapters 4 and 6, respectively.

This chapter builds on the content of earlier chapters to develop techniques for solving equations associated with processes involving rates of change. By the end of this chapter you should be able to:

- Identify a differential equation and classify it according to its order.
- Use simple examples to demonstrate the origin and nature of differential equations.
- Identify the key areas of chemistry where differential equations most often appear.
- Use the separation of variables method to find the general solutions to first order differential equations of the form $\frac{dy}{dx} = f(x)g(y)$.
- Use the integrating factor method to find the general solutions to first order differential equations linear in y.
- Find the general solutions to linear second order differential equations with constant coefficients by substitution of trial functions.
- Apply constraints (boundary conditions) to the solution(s) of differential equations.

7.1 Using the Derivative of a Function to Create a Differential Equation

Consider the function:

$$y = Be^{-2x}, \tag{7.1}$$

where B is a constant. The first derivative of this function takes the form:

$$\frac{dy}{dx} = -2Be^{-2x} \tag{7.2}$$

If we now substitute for y, using eqn (7.1), we obtain the **first order** differential equation:

$$\frac{dy}{dx} = -2y, \tag{7.3}$$

A first order differential equation is so-called because the highest order derivative is 1.

which must be solved for y as a function of x. In other words, the solution to this problem will provide us with an equation which shows quantitatively how y varies as a function of x. The solution is, of course, provided by the original eqn (7.1) but the purpose here is to explore the means by which we find that out for ourselves!

If we now differentiate eqn (7.2) with respect to x, and substitute for $\frac{dy}{dx}$ using eqn (7.3), we obtain the second order differential equation (7.4):

$$\frac{d^2y}{dx^2} = 4y \tag{7.4}$$

This differential equation is of **second order**, simply because the *highest* order derivative is two.

Problem 7.1

(a) Express the first and second derivatives of the function, $y = 1 / x$, in the form of differential equations, and give their orders.

(b) Express the second derivative of the function $y = \cos ax$ in the form of a differential equation.

(c) Show that the function $y = Ae^{4x}$ is a solution of the differential equations $\frac{dy}{dx} - 4y = 0$ and $\frac{d^2y}{dx^2} - 5\frac{dy}{dx} + 4y = 0$.

The last part of Problem 7.1 demonstrates that a given function does not necessarily correspond to the solution of only one differential equation. In later sections we shall address the question of how to determine the number of different functions (where each function differs from another by more than simply multiplication by a constant) that are solutions of a given differential equation.

If the function $y = Ae^{4x}$ is a solution to Problem 7.1(c), then so is ky, where k is a constant.

7.2 Some Examples of Differential Equations Arising in Classical and Chemical Contexts

One of the principal motivations for the development of calculus by Newton and Leibniz in the 18th Century came from the need to solve physical problems. Examples of such problems include:

- The description of a body falling under the influence of the force of gravity:

$$\frac{d^2h}{dt^2} = -g; \qquad (7.5)$$

- The motion of a pendulum, which is an example of simple harmonic motion, described by the equation:

$$\frac{d^2x}{dt^2} = -\omega^2 x. \qquad (7.6)$$

If we extend this last example to the modelling of molecular vibrations, we need to include additional terms in the differential equation to account for non-harmonic (anharmonic) forces.

In these last two example equations of motion, the objective is to determine functions of the form $h = f(t)$ or $x = g(t)$, respectively, which satisfy the appropriate differential equation. For example, the solution of the classical harmonic motion equation is an oscillatory function, $x = g(t)$, where $g(t) = \cos \omega t$, and ω defines the frequency of oscillation. This function is represented schematically in Figure 7.1 (See also Worked Problem 4.4).

In chemistry, we are mostly concerned with changing quantities. For example:

- In kinetics, the concentration of a species A may change with time in a manner described by the solution of the differential equation:

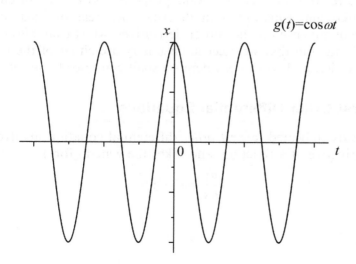

Figure 7.1 A plot of the function, $g(t) = \cos \omega t$, describing simple harmonic motion.

$$-\frac{d[\mathrm{A}]}{dt} = k[\mathrm{A}] \qquad (7.7)$$

- In quantum mechanics, the value of a wave function, ψ, changes with position. For a single particle system, ψ, is obtained as the solution of the Schrödinger equation:

$$-\frac{\hbar^2}{2m}\frac{d^2\psi}{dx^2} + V(x)\psi = E\psi, \qquad (7.8)$$

where the **Hamiltonian** operator, $\hat{H}$, given by $-\dfrac{\hbar^2}{2m}\dfrac{d^2}{dx^2} + V(x)$, is associated with the total energy, E, and $V(x)$ is the potential energy of the particle, m is the mass of the particle and $\hbar$ is the Planck constant divided by 2π.

- In spectroscopy, the response of a molecule to an oscillating electromagnetic field leads to absorption of energy, the details of which are revealed after solving an equation of the form:

i is the imaginary number $\sqrt{-1}$
(see Chapter 9)

$$i\hbar\frac{d\psi}{dt} = \left\{\hat{H} + \hat{H}'(t)\right\}\psi \qquad (7.9)$$

- In vibrational spectroscopy, where the treatment of molecular vibrations is based on the differential equation for an harmonic oscillator:

$$\frac{-\hbar^2}{2m}\frac{d^2\psi}{dx^2} + \frac{1}{2}kx^2\psi = E\psi. \qquad (7.10)$$

In all of the examples given above, we are faced with having to deal with the relationship between some property and its rate of change. The differential equations that describe such relationships contain first, second or even higher order derivatives. Most examples of this type of equation that we meet in chemistry are either of the first or second order, and so this is where we shall concentrate our efforts.

7.3 First Order Differential Equations

As already indicated, a *first order* differential equation involves the *first* derivative of a function, and takes the general form:

$$\frac{dy}{dx} = F(x,y) \qquad (7.11)$$

where y is a function of x, and $F(x, y)$ is, in general, a function of *both* x and y. The method used to solve eqn (7.11) depends upon the form of $F(x, y)$.

7.3.1 F(x, y) is Independent of y

In this simplest example, where $F(x, y) = f(x)$, the general solution is found by a simple one-step integration:

$$\frac{dy}{dx} = f(x) \quad \Rightarrow \quad y = \int f(x)dx = F(x) + C, \qquad (7.12)$$

where $F(x) + C$ is the indefinite integral and C is the constant of integration (see Chapter 6), which can, in principle, take any value. It is important to note that the solution to a *first order* equation involves:

*   *one* step of integration
*   *one* constant of integration.

The solution of an nth-order differential equation involves n steps of integration and yields n constants of integration.

Worked Problem 7.1

Q. Solve $\dfrac{dy}{dx} = x^2 + 1$.

A. Simple integration yields the general solution:

$$y = \frac{x^3}{3} + x + C,$$

which can be described in terms of a family of cubic functions, each with a different value of C (see Figure 7.2). In this example

$$F(x) = \frac{x^3}{3} + x.$$

7.3.2 Boundary Conditions

In the case of a first order differential equation, the constant of integration is usually determined by a **boundary condition**, or constraint on the solution. For example, if y is known at $x = 0$,

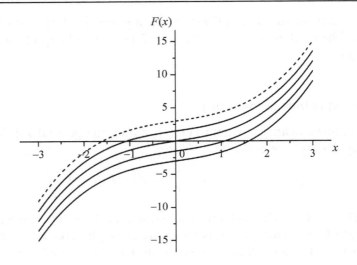

Figure 7.2 The family of solutions, $y = \dfrac{x^3}{3} + x + C$ (for $C = \ldots, 3, 1.5, 0, -1.5, -3, \ldots$) to the differential equation $\dfrac{dy}{dx} = x^2 + 1$. Note that $F(x) = \dfrac{x^3}{3} + x$. The dashed line is the solution satisfying the boundary condition $y=0, x=3$.

then this boundary condition is sufficient to determine the constant of integration, C. Thus, out of the family of possible solutions, only one solution is acceptable and this is the one satisfying the boundary condition.

For example, if the boundary condition for the solution of the differential equation in Worked Problem 7.1, is such that $y = 3$ at $x = 0$, then the solution is constrained to take the form:

$$y = F(x) + 3 = \frac{x^3}{3} + x + 3$$

since $F(0) = 0 \Rightarrow C = 3$ (see the dashed-line solution in Figure 7.2).

It should be noted that, in most chemical situations, we rarely need the general solution of a differential equation associated with a particular property, because one (or more) boundary conditions will almost invariably be defined by the problem at hand and must be obeyed; for example:

- In a first order reaction, concentration of the reacting species is specified at one particular time (usually at the start of the reaction).
- The value of the radial part of an atomic orbital wavefunction must tend to zero at very large distances from the nucleus.

As the number of boundary conditions is usually the same as the order of the differential equation for a particular chemical problem, there will be no undetermined constants of integration associated with the solution.

7.3.3 F(x, y) is in the Form f (x)g(y)

7.3.3.1 Separation of Variables Method

Suppose we are required to solve a differential equation of the form eqn (7.11), in which $\dfrac{dy}{dx}$ is equal to the product of two functions, each of which depends only on one of the variables:

$$\frac{dy}{dx} = f(x)g(y). \tag{7.13}$$

This equation is solved by first rewriting it in the form:

$$\frac{1}{g(y)} \cdot \frac{dy}{dx} = f(x), \tag{7.14}$$

which, on integration with respect to x, yields:

$$\int \frac{1}{g(y)} \cdot \frac{dy}{dx} dx = \int f(x) dx. \tag{7.15}$$

Since the differentials dx and dy are linked by the expression $dy = \dfrac{dy}{dx} dx$, the integration over x in the integral on the left-hand side of the equation can be transformed into an integration over y:

$$\int \frac{1}{g(y)} dy = \int f(x) dx \tag{7.16}$$

It now just remains to carry out separate integrations over y and x in order to obtain the required solution in the form of an expression of the general form

$$G(y) + A = F(x) + B. \tag{7.17}$$

Notice, that, although each integral yields *one constant of integration*, the two constants of integration can be combined into a *single constant, C,* by taking A over to the right-hand side of eqn (7.17).

The procedure just described is known as the separation of variables method. In some instances, it is possible to re-write eqn (7.17) in the form $y = P(x)$, to give an explicit relation between y and x [where B is contained within $P(x)$]. In other cases, the solution may have to be left in a form of an implicit relation between y and x (see Section 2.3.5).

Worked Problem 7.2

Q. Find the solution of the differential equation:

$$\frac{dy}{dx} = 3x^2 y.$$

A. This equation may be solved by the separation of variables method, as the right-hand side is in the form $f(x)g(y)$, where $f(x) = 3x^2$ and $g(y) = y$. Thus:

$$\int \frac{1}{y} dx = \int 3x^2 dx \Rightarrow \ln y = x^3 + C.$$

This expression gives y as an implicit function of x. However, from the properties of the exponential function (the inverse function of the logarithm function), we can specify y as an explicit function of x:

$$y = e^{(x^3 + C)} = e^C \cdot e^{x^3} \Rightarrow y = A e^{x^3},$$

where the constant e^C has been rewritten as A (which implies $C = \ln A$), to simplify the appearance of the solution.

Problem 7.2

Solve the differential equation:

$$\frac{dy}{dx} = -6y^2,$$

subject to the initial condition $y = 1$, $x = 0$.

> ## Problem 7.3
>
> Solve the differential equation $\dfrac{dy}{dx} = -\lambda y$, given that $y = N$ at $x = 0$. Give both implicit and explicit solutions.
>
> **Hint:** You may find it helpful to remember that $\ln A - \ln B = \ln \dfrac{A}{B}$ [see eqn (2.15)] and that $e^{\ln A} = A$ [see eqn (2.10)].

7.3.4 Separable First Order Differential Equations in Chemical Kinetics

Consider a first order rate process, with rate constant k:

$$\text{A} \xrightarrow{\ k\ } \text{B}. \qquad (7.18)$$

The rate of loss of the reactant A is proportional to its concentration, and is expressed in the form of the differential equation:

$$-\frac{d[\text{A}]}{dt} = k[\text{A}], \qquad (7.19)$$

where [A] is the concentration of the reactant at time t. Notice here that [A] is the dependent variable, and t the independent variable.

We are interested in solving eqn (7.19) to obtain an expression which describes how the concentration of A varies with time, subject to the boundary condition that the concentration of the reactant at time $t = 0$ is $[\text{A}]_0$ (note that the differential rate law above tells us only how the rate depends on [A]). Thus, using the separation of variables method, eqn (7.19) is first rearranged to:

$$-\frac{d[\text{A}]}{[\text{A}]} = k.dt, \qquad (7.20)$$

and then integrated, recognising that k is a constant:

$$-\int \frac{d[\text{A}]}{[\text{A}]} = \int k.dt = k \int dt. \qquad (7.21)$$

$$\Downarrow$$

$$-\ln[\text{A}] = kt + C. \qquad (7.22)$$

If we now impose the boundary condition above, we find that $C = -\ln [A]_0$, and the integrated rate eqn (7.22) becomes:

$$- \ln [A] = kt - \ln [A]_0, \qquad (7.23)$$

which may be expressed in the alternative forms:

$$\ln [A] = -kt + \ln [A]_0, \qquad (7.23a)$$

or

$$\ln \left(\frac{[A]}{[A]_0} \right) = -kt. \qquad (7.23b)$$

Note that in eqn (7.23a), [A] is an implicit function of t; furthermore, there is a linear relation between ln [A] and t. Thus, a plot of ln [A] against t will give a straight line of slope $-k$ and intercept ln [A]. Alternately, we can rearrange eqn (7.23b) by taking the exponential of each side, to generate an explicit function which shows the exponential decay of [A] as a function of time (see Figure 7.3 and Chapter 2):

$$\frac{[A]}{[A]_0} = e^{-kt} \qquad (7.24)$$

which rearranges to:

$$[A] = [A]_0 e^{-kt}. \qquad (7.25)$$

Figure 7.3 demonstrates clearly how the value of k determines the rate of loss of A.

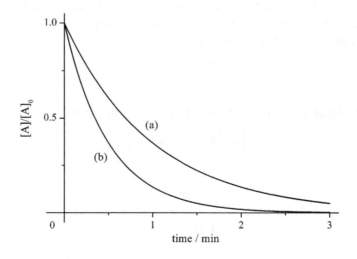

Figure 7.3   A plot of $\frac{[A]}{[A]_0}$ against t / min for (a) $k=1$ min^{-1} and (b) $k=2$ min^{-1}.

An important feature of such first order reactions is the half-life, $t_{1/2}$, which is the time taken for [A] to reduce to half of its initial value. Thus, for $t = t_{1/2}$, we have:

$$\frac{[A]_0}{2} = [A]_0 e^{-kt_{1/2}} \qquad (7.26)$$

which simplifies to:

$$\frac{1}{2} = e^{-kt_{1/2}}. \qquad (7.27)$$

Taking natural logarithms, we have:

$$\ln\frac{1}{2} = -kt_{1/2}. \qquad (7.28)$$

Remember that
$\ln\left(\frac{a}{b}\right) = \ln(a) - \ln(b)$.

Using the property of logarithms that $\ln\frac{1}{a} = -\ln a$, we can rewrite (7.28) as:

$$\ln 2 = kt_{1/2} \qquad (7.29)$$

and it follows that the half-life is expressed in terms of the rate constant, k, according to:

$$t_{1/2} = \frac{\ln 2}{k}. \qquad (7.30)$$

7.3.5 First Order Differential Equations Linear in *y*

A first-order linear differential equation has the general form:

$$\frac{dy}{dx} + yP(x) = Q(x), \qquad (7.31)$$

in which the dependent variable (here y) appears on the left-hand side with index unity. Equations of this form cannot be solved using the separation of variables method unless $Q(x) = 0$.

The general solution of a first-order linear differential equation, in the form of eqn (7.31), is:

$$y = \frac{1}{R(x)} \int R(x)Q(x)dx \qquad (7.32)$$

where $R(x)$, known as the **integrating factor**, is defined in terms of $P(x)$ as follows:

$$R(x) = e^{\int P(x)dx}. \qquad (7.33)$$

There are thus two integrations to perform, one to determine the integrating factor, and the other, which involves the product $R(x)Q(x)$ as the integrand. Since we are dealing with a first order differential equation, we expect only *one* constant of integration but, from the above discussion, it appears that *two* such constants may arise. We now describe why there is, in fact, only one undetermined constant of integration.

7.3.5.1 The Constant of Integration

In determining the integrating factor, the complete expression becomes:

$$R(x) = e^{g(x)+C} = A e^{g(x)}, \tag{7.34}$$

where $g(x) + C$ is the indefinite integral of the function $P(x)$ and $A = e^C$. Thus, if we now substitute eqn (7.34) into eqn (7.32), we obtain:

$$y = \frac{1}{A e^{g(x)}} \cdot A \int e^{g(x)} Q(x) dx. \tag{7.35}$$

Since the constant A appears in both the numerator and denominator in the right side of eqn (7.35), it can be cancelled to yield the general solution of eqn (7.31):

$$y = e^{-g(x)} \int e^{g(x)} Q(x) dx. \tag{7.36}$$

Only one constant of integration will be produced from the indefinite integral $\int e^{g(x)} Q(x) dx$ and, since the constant arising from the determination of $R(x)$ can be discarded, we see that a single constant of integration arises from the solution to a first order differential equation, as expected.

Worked Problem 7.3

Q. Solve the differential equation:

$$\frac{dy}{dx} + 2y = e^x$$

A. This is a first order linear differential equation, in which $P(x) = 2$ and $Q(x) = e^x$. The integrating factor $R(x) = e^{\int P(x)dx} = e^{\int 2dx} = e^{2x+C}$. Here $g(x) = 2x$, and so the general solution given by eqn (7.36) is:

$$y = e^{-2x} \int e^{2x} e^x dx \quad = \quad e^{-2x} \int e^{3x} dx \quad = e^{-2x} \left(\frac{e^{3x}}{3} + C \right)$$

$$\Rightarrow \quad y = \frac{1}{3} e^x + C e^{-2x}.$$

Problem 7.4

(a) Check that the solution to Worked Problem 7.3 satisfies the original differential equation.

(b) Solve the differential equation $\dfrac{dy}{dx} + \dfrac{y}{x} = x^2$, subject to the boundary condition $y = 0$, $x = 1$.

7.3.6 First Order Differential Equations in Radioactive Decay Processes

Consider the following radioactive β^- decay processes, involving two sequential first order steps, in which λ_1 and λ_2 are decay constants (analogous to rate constants in a chemical kinetic process) associated with the emission of energetic electrons:

$$^{239}_{92}\text{U} \xrightarrow{\lambda_1} {}^{239}_{93}\text{Np} \xrightarrow{\lambda_2} {}^{239}_{94}\text{Pu}. \tag{7.37}$$

The amounts of $^{239}_{92}\text{U}$, $^{239}_{93}\text{Np}$, and $^{239}_{94}\text{Pu}$ (units mol) at any given time are denoted by N_1, N_2, N_3, respectively, and we specify that, initially, $N_1 = 1$ mol. The change in the amount of $^{239}_{93}\text{Np}$ with time has two contributions: one from the decay of $^{239}_{92}\text{U}$ and the other from the decay of $^{239}_{93}\text{Np}$ itself. Thus, on the basis that these processes are first order in nature, the net rate of increase of $^{239}_{93}\text{Np}$ is given by:

$$\frac{dN_2}{dt} = \lambda_1 N_1 - \lambda_2 N_2 \tag{7.38}$$

By analogy with the first order chemical reaction [eqn (7.25)] we know that $N_1 = (N_1)_0 e^{-\lambda_1 t}$, where $(N_1)_0$ is the initial amount of $^{239}_{92}\text{U}$. Thus

^{239}U is produced when ^{238}U, the most common isotope of uranium, absorbs a neutron. The subsequent sequential decay process produces ^{239}Pu, a fissionable material that can be used as a fuel in nuclear reactors or as the core material of a nuclear bomb. 1 mol of $^{239}_{92}\text{U}$ equates to 239 g by mass, which is approximately one twentieth of the amount needed to make a nuclear device.

$$\frac{dN_2}{dt} = \lambda_1(N_1)_0 e^{-\lambda_1 t} - \lambda_2 N_2 \quad \Rightarrow \quad \frac{dN_2}{dt} + \lambda_2 N_2 = \lambda_1(N_1)_0 e^{-\lambda_1 t}. \quad (7.39)$$

If we identify t with x and N_2 with y, then we can see that eqn (7.39) is of the form of eqn (7.31), where $P(x) \equiv \lambda_2$ and $Q(x) \equiv \lambda_1(N_1)_0 e^{-\lambda_1 t}$. After determining the integrating factor, the solution is obtained using eqn (7.36). The derivation of the solution forms the basis of the next Problem.

Problem 7.5

For the radioactive two-step decay process described above:

(a) Show that the integrating factor takes the form $R(t) = Ae^{g(t)}$ where $g(t) = \lambda_2 t$.

(b) Use eqn (7.36) to show that the general solution is given by:

$$N_2 = \frac{\lambda_1(N_1)_0}{\lambda_2 - \lambda_1} e^{-\lambda_1 t} + Ce^{-\lambda_2 t}.$$

(c) Given the initial condition, $N_2 = 0$ at $t = 0$, determine an expression for C, and show that:

$$N_2 = \frac{\lambda_1(N_1)_0}{\lambda_2 - \lambda_1} \left(e^{-\lambda_1 t} - e^{-\lambda_2 t} \right). \quad (7.40)$$

(d) Given that the rate of loss of $^{239}_{92}U$, by first order decay, is expressed in the form of the differential equation $\frac{dN_1}{dt} = -\lambda_1 N_1$, deduce the solution of this differential equation from the solution to Problem 7.3, by appropriate changes of names of the independent and dependent variables.

(e) Write down the expressions for N_1 and N_2 determined in parts (d) and (c) above, and hence, from the conservation of matter (*i.e.* that at all times $N_1 + N_2 + N_3 = 1$ mol), deduce the expression for N_3 in terms of time, t.

(f) Given that the half-lives for $^{239}_{92}U$, $^{239}_{93}Np$, are 23.5 min and 2.3 days, respectively, use eqn (7.30), with the rate constant k replaced by the appropriate decay constants, to calculate the values of the decay constants λ_1 and λ_2.

(g) Use the expression for N_2 derived in part (c) to determine the time at which N_2 reaches its maximum value, and give the corresponding amount of N_2 reached at this time.

In Problem 7.5, the large disparity in the decay constants leads to a situation in which the number of $^{239}_{93}\text{Np}$ species builds up rapidly to its maximum value, and then decreases slowly (see Figure 7.4).

7.3.7 First Order Differential Equations in Chemical Kinetics Processes

Consider the following kinetic process, involving two sequential first order steps:

$$A \xrightarrow{k_1} B \xrightarrow{k_2} C,$$

This is the same model process that we described above for radioactive decay of $^{239}_{92}\text{U}$ and, if we substitute decay constants by rate constants, and amount of substance by concentration, and assume that $[A]_0 = 1 \text{ mol dm}^{-3}$, we can adapt eqn (7.40) derived in Problem 7.5(c) to describe how $[B]$ varies with time:

$$[B] = \frac{k_1}{k_2 - k_1} \left(e^{-k_1 t} - e^{-k_2 t} \right) [A]_0 \qquad (7.41)$$

If we consider initially the limiting case, in which the rate constant k_2 (governing the second step) is very much smaller than k_1 (*e.g.* $k_1 = 2 \text{ s}^{-1}$, $k_2 = 0.01 \text{ s}^{-1}$), we obtain a plot of $[B]$ over the time interval 0 to 30 s shown in Figure 7.5.

We can see in Figure 7.5 that, at the start of the reaction, the concentration of the intermediate, B, initially rises quite rapidly to a maximum, and thereafter declines slowly. The level to which the

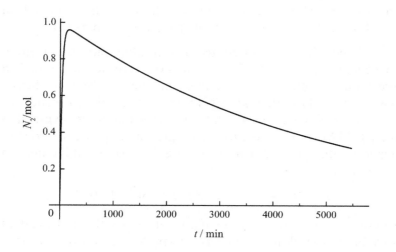

Figure 7.4 A plot of the variation in the amount of $^{239}_{93}\text{Np}$, given by N_2, as a function of time in the radioactive decay of $^{239}_{92}\text{U}$.

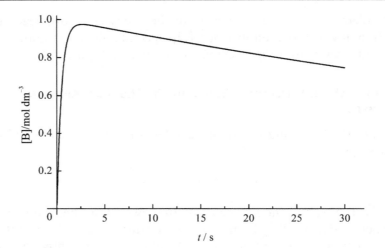

Figure 7.5 A plot of the variation in the concentration of the reaction intermediate, [B], with time in a two sequential first order reaction mechanism.

concentration of B builds up will depend on the rate constants governing the two steps. For reactions where the rate constant k_2 is very much larger than for k_1, the concentration of B never has the opportunity to build up and remains essentially constant throughout most of the reaction. Under such conditions it is said to have reached a steady state.

Problem 7.6

(a) Use eqn (7.41) to plot [B] / mol dm^{-3} against t / s (see hint below), using first $k_1 = 2$ s^{-1}, $k_2 = 3$ s^{-1} and second $k_1 = 2$ s^{-1}, $k_2 = 10$ s^{-1}.

(b) Comment on the maximum values of [B] obtained in Figure 7.5, and in your plots generated in part (a).

Hint: If possible, you should use a computing plotting program to generate your plots. If you do not have access to such a program, a plotter is available on the Internet at http://www.ucl.ac.uk/mathematics/geomath/plot.html. In order to use such a plotter, you will need to enter the model expression for [B] / [A]$_0$ into a function descriptor box. Thus, in order to generate the plot shown in Figure 7.5, we would enter an expression such as:

$$(2/(0.01-2))*(\exp(-2*x)-\exp(-0.01*x)),$$

where x is the appropriate name given to the independent variable in the plotter, and * is interpreted as 'multiplication'.

7.4 Second Order Differential Equations

In Section 7.1, we saw how differential equations of second order can be generated from a particular function. Thus, for example, if we differentiate the function $y=e^{2x}$ twice, we obtain:

$$\frac{dy}{dx} = 2y \qquad (7.42)$$

from the first differentiation and then:

$$\frac{d^2y}{dx^2} = 4y, \qquad (7.43)$$

from the second. Eqn (7.43) is an example of a second order differential equation. If we now reverse this process, we need two acts of integration to yield an expression for y. However, as each integration step leads to a constant of integration, the resulting expression for y, which now contains two undetermined constants, represents the general solution of the second order differential eqn (7.43).

In general, a second order differential equation will take the form:

$$\frac{d^2y}{dx^2} = G\left(\frac{dy}{dx}, y, x\right) \qquad (7.44)$$

where $G\left(\frac{dy}{dx}, y, x\right)$ can, in principle, represent any function of $\frac{dy}{dx}$, y and x. In the example above, eqn (7.43) is not, in fact, a function of $\frac{dy}{dx}$ or x, and so we can write instead that eqn (7.43) has the form

$$\frac{d^2y}{dx^2} = G(y).$$

Most of the problems involving second order differential equations which we encounter in chemistry, are second order linear differential equations which take the general form:

$$\frac{d^2y}{dx^2} + P(x)\frac{dy}{dx} + S(x)y = Q(x) \qquad (7.45)$$

Unfortunately, unlike the general linear *first order* differential equation (7.31), there is no simple template which provides the solution, and we need therefore to apply different methods to suit the equation we meet in the chemical context. Equations of the general form given in eqn (7.45) crop up in all branches of the physical sciences in which a system is under the influence of an oscillatory or periodic change. In chemistry, some of the most important examples can be found in modelling:

- Vibrational motions of molecules.
- The interaction of molecules with electromagnetic radiation (light).
- The radial motion of the electron in a hydrogen-like species.

As a first example, we will consider the differential equation describing the dynamics of simple harmonic motion, and demonstrate how the general solution is found.

7.4.1 Simple Harmonic Motion

The special case of eqn (7.45) with $P(x) = Q(x) = 0$ and $S(x)$ equal to a positive constant, n^2 (the choice of n^2 as the constant ensures that it is positive quantity for any real value for n), gives rise to eqn (7.46) for simple harmonic motion, the solution of which can be used to model nuclear motion in molecules:

$$\frac{d^2y}{dx^2} = -n^2y \qquad (7.46)$$

Thus, for example, if we apply eqn (7.46) to describe the periodic vibrational motion in a diatomic molecule, x represents time, and positive and negative values for y correspond to bond extension and compression, respectively. Finally, we can see that eqn (7.46) is an eigenvalue equation in which y is the eigenfunction and $-n^2$ the eigenvalue (see Section 4.3.1).

Worked Problem 7.4

Q. The form of eqn (7.46) implies that y is such that its second derivative is a negative multiple of itself (since n^2 is positive real number).

(a) Which of the following functions are solutions of eqn (7.46)?

(i) $y = \sin nx$; (ii) $y = \cos nx$; (iii) $y = e^{nx}$; (iv) $y = e^{-nx}$.

(b) Explain why none of these functions corresponds to the general solution of eqn (7.46).

(c) Give the form of the general solution to eqn (7.46).

(d) Find the solution to eqn (7.46), given the boundary condition $y = 0$ when $x = 0$.

(e) If we additionally impose the boundary condition $y = 0$ when $x = $ L, show that the solution becomes $y = A \sin \dfrac{n\pi x}{L}$, where A is a constant.

A. (a) (i) $\dfrac{dy}{dx} = n \cos nx \quad \Rightarrow \quad \dfrac{d^2y}{dx^2} = -n^2 \sin nx = -n^2 y;$

(ii) $\dfrac{dy}{dx} = -n \sin nx \quad \Rightarrow \quad \dfrac{d^2y}{dx^2} = -n^2 \cos nx = -n^2 y;$

(iii) $\dfrac{dy}{dx} = n e^{nx} \quad \Rightarrow \quad \dfrac{d^2y}{dx^2} = n^2 e^{nx} = n^2 y;$

(iv) $\dfrac{dy}{dx} = -n e^{-nx} \quad \Rightarrow \quad \dfrac{d^2y}{dx^2} = n^2 e^{-nx} = n^2 y.$

(b) Functions (i) and (ii) are both solutions of the differential eqn (7.46), but neither of them forms the general solution, which must contain two, as yet undetermined, constants of integration. Neither of functions (iii) nor (iv) are solutions because they lead to the constant n^2 rather than $-n^2$.

(c) We can obtain the general solution from the sum of constant multiples of the two solutions obtained in (a) (i) and (ii) above to give $y = A \cos nx + B \sin nx$. We can cross-check our answer by substituting the function and its second derivative into eqn (7.46):

$$\dfrac{dy}{dx} = -nA \sin nx + nB \cos nx$$

$$\Rightarrow \quad \dfrac{d^2y}{dx^2} = -n^2(A \cos nx + B \sin nx) = -n^2 y$$

(d) On substituting the boundary condition values $y = 0$, $x = 0$ into the general solution, we find that $0 = A$, and hence the solution becomes $y = B \sin nx$.

(e) We can further refine our solution by applying the second of our two boundary conditions. If $y = 0$ when $x = L$, then:

$$B \sin nL = 0$$

which is satisfied when $nL = m\pi$ (where m is an integer). Thus, $n = \dfrac{m\pi}{L}$ and the required solution is:

Neither of equations (iii) and (iv) are solutions to eqn (7.46). However, if n was such that n^2 was negative, then both functions would be solutions to the equation. This would require us to define the square root of a negative number, which is at odds with our understanding of what constitutes a real number. In Chapter 9, we extend the concept of the number to include so called imaginary and complex numbers which embrace the idea that the square root of a negative number can be defined.

$$y = B \sin\left(\frac{m\pi}{L} x\right).$$

Note: This same solution with $m = 1$ arose in the context of the particle in the box problem in Problem 4.11.

Problem 7.7

(a) Use the results from (a) of Worked Problem 7.4 to express the general solution of the differential equation $\frac{d^2 y}{dx^2} = n^2 y$ in terms of exponential functions.

(b) Use the definitions for the hyperbolic functions, cosh x and sinh x, defined in Section 2.3.4, to show that the general solution obtained in (a) can be rewritten as:

$$y = (A + B)\cosh nx + (A - B)\sinh nx.$$

(c) Give the form of this solution when constrained by the boundary condition $y = 0$ when $x = 0$.

7.4.2 Second Order Differential Equations with Constant Coefficients

Linear second order differential equations of the general form given in eqn (7.45) are quite tricky to solve but, fortunately, we are usually interested in situations where $P(x)$ and $S(x)$ are both constant functions, say c_1 and c_2:

$$\frac{d^2 y}{dx^2} + c_1 \frac{dy}{dx} + c_2 y = Q(x) \tag{7.47}$$

A number of possible variants of this equation can result from the different choices for c_1 and c_2:

(a) $c_1 = 0$ and $c_2 = 0$:

This simple case results in the differential equation:

$$\frac{d^2 y}{dx^2} = Q(x) \tag{7.48}$$

which simply requires two steps of integration to yield an expression for y.

Worked Problem 7.5

Q. Solve the second order differential equation $\dfrac{d^2y}{dx^2} = 2x$.

A. The equation is solved by integrating once to give:

$$\frac{dy}{dx} = x^2 + C,$$

and a second time to give:

$$y = \frac{x^3}{3} + Cx + D.$$

(b) $c_1 = 0$, $Q(x) = 0$ (the null function), c_2 positive.

The differential equation now adopts the form of an eigenvalue problem [see eqn (7.46)]:

$$\frac{d^2y}{dx^2} = -c_2y, \tag{7.49}$$

which is solved using the procedure described in Worked Problem 7.4. For $c_2 < 0$, the solution is obtained using the same procedure.

(c) c_1 and c_2 are both positive and $Q(x) = 0$.

Eqn (7.47) now becomes a homogeneous linear second order differential equation having the form:

$$\frac{d^2y}{dx^2} + c_1\frac{dy}{dx} = -c_2y. \tag{7.50}$$

Eqn (7.50) is also an example of an eigenvalue problem (see Section 4.3.1), of the type commonly encountered in chemistry when modelling electronic and nuclear motions.

7.4.3 How is an Eigenvalue Problem Recognised?

The simplest way of thinking about an eigenvalue problem is to consider the result of some operator, $\hat{A}$, acting on a suitable function, $f(x)$, to yield a constant, λ, multiplied by the original function, $f(x)$:

The choice of $f(x)$ to label a function is entirely arbitrary – in principle, any label will do. We could just as easily have labelled it $\psi(x)$, $G(x)$, ϕ, γ or y. Later on we do use the label y to represent the eigenfunction of a given operator, but y is also routinely used to name the value of the dependent variable for some given function, and as a result it can be confusing sometimes to distinguish between the label applied to the function itself and the label applied to the value of the function for a particular choice of independent variable.

$$\hat{A}f(x) = \lambda f(x). \tag{7.51}$$

The objective in this type of problem is to find the eigenfunctions, $f(x)$, and associated eigenvalues, λ, for a given operator, $\hat{A}$. The solution will generally yield a number of different eigenfunctions, and associated eigenvalues, all of which emerge from a single general solution.

The procedure used to solve second order differential equations of the form of eqn (7.50) is essentially the same as that described in Worked Problem 7.4 and involves the construction of trial eigenfunctions from some of the functions introduced in Chapter 2.

7.4.3.1 The Search for Eigenfunctions

We first re-write eqn (7.49), using differentiation operators $\hat{D}^2 = \dfrac{d^2}{dx^2}$ and $\hat{D} = \dfrac{d}{dx}$ to give:

$$\hat{D}^2 y + c_1 \hat{D} y = -c_2 y. \tag{7.52}$$

Factorising yields:

$$(\hat{D}^2 + c_1 \hat{D}) y = -c_2 y, \tag{7.53}$$

which is of the form of an eigenvalue equation:

$$\hat{A} y = -c_2 y,$$

where:

$$\hat{A} = \hat{D}^2 + c_1 \hat{D} \tag{7.54}$$

is the operator; y is the eigenfunction, and $-c_2$ the eigenvalue. The form of the operator $\hat{A}$ is such that its eigenfunctions must be functions whose first and second derivatives differ only by a constant.

Worked Problem 7.6

Q. Which of the following functions:

$$y = \cos nx, \; y = \sin nx, \; y = e^{nx}, \; y = e^{-nx},$$

are eigenfunctions of:

$$\text{(a) } \hat{D}, \text{ (b) } \hat{D}^2, \text{ (c) } \hat{D}^2 + \hat{D}?$$

For each eigenfunction, give the associated eigenvalue.

A. (a) $\hat{D} \cos nx = -n \sin nx$; $\hat{D} \sin nx = n \cos nx$; $\hat{D} e^{nx} = n e^{nx}$, $\hat{D} e^{-nx} = n e^{-nx}$. Only the two exponential functions are eigenfunctions of $\hat{D}$, with associated eigenvalues n and $-n$, respectively.

(b) $\hat{D}^2 \cos nx = -n^2 \cos nx$; $\hat{D}^2 \sin nx = -n^2 \sin nx$; $\hat{D}^2 e^{nx} = n^2 e^{nx}$, $\hat{D}^2 e^{-nx} = n^2 e^{-nx}$. All four functions are eigenfunctions of $\hat{D}^2$, with respective eigenvalues of $-n^2$, $-n^2$, n^2 and n^2.

(c) $(\hat{D}^2 + \hat{D}) \cos nx = -n^2 \cos nx - n \sin nx$;
$(\hat{D}^2 + \hat{D}) \sin nx = -n^2 \sin nx - n \cos nx$;
$(\hat{D}^2 + \hat{D}) e^{nx} = n^2 e^{nx} + n e^{nx} = (n^2 + n) e^{nx}$;
$(\hat{D}^2 + \hat{D}) e^{-nx} = n^2 e^{-nx} - n e^{-nx} = (n^2 - n) e^{-nx}$. Only the two exponential functions are eigenfunctions of both operators, with eigenfunctions $n^2 + n$ and $n^2 - n$, respectively.

The key feature emerging from Worked Problem 7.6 is that if two functions, f_1 and f_2, are eigenfunctions of an operator $\hat{A}$, and have the same eigenvalues, λ, then an arbitrary linear combination of the two functions is also an eigenfunction of $\hat{A}$ with eigenvalue, λ. In this example, the two functions $y = \cos nx$ and $y = \sin nx$ are both eigenfunctions of the $\hat{D}^2$ operator with eigenvalue $-n^2$. Consequently, it follows that an arbitrary linear combination of the two functions is also an eigenfunction of this operator with eigenvalue $-n^2$. This concept is expressed formally in the following expression:

$$\hat{A}(bf_1 + cf_2) = b\lambda f_1 + c\lambda f_2 = \lambda(bf_1 + cf_2) \tag{7.55}$$

Problem 7.8

Show that the function $y = A\cos nx + B\sin nx$ is an eigenfunction of the $\hat{D}^2$ operator with eigenvalue $-n^2$.

Worked Problem 7.7

Q. The differential equation $\dfrac{d^2y}{dx^2} - 3\dfrac{dy}{dx} + 2y = 0$ has the same form as eqn (7.50) and, as such, can be written as an eigenvalue equation, $\hat{A}y = \lambda y$.

(a) Give the form of the operator $\hat{A}$, expressed in terms of the $\hat{D}^2$ and $\hat{D}$ operators and the value of λ, the eigenvalue.

(b) Find the eigenfunctions of $\hat{A}$, and hence deduce the general solution.

(c) Give the form of the solution for the boundary conditions $y = 0$ when $x = 0$ and $y = e$ when $x = 1$.

A. (a) The operator $\hat{A} = \hat{D}^2 - 3\hat{D}$ and has the same form as that given in eqn (7.54) with $c_1 = -3$. The eigenvalue $\lambda = -2$.

(b) We have seen already in Worked Problem 7.6 that e^{nx} is an eigenfunction of the operators $\hat{D}^2$ and $\hat{D}$. It follows that e^{nx} is also an eigenfunction of the operator $\hat{D}^2 + c_1\hat{D}$ and hence of $\hat{A} = \hat{D}^2 - 3\hat{D}$. We can now tailor this function by finding the appropriate values of n that are consistent with this operator, and an eigenvalue $\lambda = -2$. Thus, since:

$$\hat{D}^2 e^{nx} = n^2 e^{nx} \text{ and } -3\hat{D}e^{nx} = -3ne^{nx},$$

it follows that:

$$\hat{A}e^{nx} = (n^2 - 3n)e^{nx} = -2e^{nx},$$

and, on cancelling the e^{nx} terms, we obtain $(n^2 - 3n) = -2$, which may be written as:

$$n^2 - 3n + 2 = 0. \tag{7.56}$$

The two roots of the quadratic eqn (7.56) are $n = 2$ and $n = 1$, and it follows that e^{2x} and e^x are two solutions of the differential equation which we may use to build the general solution in the form, $y = Be^{2x} + Ce^x$.

(c) If we now apply the two boundary conditions, $y = 0$, $x = 0$ and $y = e$, $x = 1$, we obtain the two equations, $B + C = 0$, and $Be^2 + Ce = e$. Solving for A and B yields:

$$C = \frac{1}{1-e}, \text{ and } B = -\frac{1}{1-e}$$

and the final solution then takes the form:

$$y = -\frac{1}{e-1}e^{2x} + \frac{1}{e-1}e^x.$$

$$\Rightarrow \quad y = \frac{1}{e-1}\left(e^x - e^{2x}\right).$$

Problem 7.9

Solve the differential equation $\dfrac{d^2y}{dx^2} - 5\dfrac{dy}{dx} + 6y = 0$, subject to the boundary conditions $y = 0$ when $x = 0$ and $dy / dx = 1$ when $x = 0$.

Summary of Key Points

This chapter has been concerned with bringing differential and integral calculus together, in order to solve a number of differential equations that are used widely in chemistry. The key points discussed in this chapter include:

1. The order of a differential equation.
2. Creating a differential equation using the first or second derivative of a function.
3. Examples of differential equations in a chemical context.
4. The solution to first order differential equations of the form $\dfrac{dy}{dx} = f(x)g(y)$ using the separation of variables method.
5. The application of boundary conditions to determine the value of constant(s) of integration.

6. Finding general solutions to linear first order differential equations using the integrating factor method.
7. Finding general solutions to linear second order differential equations by substitution of trial functions.
8. A revision of the eigenvalue problem and the construction of trial functions to provide a solution of second order differential equations with constant coefficients.

8
Power Series

In Chapter 2, we saw that a **polynomial function**, $P(x)$, of degree n in the independent variable, x, is a finite sum of the form:

$$P(x) = c_0 + c_1 x + c_2 x^2 + \ldots + c_n x^n. \tag{8.1}$$

Such polynomial functions have as a **domain** the set of all real (finite) numbers; in other words they yield a finite result for any real value of the independent variable x. For example, the polynomial function:

$$P(x) = 1 + x + x^2 + x^3 \tag{8.2}$$

will have a finite value for any real number x because each term in the polynomial will also have a finite value.

An **infinite series** is very similar in form to a polynomial, except that it does not terminate at a particular power of x and, as a result, is an example of a **power series**:

$$p(x) = c_0 + c_1 x + c_2 x^2 + \ldots + c_n x^n + \ldots \tag{8.3}$$

One important consequence of this lack of termination is that we need to specify a domain which includes only those real numbers, x, for which $p(x)$ is finite. For example, the power series:

$$p(x) = 1 + x + x^2 + x^3 + \ldots \tag{8.4}$$

does not yield a finite result for $x > 1$, or for $x < -1$, because in the former case, the summation of terms increases without limit and, in the latter, it oscillates between increasingly large positive and negative numbers as more and more terms are included. Try this for yourself by substituting the numbers $x = 2$ and $x = -2$ into eqn (8.4) and then $x = 0.5$ and $x = -0.5$ and observe what happens to the sum as more and more terms are included.

Power series are useful in chemistry (as well as in physics, engineering and mathematics) for a number of reasons.

- Firstly, they provide a means to formulate alternative representations of transcendental functions such as the exponential, logarithm and trigonometric functions introduced in Chapter 2.
- Secondly, as a direct result of the above, they also allow us to investigate how an equation describing some physical property behaves for small (or large) values for one of the independent variables.

For example, the radial part of the 3s atomic orbital for hydrogen has the same form as the expression:

$$R(x) = N(2x^2 - 18x + 27)e^{-x/3}. \tag{8.5}$$

If we replace the exponential part of the function, $e^{-x/3}$, with the first two terms of its power series expansion $(1 - x/3)$, we obtain a polynomial approximation to the radial function given by:

$$R(x) = N(2x^2 - 18x + 27)(1 - x/3)$$
$$\approx N(8x^2 - 2x^3/3 - 27x + 27). \tag{8.6}$$

We can see how well eqn (8.6) approximates eqn (8.5), by comparing plots of the two functions in the range $0 \leqslant x \leqslant 20$, as shown in Figure 8.1. In this example, the polynomial approximation to the form of the radial wave function gives an excellent fit for small values of x (*i.e.* close to the nucleus), but it fails to reproduce even one radial node [a value of x for which $R(x) = 0$].

- Thirdly, power series are used when we do not know the formula of association between one property and another. It is usual in such situations to use a power series to describe the formula of

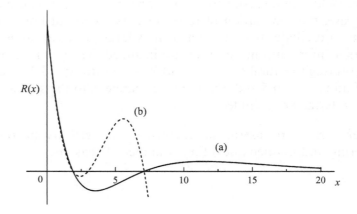

Figure 8.1 A comparison of (a) the 3s radial function, $R(x)$, of the hydrogen atom with (b) an approximation obtained by substituting a two term expansion of the exponential part of the function.

association, and to fit the series to known experimental data by varying the appropriate coefficients in an iterative manner. For example, the parameters, α, β and γ, in the polynomial expression:

$$C_P = g(T), \text{ where } g(T) = \alpha + \beta T + \gamma T^2,$$

describing the temperature dependence of the specific heat capacity of a substance at constant pressure, C_P, may be found by fitting measured values of C_P over a range of temperatures to this equation.

Much of this chapter is concerned with a discussion of power series but, before we go into detail, we consider the general concepts of sequences and series, both finite and infinite.

Aims:

By the end of this chapter, you should be comfortable with the idea that functions can be represented in series form and be able to:

- Understand the distinction between Maclaurin and Taylor series expansions and appreciate when one or the other is the more appropriate.
- Understand what factors influence the accuracy of a given power series expansion;
- Determine the values of x for which the power series is useful (the interval of convergence).
- Understand why the interval of convergence may differ from the domain of the original function.
- Manipulate power series to obtain series for new functions.
- Apply some of the ideas explored in this chapter to probe the limiting behaviour of functions for increasingly large or small values of the independent variable.

8.1 Sequences

A **sequence** is simply a list of terms:

$$u_1, u_2, u_3, \ldots \tag{8.7}$$

each of which is defined by a formula or prescription. The sequence may be finite or infinite, depending on whether it terminates at u_n, or continues indefinitely. Furthermore, the sequence u_1, u_2, u_3, ... represents a function, with a domain specified either as a subset of the positive integers or all positive integers.

8.1.1 Finite Sequences

8.1.1.1 The Geometric Progression

The numbers:

$$1, 2, 4, 8, \ldots, 256$$

form a finite sequence generated by the general term:

$$u_r = 2^r, \text{ where } r = 0, 1, 2, 3, \ldots, 8. \tag{8.8}$$

Here, the formula is only 2 raised to a power, the value of which is defined by each element of the domain. Notice that the use of r as a counting index is arbitrary: any other appropriate letter (with the exception of u which we have used already) would do. A counting index such as r is often termed a dummy index. An alternative way of generating this sequence is accomplished using a recurrence relation as the prescription, where each successive term is obtained from the previous term. For example, the sequence given in eqn (8.8) can alternatively be expressed as:

$$u_r = \begin{cases} 1, r = 0 \\ 2u_{r-1}, r = 1, 2, 3, \ldots, 8 \end{cases} \tag{8.9}$$

which simply means that, starting from 1 as the first term, each successive term is obtained by multiplying the previous term by 2.

The finite sequence in eqn (8.8) is an example of a geometric progression, having the general form:

$$a, ax, ax^2, ax^3, \ldots, ax^8 \tag{8.10}$$

In the case of the geometric progression defined by eqn (8.8), $a = 1$, $x = 2$ and $u_r = ax^r$ for $r = 0, 1, 2, \ldots$

8.1.1.2 Arithmetic Progression

Consider the sequence of odd positive numbers 1, 3, 5, 7, ..., 31, which can be expressed either in terms of the general term:

$$u_r = 2r - 1, r = 1, 2, 3, \ldots, 16 \qquad (8.11)$$

or as a recurrence relation expressed in prescription form:

$$u_r = \begin{cases} 1, r = 1 \\ u_{r-1} + 2, r = 2, 3, \ldots, 8 \end{cases} \qquad (8.12)$$

This finite sequence is an example of an arithmetic progression, because each successive term is given by a sum having the general form:

$$a, a+d, a+2d, a+3d, \ldots, a+nd \qquad (8.13)$$

where, in this example, $a = 1$, $d = 2$.

8.1.2 Sequences of Indefinite Length

In the sequence given in eqn (8.13), the magnitudes of successive terms progressively increase. Some sequences, however, have the property that as the number of terms increases, the values of successive terms appear to be approaching a limiting value. For example, the terms in the harmonic sequence:

$$1, \frac{1}{2}, \frac{1}{3}, \frac{1}{4}, \ldots, \frac{1}{n}, \ldots, \qquad (8.14)$$

where $u_r = 1/r$, $r = 1, 2, 3,\ldots$, decrease in magnitude as r increases, and approach zero as r tends to infinity. Thus, we can define the limit of the sequence as:

$$\lim_{r \to \infty}(u_r) = \lim_{r \to \infty}\frac{1}{r} = 0. \qquad (8.15)$$

If the limit of a sequence is a single finite value, say, m, then:

$$\lim_{r \to \infty}(u_r) = m \qquad (8.16)$$

and the sequence is said to converge to the limit m. However, if this is not the case, then the sequence is said to diverge. Thus, for the arithmetic progression defined in eqn (8.13), the magnitudes of successive values in the sequence increase without limit and the sequence diverges. In contrast, the geometric progression in eqn (8.10) will converge to a limiting value of zero if $-1 < x < 1$.

Problem 8.1

Test each of the following sequences for convergence. Where convergence occurs, give the limiting value.

(a) $u_r = \dfrac{1}{2^r}$, $r = 0, 1, 2, \ldots$;

(b) $u_n = \dfrac{n-1}{2n}$, $n = 1, 2, \ldots$;

(c) $u_r = \cos r\pi$, $r = 0, 1, 2, \ldots$

8.1.3 Functions Revisited

In our discussion of algebraic manipulation in Chapter 1, we used the three-spin model for counting the various permitted orientations of three spin-½ nuclei. If we focus on the number of arrangements where r nuclei are in the spin-up state, then we see that there is only one arrangement where none of the nuclei has spin up; three where one nucleus has spin up; three where two nuclei have spin up; and one where all three nuclei have spin up. Thus, we can define the sequence 1, 3, 3, 1, where the general term is given by $u_r = 3!/(3-r)!r!$, ($r = 0, 1, 2, 3$). In general, the number of ways of selecting r specified objects from n objects is given by the expression $^nC_r = n!/(n-r)!r!$. In this example, there are three nuclei and so $n = 3$, and there are $^3C_r(r = 0, 1, 2, 3)$ ways in which 0, 1, 2 and 3 nuclei have spin up.

Problem 8.2

Find the sequence that represents the number of arrangements of six spin-½ nuclei, with r spin–up arrangements, where r now runs from 0 to 6.

8.2 Finite Series

For any sequence of terms $u_1, u_2, u_3, \ldots$ we can form a **finite series** by summing the terms in the sequence up to and including the nth term:

$$S_n = u_1 + u_2 + u_3 + \cdots + u_n = \sum_{r=1}^{n} u_r \qquad (8.17)$$

We first met the summation notation Σ in Chapter 1. In this example, the counting index, r, takes values from 1 to n.

For example, the sum of the first n terms in the series obtained from the sequence defined by eqn (8.8) is given by:

$$S_n = 1 + 2 + 2^2 + \ldots + 2^{n-1}. \qquad (8.18)$$

Evaluating this sum for $n = 1, 2, 3, 4, 5$ yields the sequence of **partial sums**:

$$S_1 = 1,\ S_2 = 3,\ S_3 = 7,\ S_4 = 15 \text{ and } S_5 = 31. \qquad (8.19)$$

If we now look closely at this new sequence of partial sums, we may be able to deduce that the sum of the first n terms is $S_n = 2^n - 1$. In general, for a **geometric series** obtained by summing the members of the geometric progression, defined by eqn (8.10), the sum of the first n terms is given by:

$$S_n = a + ax + ax^2 + \cdots ax^{n-1}$$
$$= a\left(\frac{1 - x^n}{1 - x}\right) \qquad (8.20)$$

Problem 8.3

For the geometric series obtained by summing the first n terms of the geometric progression in eqn (8.8), use eqn (8.20) and appropriate values of a and x given in eqn (8.10) to confirm that the sum of the first n terms is $2^n - 1$.

8.3 Infinite Series

We can also form an **infinite series** from a sequence by extending the range of the dummy index to an infinite number of terms:

$$S = u_1 + u_2 + u_3 + \cdots = \sum_{r=1}^{\infty} u_r. \qquad (8.21)$$

The summation of a finite series will always yield a finite result, but the summation of an infinite series needs careful examination to confirm that the addition of successive terms leads to a finite result, *i.e.* the series converges. It is important not to confuse the notion of **convergence** as applied to a series with that applied to a sequence. For example, the harmonic *sequence* given by eqn (8.14) converges to the limit zero. However, somewhat surprisingly, the harmonic *series*:

$$S = \sum_{r=1}^{\infty} \frac{1}{r} = 1 + \frac{1}{2} + \frac{1}{3} + \frac{1}{4} + \cdots \qquad (8.22)$$

does not yield a finite sum, S, and consequently does not converge. In other words the sum of the series increases without limit as the number of terms in the series increases, even though the values of successive terms converge to zero. We can see more easily how this is true by breaking down the series into a sum of partial sums:

$$S = 1 + \frac{1}{2} + \left(\frac{1}{3} + \frac{1}{4}\right) + \left(\frac{1}{5} + \frac{1}{6} + \frac{1}{7} + \frac{1}{8}\right) + \cdots \qquad (8.23)$$

Here, each successive sum of terms in parentheses will always be greater than ½; for example,

$$\left(\frac{1}{3} + \frac{1}{4}\right) > \left(\frac{1}{4} + \frac{1}{4}\right) \text{ and } \left(\frac{1}{5} + \frac{1}{6} + \frac{1}{7} + \frac{1}{8}\right) > \left(\frac{1}{8} + \frac{1}{8} + \frac{1}{8} + \frac{1}{8}\right), \quad (8.24)$$

and, because this is an infinite series, it follows that the sum increases without limit

$$S > 1 + \frac{1}{2} + \frac{1}{2} + \frac{1}{2} + \cdots \qquad (8.25)$$

8.3.1 π Revisited – The Rate of Convergence of an Infinite Series

In Chapter 1 we saw that the irrational number, π, can be calculated from the sum of an infinite series. One example given involved the sum of the inverses of the squares of all positive integers:

$$\frac{\pi^2}{6} = \sum_{r=1}^{\infty} \frac{1}{r^2} = 1 + \frac{1}{2^2} + \frac{1}{3^2} + \frac{1}{4^2} + \frac{1}{5^2} + \cdots + \frac{1}{n^2} + \cdots \qquad (8.26)$$

This series converges extremely slowly, requiring well over 600 terms to provide precision to the second decimal place. In order to achieve 100 decimal places for π, we would need more than 10^{50} terms! However, the alternative series:

$$\frac{\pi}{2} = \frac{1}{1} + \frac{1 \cdot 1}{1 \cdot 3} + \frac{1 \cdot 1 \cdot 2}{1 \cdot 3 \cdot 5} + \frac{1 \cdot 1 \cdot 2 \cdot 3}{1 \cdot 3 \cdot 5 \cdot 7} + \cdots \qquad (8.27)$$

converges more rapidly, achieving a precision to the second decimal place in a relatively brisk 10 terms.

8.3.2 Testing a Series for Convergence

The non-convergence of the harmonic series, discussed above, highlights the importance of testing whether a particular series is convergent or divergent. For a series given by:

$$\sum_{r=1}^{\infty} u_r = u_1 + u_2 + u_3 + \cdots \tag{8.28}$$

the first, and necessary, condition needed to ensure convergence is that $\lim_{r \to \infty} u_r = 0$. If this condition is satisfied (as it is in the series above for determining π), we can then proceed to test the series further for convergence. It should be emphasised, however, that satisfying this first condition does not necessarily imply that the series converges (*i.e.* we say that the first condition is not sufficient). For example, as we have seen, the harmonic sequence is an example of one for which u_r tends to zero as $r \to \infty$, but the corresponding harmonic series is not convergent.

8.3.2.1 The Ratio Test

A number of tests are available for confirming the convergence, or otherwise, of a given series. The test for absolute convergence is the simplest, and is carried out using the ratio test.

For successive terms in a series, u_r and u_{r+1}, the series:

- converges if $\lim_{r \to \infty} \left| \dfrac{u_{r+1}}{u_r} \right| < 1;$ $\qquad$ (8.29)

- diverges if $\lim_{r \to \infty} \left| \dfrac{u_{r+1}}{u_r} \right| > 1.$ $\qquad$ (8.30)

If, however, $\lim_{r \to \infty} \left| \dfrac{u_{r+1}}{u_r} \right| = 1$, then the series may either converge or diverge, and further tests are necessary.

8.3.2.2 The Infinite Geometric Series

The form of the geometric series in eqn (8.20) generalises to the form of eqn (8.21) where, now:

$$S_n = a + ax + ax^2 + \cdots = \sum_{r=1}^{\infty} ax^{r-1},$$

and $u_r = ax^{r-1}$. On using the ratio test in eqn (8.29), we find that:

$$\lim_{r \to \infty} \left| \frac{ax^r}{ax^{r-1}} \right| < 1.$$

but only when $|x| < 1$. This constraint on the permitted values of x, for which the series converges, defines the **interval of convergence**.

In general, it may not be possible to specify the value of the sum, S, in terms of x: instead, we chose a value of x, and compute the sum to a given number of significant figures or decimal places.

Worked Problem 8.1

Q. (a) Give the forms of u_{r+1} and u_r for the geometric series $1+x+x^2+x^3+\ldots$

(b) Use the ratio test to establish that the series converges and find the interval of convergence.

(c) Given $x = 0.27$ calculate the sum of the series to two decimal places.

A. (a) Since the first term in the series is a constant, we define the rth term as $u_r = x^{r-1}$ and the $(r+1)^{\text{th}}$ term as $u_{r+1} = x^r$.

(b) The ratio test yields:

$$\lim_{r \to \infty} \left| \frac{u_{r+1}}{u_r} \right| = \lim_{r \to \infty} \left| \frac{x^r}{x^{r-1}} \right| = |x|,$$

and so the series converges if $|x| < 1$ and diverges if $|x| > 1$. If $x = \pm 1$, then, as we saw earlier, further tests are required to establish whether the series converges or diverges at these end points. However, in this case, we can see by inspection that for $x = +1$ the sum of the first r terms will be r and thus increases without limit as $r \to \infty$. For $x = -1$, the sum oscillates between 0 and 1, depending on whether r is even or odd. In both cases, a finite sum is not obtained as $r \to \infty$, and we can say that the series fails to converge for $x = \pm 1$. Consequently, the series converges if x takes the values $-1 < x < 1$; an inequality that defines the interval of convergence.

(c) Let S_n designate the sum of the first n terms of the geometric series. Table 8.1 summarises the values of S_n, and the incremental changes for $n = 1, 2, \ldots, 8$, using

$x = 0.27$. We can see from the Table that in order to specify the sum to a given number of decimal places, we have to compute its value to one more decimal place than required, in case rounding up is necessary. In this case, convergence to two decimal places yields a sum of 1.370 at $n = 6$.

Table 8.1 Numerical summation of the geometric series $\sum_{r=1}^{n} x^{r-1}$.

n	1	2	3	4	5	6	7	8
S_n	1.0	1.27	1.3429	1.3626	1.3679	1.3693	1.3697	1.3698
ΔS_n	-	0.27	0.0729	0.0197	0.0053	0.0014	0.0004	0.0001

Problem 8.4

For each of the following infinite series, use the ratio test to establish the interval of convergence.

(a) $S = 1 + 2x + 3x^2 + 4x^3 + \ldots$

(b) $S = 1 - x + \dfrac{x^2}{2!} - \dfrac{x^3}{3!} + \dfrac{x^4}{4!} - \cdots + (-1)^{r-1} \dfrac{x^{r-1}}{(r-1)!} + \cdots$

(c) $S = 1 + \dfrac{x^2}{2} - \dfrac{x^4}{4} + \cdots + (-1)^{r-1} \dfrac{x^{2r-2}}{2r-2}$.

Hint: For part (a), you will need to find the general term first.

8.4 Power Series as Representations of Functions

We have seen above that, for a geometric progression of the type given in eqn (8.10), the sum of the first n terms is given by eqn (8.20). Furthermore, for $a = 1$, we can see that:

$$\frac{1-x^n}{1-x} = 1 + x + x^2 + \cdots + x^{n-1}. \qquad (8.31)$$

This is an important expression because it allows us to see how a function such as $\dfrac{1-x^n}{1-x}$ can be represented by a polynomial of degree $n - 1$. However, if we now extend the progression indefinitely to form the infinite geometric series $1+x+x^2+\ldots+x^{n-1}+\ldots$, we obtain an expansion of a function $\lim\limits_{n\to\infty}\dfrac{1-x^n}{1-x}$ which converges only for values of x in the range $-1 < x < 1$ (see Worked Problem 8.1). If we now evaluate the limit as $n \to \infty$, for any x in the interval of convergence $-1 < x < 1$, we obtain:

$$\lim_{n\to\infty}\frac{1-x^n}{1-x}=\frac{1-0}{1-x}=\frac{1}{1-x}. \tag{8.32}$$

Note that in the limit $n \to \infty$, the term $x^n \to 0$ because we are restricting the values of x to the interval of convergence $-1 < x < 1$. We now see that the function $f(x)=\dfrac{1}{1-x}$ can be represented by the infinite series expansion $1+x+x^2+\ldots+x^{n-1}+\ldots$, which converges for $-1 < x < 1$. For all other values of x the expansion diverges.

The infinite geometric series is an example of a **power series** because it contains a sum of terms involving a systematic pattern of change in the power of x. In general, the simplest form of a power series is given by:

$$f(x)=c_0+c_1x+c_2x^2+c_3x^3+\cdots+c_nx^n+\ldots \tag{8.33}$$

where c_0, c_1, c_2, ... are coefficients and successive terms involve an increasing power of the independent variable, x. Such series involving simple powers of x are termed **Maclaurin series**. The more general **Taylor series** are similar in form, but involve powers of $(x - a)$:

$$f(x)=c_0+c_1(x-a)+c_2(x-a)^2+c_3(x-a)^3$$
$$+\cdots+c_n(x-a)^n+\ldots \tag{8.34}$$

where a is any number other than 0 (in which case, we revert to a Maclaurin series). The significance of the value of a is that it represents the point about which the function is expanded. Thus the Taylor series are expanded about the point $x = a$, whereas the Maclaurin series are simply expanded about the point $x = 0$. Maclaurin and Taylor series are used most frequently to provide alternative ways of representing many types of function. In addition, such series in truncated polynomial form provide an excellent tool for fitting experimental data when there is no model formula available.

There are two important features associated with the generation of power series representations of functions: first, a value of x lying in the domain of the function must be chosen for the expansion point, a; second, the function must be infinitely differentiable at the chosen point in its domain: in other words, differentiation of the function must never yield a constant function because subsequent derivatives will be zero, and the series will be truncated to a polynomial of finite degree. The question as to whether the power series representation of a function has the same domain as the function itself is discussed in a later subsection. The next subsection is concerned with determining the coefficients, c_i, for the two kinds of power series used to represent some of the functions introduced in Chapter 2.

8.4.1 The Maclaurin Series: Expansion about the Point $x = 0$

Let us start by using eqn (8.33) as a model expression to generate a power series expansion for a function $f(x)$, assuming that the requirements given in the paragraph above are satisfied. In order to obtain the explicit form of the series, we need to find values for the coefficients c_0, c_1, c_2, ... This is achieved in the following series of steps:

The original function, and its first, second and third derivatives are:

$$f(x) = c_0 + c_1 x + c_2 x^2 + c_3 x^3 + c_4 x^4 + \cdots + c_n x^n + \cdots \tag{8.35}$$

$$f^{(1)}(x) = c_1 + 2c_2 x + 3c_3 x^2 + 4c_4 x^3 + \cdots + nc_n x^{n-1} + \cdots \tag{8.36}$$

$$f^{(2)}(x) = 2c_2 + 2\cdot 3c_3 x + 3\cdot 4c_4 x^2 \cdots + n(n-1)c_n x^{n-2} + \cdots \tag{8.37}$$

$$f^{(3)}(x) = 2\cdot 3c_3 + 2\cdot 3\cdot 4c_4 x \cdots + n(n-1)(n-2)c_n x^{n-3} + \cdots \tag{8.38}$$

If we now substitute the expansion point, $x = 0$, into each of the above equations we obtain:

$$f(0) = c_0 \tag{8.39}$$

$$f^{(1)}(0) = c_1 \tag{8.40}$$

$$f^{(2)}(0) = 2c_2 = 2!c_2 \tag{8.41}$$

$$f^{(3)}(0) = 2\cdot 3c_3 = 3!c_3 \tag{8.42}$$

For clarity, we use a superscript containing a counting number in parentheses to denote a particular order of derivative.

and, by inspection, the nth derivative has the form:

$$f^{(n)}(0) = n!c_n. \tag{8.43}$$

If we now substitute the coefficients obtained from each of the expressions (8.39) to (8.43) into (8.35) we obtain the **Maclaurin series** for $f(x)$:

$$f(x) = f(0) + f^{(1)}(0)x + \frac{f^{(2)}(0)}{2!}x^2 +$$

$$\frac{f^{(3)}(0)}{3!}x^3 + \cdots + \frac{f^{(n)}(0)}{n!}x^n + \cdots \qquad (8.44)$$

This series, which is generated by evaluating the function and its derivatives at the point $x=0$, is valid only when the function and its derivatives exist at the point $x=0$ and, furthermore, for functions that are infinitely differentiable.

8.4.1.1 The Maclaurin Series Expansion for e^x

The exponential function $f(x) = e^x$ is unique insofar as the function and all its derivatives are the same. Thus, since $f^{(n)}(x) = e^x$, for all n, we have:

$$f(0) = f^{(1)}(0) = f^{(2)}(0) = f^{(3)}(0) = \cdots f^{(n)}(0) = e^0 = 1 \qquad (8.45)$$

and, using eqn (8.44), we obtain:

$$f(x) = e^x = 1 + x + \frac{x^2}{2!} + \frac{x^3}{3!} + \cdots + \frac{x^{n-1}}{(n-1)!} + \cdots, \qquad (8.46)$$

in which the nth term is given explicitly.

8.4.1.2 Truncating the Exponential Power Series

For any power series expansion, the accuracy of a polynomial truncation depends upon the number of terms included in the expansion. Since it is impractical to include an infinite number of terms (at which point the precision is perfect), a compromise has to be made in choosing a sufficient number of terms to achieve the desired accuracy. However, in truncating a Maclaurin series, the chosen degree of polynomial is always going to best represent the function close to $x = 0$. The further away from $x = 0$, the worse the approximation becomes, and more terms are needed to compensate; a feature which is demonstrated nicely in Figure 8.2 and Table 8.2.

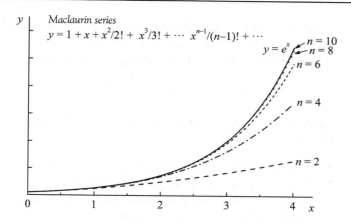

Figure 8.2 A comparison of the accuracy of polynomial approximations to the function $y = e^x$ ($x \geqslant 0$), using polynomials of degrees two to ten.

Table 8.2 The accuracy of first, second and third degree polynomial approximations to the function $f(x) = e^x$.

x	$1+x$	$1+x+\frac{x^2}{2!}$	$1+x+\frac{x^2}{2!}+\frac{x^3}{3!}$	e^x
0	1	1	1	1
0.0001	1.0001	1.000100005	1.000100005	1.000100005
0.001	1.001	1.0010005	1.0010005	1.0010005
0.01	1.01	1.01005	1.01005017	1.01005017
0.1	1.1	1.105	1.105167	1.1051709
0.2	1.2	1.22	1.22133	1.221403
1.0	2	2.5	2.6667	2.718282

8.4.1.3 The Maclaurin Expansions of Trigonometric Functions

The trigonometric functions $\sin x$, $\cos x$, $\tan x$ have derivatives which exist at $x = 0$, and so can be represented by Maclaurin series.

Worked Problem 8.2

Q. (a) Find the first three non-zero terms of the Maclaurin series expansion of the sine function.

(b) Deduce the form of the general term, and give the interval of convergence for the series.

A. (a) Proceeding in the same manner used for the exponential function:

$$f(x) = \sin x \quad f^{(1)}(x) = \cos x \quad f^{(2)}(x) = -\sin x \quad f^{(3)}(x) = -\cos x \quad f^{(4)}(x) = \sin x \quad f^{(5)}(x) = \cos x$$
$$f(0) = 0 \quad f^{(1)}(0) = 1 \quad f^{(2)}(0) = 0 \quad f^{(3)}(0) = -1 \quad f^{(4)}(0) = 0 \quad f^{(5)}(0) = 1$$

As every other derivative is zero at $x = 0$, we need to go as far as the 5th derivative in order to obtain the first three non-zero terms. Thus, using eqn (8.44) we have:

$$f(x) = \sin x = x - \frac{1}{3!}x^3 + \frac{1}{5!}x^5 - \cdots$$

(b) Finding the general term requires some trial and error. In this case:

- The coefficients of even powers of x are zero.
- The denominators in the coefficients of the odd powers of x are odd numbers (formed by adding or subtracting 1 to or from an even number, $2n$).
- As the signs of the coefficients for $c_1, c_3, c_5, \ldots$ alternate, starting with a positive value for c_1, the factor $(-1)^{n-1}$ takes care of the sign alternation.
- Only odd powers of x appear, suggesting that the index can be generated by subtracting 1 from an even number; thus the power of x in the nth term can be written as x^{2n-1}. (Check that this generates the terms x, x^3 and x^5 by substituting $n = 1$, 2 and 3, respectively). Thus, the general term in this case is given by: $\frac{(-1)^{n-1}}{(2n-1)!}x^{2n-1} + \cdots$, where $n = 1, 2, 3, \ldots$

Problem 8.5

Use eqn (8.44) to find the first four non-zero terms, as well as the general term, in the Maclaurin series expansions of each of the following functions:

(a) e^{-x}; (b) $\cos x$; (c) $(1-x)^{-1}$.

8.4.1.4 The Problem with Guessing the General Term: A Chemical Counter Example

In our discussion of the geometric series and the Maclaurin series for $\sin x$, we made the assumption, from the pattern emerging from the first few terms, that we could predict how the series will continue *ad infinitum*. In most cases, this confidence is justified, but sometimes we encounter problems where finding the general term requires a knowledge of the physical context of the problem. An example of such a problem in chemistry involves the computation of the ion–ion interaction energy in an ionic solid, such as NaCl. If we compute the interaction energy arising from the interaction between one ion (positive or negative) and all the other ions in the NaCl lattice structure, then we obtain the Madelung energy in the form:

$$V = -\frac{e^2}{4\pi\varepsilon_0 R}\left\{\frac{6}{\sqrt{1}} - \frac{12}{\sqrt{2}} + \frac{8}{\sqrt{3}} - \frac{6}{\sqrt{4}} + \frac{24}{\sqrt{5}} - \frac{24}{\sqrt{6}} + \cdots\right\}$$

$$= -\frac{e^2 A}{4\pi\varepsilon_0 R}. \tag{8.47}$$

Here A is the Madelung constant for the NaCl structure, and R is the distance between any adjacent Na^+ and Cl^- ions. If we inspect the terms in the series, we can see not only that the sign alternatives but also what appears to be a pattern in the square root values given in the denominators of successive terms. However, in contrast, it is very difficult to see any pattern to the values of the numerators; the reason being that there is none: we can only determine their values from knowledge of the NaCl structure. In this example, the first term arises from the interaction between a Na^+ ion and the six nearest neighbour Cl^- anions at a distance $\sqrt{1}R$; the second term arises from the interaction of nearest neighbour ions of the same charge, which in this case involves an Na^+ ion and 12 second nearest neighbour Na^+ ions at a distance $\sqrt{2}R$; the third term is then the interaction between an Na^+ and eight Cl^- at a distance $\sqrt{3}R$, and so on. The general term in this case is $\frac{m}{R\sqrt{n}}$, where m is the number of nth neighbours at a distance of $R\sqrt{n}$. The next term in the series is, somewhat unexpectedly, $+\frac{12}{\sqrt{8}}$ because the number of 7th nearest neighbours at a distance $R\sqrt{7}$ is zero!

8.4.2 The Taylor Series: Expansion about Points Other Than Zero

In many situations, we need to find the power series expansion of a function in terms of the values of the function and its derivatives at some point other than $x = 0$. For example, in the case of a vibrating diatomic molecule, the natural choice of origin for describing the energy of the molecule is the equilibrium internuclear separation, R_e, and not $R = 0$ (where the nuclei have fused). We can determine the expansion of a function $f(x)$ about an origin $x = a$ (where a is now, by definition, generally not zero) using the Taylor series which is given by the expression:

$$f(x) = f(a) + f^{(1)}(a)(x-a) + \frac{f^{(2)}(a)}{2!}(x-a)^2 + \cdots +$$

$$\frac{f^{(n)}(a)}{n!}(x-a)^n + \cdots. \tag{8.48}$$

Here, $f^{(n)}(a)$ is the value of the nth derivative of $f(x)$ at the point $x = a$. The special case where $a = 0$, as discussed above, generates the Maclaurin series.

Worked Problem 8.3

Q. Find the Taylor series expansion for the function $f(x) = e^x$ about the point $x = 1$.

A. Since all the derivatives of e^x at $x = 1$ have the value e, the Taylor series takes the form:

$$e^x = e\left\{1 + (x-1) + \frac{(x-1)^2}{2!} + \frac{(x-1)^3}{3!} + \cdots \right. $$
$$\left. + \frac{(x-1)^{n-1}}{(n-1)!} + \cdots \right\}.$$

$$(8.49)$$

where the last term in the brackets is the nth term. We can see from a plot of the Taylor series expansion of the exponential function shown in Figure 8.3, that far fewer terms are necessary to achieve a good degree of accuracy in the region around $x = 1$ than is the case with the MacLaurin series. However, we also see that the further away we are from the point $x = 1$, the poorer the approximation and the more terms we will need to achieve a given accuracy. Although it is not obvious from Figure 8.3, the Maclaurin series will be better than this Taylor series expansion for values of x close to $x = 0$.

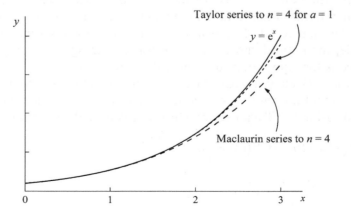

Figure 8.3 An illustration of the improved accuracy achieved with the Taylor series expansion of $f(x) = e^x$ about $x = 1$, compared with the Maclaurin series (expanded about $x = 0$).

Problem 8.6

Find the first four non-zero terms in the Taylor series expansions of the following functions, expanded about the given point, and deduce the form of the general term for each series:

(a) $(1-x)^{-1}$, at $x = -1$; (b) $\sin x$, at $x = \pi/2$; (c) $\ln x$, at $x = 1$.

Worked Problem 8.4

The variation of potential energy, $E(R)$, with internuclear separation, R, for a diatomic molecule can be approximated by the **Morse potential** $E(R) = D_e\left\{1 - e^{-\alpha(R-R_e)}\right\}^2$, shown schematically in Figure 8.4. The dissociation energy, D_e and α are both constants for a given molecule.

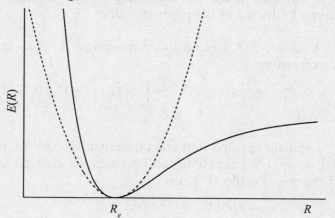

Figure 8.4 Schematic plot the Morse potential energy function (continuous line). The minimum energy is at $R=R_e$. The harmonic approximation (see text) is shown as a dashed line.

However, one of the limitations of the Morse potential energy is that, in contrast to the "experimental" curve, the value of the energy at $R = 0$ (corresponding to nuclear fusion) is finite, rather than infinite. The minimum in the Morse potential energy curve occurs at $R = R_e$, which represents the equilibrium bond length.

Q. Find the first three terms in the Taylor series expansion for the Morse function about the point $R = R_e$.

A. The general expression for the Taylor series expanded about an arbitrary point $x = a$ is

$$f(x) = f(a) + f^{(1)}(a)(x-a) + \frac{f^{(2)}(a)}{2!}(x-a)^2 + \cdots +$$

$$\frac{f^{(n)}(a)}{n!}(x-a)^n + \cdots$$

Step 1: Identify R with x; R_e with a, and $E(R)$ with $f(x)$. We can then see that the three terms $f(a)$, $f^{(1)}(a)$, and $f^{(2)}(a)$ are equivalent to $E(R_e)$, $E^{(1)}(R_e)$ and $E^{(2)}(R_e)$, which enables us to re-express the Taylor series in the form:

$$E(R) = E(R_e) + e^{(1)}(R_e)(R - R_e)$$

$$+ \frac{e^{(2)}(R_e)}{2!}(R - R_e)^2 + \cdots \tag{1.50}$$

Thus, we need to evaluate each of the terms $E(R_e)$, $E^{(1)}(R_e)$ and $E^{(2)}(R_e)$, corresponding to the energy and its first two derivatives, evaluated at the point $R = R_e$.

Step 2: Evaluate $E(R_e)$ using the substitution $R = R_e$ in the energy expression:

$$E(R_e) = D_e\left\{1 - e^{-\alpha(R_e - R_e)}\right\}^2 = D_e\{1 - 1\}^2 = 0,$$

since $e^0 = 1$.

Step 3: Evaluate the first and second derivatives, $E^{(1)}(R_e)$ and $E^{(2)}(R_e)$, by applying the chain rule (the detailed working forms part of the next Problem). Thus:

$$e^{(1)}(R) = 2D_e\left\{1 - e^{-\alpha(R - R_e)}\right\} \times \alpha e^{-\alpha(R - R_e)}$$

$$e^{(2)}(R) = 2\alpha^2 D_e\left\{2e^{-2\alpha(R - R_e)} - e^{-\alpha(R - R_e)}\right\}$$

Substituting $R = R_e$ into the above equations gives:

$$E^{(1)}(R_e) = 2D_e\{1 - 1\} \times \alpha = 0$$
$$E^{(2)}(R_e) = 2\alpha^2 D_e\{2 - 1\} = 2\alpha^2 D_e$$

Step 4: Substitute the expressions for $E(R_e)$, $E^{(1)}(R_e)$ and $E^{(2)}(R_e)$, into the Taylor expansion [eqn (8.50)] yields:

$$E(R) = 0 + 0 \times (R - R_e) + \frac{2\alpha^2 D_e}{2!}(R - R_e)^2 + \cdots$$

$$\Rightarrow E(R) = \alpha^2 D_e (R - R_e)^2 + \cdots \tag{8.51}$$

This is the total energy, $E(R)$, to second order in R, and is commonly known as the **harmonic approximation**. The expression for $E(R)$ gives a good approximation to the potential energy for small displacements of the nuclei, but a somewhat poorer one as the displacements from the equilibrium bond length increase, or decrease, as is seen in Figure 8.4.

Problem 8.7

(a) Use the chain rule (see Section 4.2.4) to find (i) the first and (ii) the second derivative of the Morse function $E(R) = D_e\left\{1 - e^{-\alpha(R - R_e)}\right\}^2$, checking your answers with those given in Worked Problem 8.4.

(b) Verify, by checking the values of $E^{(1)}(R_e)$ and $E^{(2)}(R_e)$, that $E(R_e)$ corresponds to a minimum energy.

(c) Given that the force acting between the nuclei of the molecule is given by:

$$F = -\frac{dE}{dR},$$

use eqn (8.51) to find an expression for F (for small displacements of the nuclei).

(d) The restoring force acting on a simple harmonic oscillator is given by the expression $F = -kx$. Comment on any similarity between the form of this expression and the one obtained in (c), assuming that the displacement x is equivalent to $(R - R_e)$. What conclusions do you draw about applicability of the harmonic approximation for diatomic molecules?

8.4.3 Manipulating Power Series

8.4.3.1 Combining Power Series

If two functions are combined by some operation (for example addition, multiplication, differentiation or integration) then we can find the power series expansion of the resulting function by applying the appropriate operation to the reference series. However, the outcome will be valid only within a domain common to both power series. So, for example, if the Maclaurin series for e^x (interval of convergence: all x in R) is multiplied by that for $\ln(1 + x)$ (interval of convergence: $-1 < x \leqslant 1$), the resulting series only converges in the common interval of convergence $-1 < x \leqslant 1$.

Worked Problem 8.5

Q. Given that the hyperbolic cosine function $\cosh x$ is defined by:

$$\cosh x = \frac{1}{2}\{e^x + e^{-x}\},$$

use the Maclaurin series for e^x and e^{-x} to obtain a power series expansion for the $\cosh x$ function. Give the form of the general term.

A. If we substitute the Maclaurin series for e^x and e^{-x} in the defining equation for $\cosh x$, we obtain:

$$\cosh x = \frac{1}{2}\left\{ 1 + x + \frac{x^2}{2!} + \frac{x^3}{3!} + \cdots + \frac{x^{n-1}}{(n-1)!} \right.$$

$$+ \cdots + 1 - x + \frac{x^2}{2!} - \frac{x^3}{3!} + \frac{x^4}{4!}$$

$$\left. - \cdots (-1)^{n-1} \frac{x^{n-1}}{(n-1)!} + \cdots \right\}$$

$$\Rightarrow \cosh x = 1 + \frac{x^2}{2!} + \frac{x^4}{4!} + \cdots + \frac{x^{2(r-1)}}{[2(r-1)]!} \qquad (8.52)$$

$$+ \cdots r = 1, 2, 3, \ldots$$

Since both e^x and e^{-x} converge for all x, the above series for $\cosh x$ will also converge for all x.

Problem 8.8

(a) Give the form of the Maclaurin series for the function sinh x, where sinh $x = \frac{1}{2}(e^x - e^{-x})$.

(b) Deduce the first three terms of the Maclaurin series for the function $f(x) = \dfrac{e^{-x}}{(1-x)}$ using the series for e^{-x} and $\dfrac{1}{(1-x)}$ taken from your answers to Problem 8.5. Give the interval of convergence for $f(x)$.

8.4.3.2 A Shortcut for Generating Maclaurin Series

Sometimes we can generate Maclaurin series for a given function by a simple substitution. For example, the Maclaurin series for the function e^{-x} can be found as in Problem 8.5. However, an alternative, and much less labour intensive, approach involves writing $X = -x$ and then using the existing series for e^x, with X replacing x.

Worked Problem 8.6

Q. Use the substitution $X = -x$ for the Maclaurin series for e^X, to find the related series for e^{-x}. How is the interval of convergence for e^{-x} related to that for e^x?

A. The Maclaurin expansion for the exponential function e^X is:

$$e^X = 1 + X + \frac{X^2}{2!} + \frac{X^3}{3!} + \frac{X^4}{4!} + \cdots \frac{X^{n-1}}{(n-1)!} + \cdots$$

where X is the independent variable. If we now write $X = -x$ we obtain the series for e^{-x} as required:

$$e^{-x} = 1 - x + \frac{x^2}{2!} - \frac{x^3}{3!} + \frac{x^4}{4!} - \cdots (-1)^{n-1}\frac{x^{n-1}}{(n-1)!} + \cdots \quad (8.53)$$

The test for absolute convergence shows that the interval of convergence is the same as for e^x.

Problem 8.9

(a) Use the substitution $X = ax$ and the Maclaurin series for e^X to find the series for e^{ax}.

(b) (i) Use the equality $\sin 2x = 2 \sin x \cos x$, and Maclaurin series for $\sin x$ and $\cos x$, to find the first three terms in the related series for $\sin 2x$.

   (ii) Use the substitution $X = 2x$ and the Maclaurin series for $\sin X$, to find the first three terms in the related series for $\sin 2x$. Compare your answer to (b)(i) above.

8.4.4 The Relationship Between Domain and Interval of Convergence

We saw earlier that the Maclaurin series expansion of the function $f(x) = (1 - x)^{-1}$ takes the form $1+x+x^2+\dots$. Although the domain of $f(x)$ includes *all* x values, with the exclusion of $x = 1$, where the function is undefined, the domain of the Maclaurin series, determined by applying the ratio test, is restricted to $-1 < x < 1$. The point $x = 1$ is excluded from the domains of both the function, and the series. However, although the point $x = -1$ is clearly included in the domain of the function, since $f(-1)=1/2$, it is *excluded* from the domain of the series. We can further illustrate this by comparing a plot of the function $y = f(x) = (1 - x)^{-1}$ with the MacLaurin series expansion of this function up to the third, fourth, fifth and sixth terms (see Figure 8.5). Clearly the three plots match quite well for $-1 < x < 1$ but differ dramatically for all other values of x. We also see at $x = -1$ that the series representation oscillates between zero and +1 as each new term is added to the series, thus indicating divergence at this point.

8.4.5 Limits Revisited: Limiting Forms of Exponential and Trigonometric Functions

In Chapter 3 we discussed the behaviour of a function close to some limiting value of the independent variable. Some of the examples concern finite limiting values, but more often we are interested in how functions behave for increasingly small or large values in the independent variable. It is usually straightforward to evaluate the limit of simple functions for increasingly large or small values of x but, for some of the transcendental functions, we need to use power series expansions to probe their asymptotic behaviour.

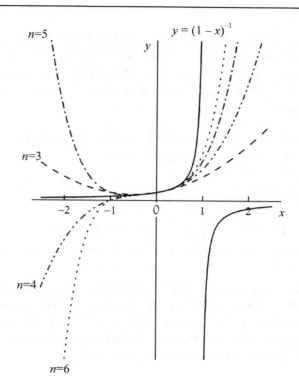

Figure 8.5 A plot of $f(x) = (1-x)^{-1}$ (continous line), compared with plots of the polynomial truncations of the Maclaurin series expansion $1+x+x^2+\ldots+x^{n-1}+$ for $n = 3$ to 6.

8.4.5.1 Exponential Functions

The behaviour of the function e^{ax}, as x tends to large or small values, depends upon the signs and magnitudes of x and a. Thus:

- for $x = 0$, $e^{ax} = 1$, irrespective of the value of a;
- for large positive x, e^{ax} increases without limit as x increases for $a > 0$, but e^{ax} becomes increasingly small as x increases for $a < 0$;
- regardless of the signs of x or a, e^{ax} approaches 1, for increasingly small values of x, according to the MacLaurin power series expansion [as seen in Problem 8.9(a)]:

$$e^{ax} = 1 + ax + \frac{(ax)^2}{2!} + \frac{(ax)^3}{3!} + \frac{(ax)^4}{4!} + \cdots \frac{(ax)^{n-1}}{(n-1)!} + \cdots \quad (8.54)$$

Worked Problem 8.7

Q. For the radial function of a 3d hydrogen atomic orbital, $R(r) = Nr^2 e^{-r/3a_0}$ (N is a normalising constant, and a_0 is the Bohr radius), find:

    (a) the form of $R(r)$ at small r, using the expansion of the exponential function given above;

    (b) $\lim\limits_{r \to 0} R(r)$;

    (c) $\lim\limits_{r \to \infty} R(r)$.

A. (a) $R(r) \approx Nr^2 \left(1 - \dfrac{r}{3a_0} + \ldots \right) = Nr^2$ for small r.

    (b) Using the approximation from (a), we see that $\lim\limits_{r \to 0} R(r) = 0$.

    (c) For large r we see that the limiting value of the function will be determined by the outcome of the competition between the Nr^2 term and the $e^{-r/3a_0}$ term. As we saw in Section 2.3.4, the exponential term will always overcome the power term, and so $\lim\limits_{r \to \infty} R(r) = 0$.

Problem 8.10

The Einstein model for the molar heat capacity at constant volume, C_V, of a solid yields the formula:

$$C_V = 3R \left(\frac{hv}{kT} \right)^2 \left\{ \frac{e^{\frac{hv}{2kT}}}{e^{\frac{hv}{kT}} - 1} \right\}^2 .$$

Show that at high values of T, when we can justifiably substitute the exponential terms by their two-term series approximations, C_V tends to the limit $3R$.

8.4.5.2 Trigonometric Functions

In an analogous way, the series expansions for the sine and cosine functions have the forms:

$$\cos ax = 1 - \frac{(ax)^2}{2!} + \cdots; \quad \sin ax = ax - \frac{(ax)^3}{3!} + \cdots, \tag{8.55}$$

as $x \rightarrow 0$. For very small values of x, $\cos ax$ and $\sin ax$ may be approximated by 1 and ax, respectively. However, as x increases without limit, in both positive and negative senses, the values of the sine or cosine functions oscillate between ± 1.

Problem 8.11

Consider a particle confined to move in a constant potential between the points $x = 0$ and $x = L$ at which the potential is infinite. The associated wavefunction has the form:

$$\psi = \sqrt{\frac{2}{L}} \sin \frac{n\pi x}{L},$$

where n is the quantum number defining the state of the particle, and has values 1, 2, 3, ...

    Find the expression for ψ:

(a) at $x = 0$;
(b) at $x = L$;
(c) when x is very small.

Summary of Key Points

The key points discussed in this chapter include:

1. The definition of finite sequences with examples including the geometric and arithmetic progressions.
2. The definition of indefinite sequences and the concepts of convergence and divergence of a sequence of numbers.
3. The distinction between a finite series having a finite sum and an infinite series where a finite sum exists only if the series converges.
4. Testing an infinite series for convergence: the ratio test for absolute convergence and the interval of convergence.
5. Power series: the Maclaurin and Taylor series.
6. Power series expansions of functions and the appropriate choice of expansion point.
7. Truncation of power series.

8. Determining the general term in a power series.
9. The Taylor series expansion of the Morse potential leading to the harmonic approximation.
10. Manipulating power series.
11. Using power series expansions of functions to probe limiting behaviour for increasingly large or small values of the independent variable.

9

Numbers Revisited: Complex Numbers

In Chapter 2, we saw that the solution of a quadratic equation of the form:

$$ax^2 + bx + c = 0 \qquad (9.1)$$

can yield up to two real roots depending on the values of the coefficients, a, b and c. The general solution to quadratic equations of this form is given by the formula:

$$x = \frac{-b \pm \sqrt{b^2 - 4ac}}{2a}, \qquad (9.2)$$

where the quantity $b^2 - 4ac$ is known as the **discriminant** (see Section 2.4). If the discriminant is positive, then the equation has two real and different roots; if it is zero then the equation will have two identical roots, and if it is negative, there are no real roots, as the formula involves the square root of a negative number. For example, the equation:

$$x^2 - 4x + 3 = 0$$

yields two real roots, $x = 3$ and $x = 1$, according to:

$$x = \frac{4 \pm \sqrt{16 - (4 \times 3)}}{2} = 2 \pm \frac{\sqrt{4}}{2} = 2 \pm 1 = 3, 1.$$

We can represent this solution graphically (see Figure 9.1) in terms of where the function $y = x^2 - 4x + 3$ cuts the x-axis (where $y = 0$).

However, if we use eqn (9.2) to find the roots of the quadratic equation:

$$x^2 - 4x + 6 = 0$$

we find that the solution yields:

$$x = \frac{4 \pm \sqrt{16 - (4 \times 6)}}{2} = 2 \pm \frac{\sqrt{-8}}{2}, \qquad (9.3)$$

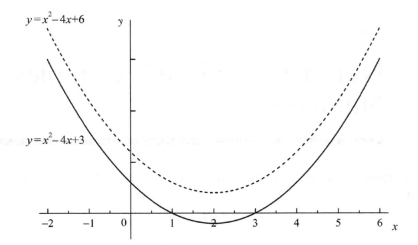

Figure 9.1 Plots of the two functions $y = x^2 - 4x + 3$ and $y = x^2 - 4x + 6$, showing the presence of two and zero real roots of the respective equations $x^2 - 4x + 3 = 0$ and $x^2 - 4x + 6 = 0$.

which requires us to find the square root of -8. Graphically, we see in Figure 9.1 that a plot of the function $y = x^2 - 4x + 6$ does not cut the x-axis at all. Logic would seem to dictate that any solution to the second of these two equations is non-sensical, and that the result cannot possibly be real – especially when we view the plot of the function, which clearly does not cut the x-axis. However, there is a way of circumventing this problem by simply extending the number system to include so-called **complex numbers**, which incorporate $\sqrt{-1}$ as a legitimate number. This concept can naturally seem somewhat bemusing but, once we get over the shock, we find that the treatment of complex numbers is really quite straightforward and, more importantly, we find that they allow us to tackle real problems in chemistry in a way that would otherwise be impossible.

Aims:

This chapter extends the familiar number system to include complex numbers containing the imaginary number i. By the end of this chapter, you should be able to:

- Recognise the real and imaginary parts of a complex number expressed in either Cartesian or plane polar coordinates.
- Determine the modulus and argument of a complex number, and denote its location on an Argand diagram.

- Perform arithmetical operations on complex numbers.
- Use the Euler formula and the De Moivre theorem to evaluate powers of complex numbers, to determine n^{th} roots of a complex number, and to identify real and imaginary parts of functions of a complex variable.

9.1 The Imaginary Number i

As we saw above, the solutions to algebraic equations do not always yield real numbers. The solution of the equation $x^2 + 1 = 0$ yields the apparently meaningless result

$$x = \pm\sqrt{-1}, \tag{9.4}$$

because the square root of a negative number is not defined in terms of a real number. However, if we now *define* the imaginary number $i = \sqrt{-1}$, then the two roots may be specified as $x = \pm i$. In general, an imaginary number is defined as any real number multiplied by i. Thus, for example, the number $\sqrt{-8}$, which emerged from the solution to eqn (9.3) above, can be written as $\sqrt{8}\sqrt{-1} = \sqrt{8}i$.

Worked Problem 9.1

Q. Solve the quadratic equation $x^2 + 2x + 5 = 0$.
A. The formula given in eqn (2.2) for the roots of a quadratic equation yields:

$$x = -1 \pm \frac{\sqrt{4-20}}{2} = -1 \pm \frac{\sqrt{-16}}{2}$$

$$= -1 \pm \frac{4}{2}\sqrt{-1} = -1 \pm 2i.$$

Problem 9.1

(a) Draw plots of the following functions:
  (i) $y = x^2 - 2x - 3$ and (ii) $y = x^2 - 2x + 2$.

In each case, comment on whether the plot cuts the x-axis and, if so, where.

(b) Use eqn (9.2) to find the roots of each of the quadratic equations $x^2 - 2x - 3 = 0$ and $x^2 - 2x + 2 = 0$. Comment on your answers with respect to your plots from part (a).

9.2 The General Form of Complex Numbers

In the answer to Worked Problem 9.1, we obtained the required roots of the quadratic equation in the form of a sum of a real number (-1) and an imaginary number ($2i$ or $-2i$). Such numbers are termed **complex numbers**, and have the general form:

$$z = x + iy \tag{9.5}$$

The real and imaginary parts of a complex number, z, are often expressed as Re z, and Im z, respectively.

where x and y are real numbers, termed the **real** and **imaginary** parts of z, respectively. Clearly, if $x = 0$, $y \neq 0$, then z is an imaginary number (because the real part vanishes); likewise, if $x \neq 0$, $y = 0$, then z is a real number (because the imaginary part vanishes).

9.3 Manipulation of Complex Numbers

The algebraic manipulation of pairs of complex numbers is really quite straightforward, so long as we remember that, since $i = \sqrt{-1}$, it follows that $i^2 = -1$.

Problem 9.2

Evaluate (a) i^3, (b) i^4 and (c) i^5.

9.3.1 Addition, Subtraction and Multiplication

For addition or subtraction of complex numbers, the appropriate operation is carried out separately on the real and imaginary parts of the two numbers.

Multiplication of a complex number by a **scalar** (real number), is achieved by simply multiplying the real and imaginary parts of the complex number by the scalar quantity. Multiplication of two complex numbers is performed by expanding the expression

$(a + ib)(c + id)$ as a sum of terms, and then collecting the real and imaginary parts to yield a new complex number.

Worked Problem 9.2

Q. If $z_1 = 1 + 2i$ and $z_2 = -2 + i$, write down (a) $z_1 + z_2$, (b) z_1z_2, (c) $z_1 - z_2$ and (d) $2(z_1 - z_2)$.

A. (a) $z_1 + z_2 = (1 + 2i) + (-2 + i) = 1 + 2i - 2 + i = -1 + 3i$;

(b) $z_1z_2 = (1 + 2i)(-2 + i) = -2 + i - 4i - 2 = -4 - 3i$;

(c) $z_1 - z_2 = (1 + 2i) - (-2 + i) = 3 + i$;

(d) $2(z_1 - z_2) = 2(3 + i) = 6 + 2i$.

Note that all of the answers are in the form $x + iy$.

Problem 9.3

If $z_1 = 2 + 3i$, $z_2 = -1 + i$, $z_3 = 3 - 2i$, give expressions for:

(a) $z_1 + z_2 - 2z_3$;

(b) $z_1z_2 + z_3^2$.

9.3.2 The Complex Conjugate

The complex conjugate, z^*, of the complex number $z = x + iy$, is obtained by changing the sign of the imaginary part of z to yield $z^* = x - iy$. Thus, for example, the complex conjugate of the number $z_2 = -1 + i$, given in Problem 9.3, is $z_2^* = -1 - i$.

The two numbers z and z^* have the properties that their sum and product are both real, but their difference is imaginary:

$$z + z^* = 2x, \qquad (9.6)$$

$$zz^* = (x + iy)(x - iy) = x^2 + y^2, \qquad (9.7)$$

$$z - z^* = 2iy. \qquad (9.8)$$

Problem 9.4

Express the following in the form $x + iy$, and write down the complex conjugate in each case.

(a) $(-1 - 2i) + (2 + 7i)$,
(b) $(3 - i) - (4 - 2i)$,
(c) $i(1 + 3i)$,
(d) $(1 + 3i)(3 + 2i)$.

9.3.3 Division of Complex Numbers

As we have seen, addition, subtraction and multiplication of complex numbers is generally quite straightforward, requiring little more than the application of elementary algebra. However, the division of one complex number by another requires that a quotient such as $\frac{z_1}{z_2} = \frac{x_1 + iy_1}{x_2 + iy_2}$, be transformed into a complex number in the form of eqn (9.5). The solution to this conundrum is not immediately obvious, until we remember that the product of a complex number with its own complex conjugate zz^* is a real number [see eqn (9.7)]. This suggests that we could achieve the required form for the quotient by multiplying both numerator and denominator by z_2^*:

$$\frac{z_1 z_2^*}{z_2 z_2^*} = \frac{x_1 + iy_1}{x_2 + iy_2} \times \frac{x_2 - iy_2}{x_2 - iy_2} \qquad (9.9)$$

This has the same effect as multiplying by unity since $z_2^*/z_2^* = 1$, but it allows us to express the quotient in the required form. Thus, multiplying out the numerator and denominator on the right side of eqn (9.9), and collecting terms, gives:

$$\frac{z_1}{z_2} = \frac{x_1 x_2 + y_1 y_2}{x_2^2 + y_2^2} + i\frac{y_1 x_2 - x_1 y_2}{x_2^2 + y_2^2}. \qquad (9.10)$$

The net result is that the original quotient is transformed into the form of a complex number with the real and imaginary parts $\frac{x_1 x_2 + y_1 y_2}{x_2^2 + y_2^2}$ and $\frac{y_1 x_2 - x_1 y_2}{x_2^2 + y_2^2}$, respectively.

Worked Problem 9.3

Q. Using z_1 and z_2 as defined in Worked Problem 9.2, express z_1 / z_2 in the form $x + iy$.

A. $\dfrac{z_1}{z_2} = \dfrac{1+2i}{-2+i} = \dfrac{1+2i}{-2+i} \times \dfrac{-2-i}{-2-i} = \dfrac{-2-i-4i+2}{4+2i-2i+1} = \dfrac{-5i}{5} = -i.$

In this example, the answer is an imaginary number, since $x = 0$ and $y = -1$.

Problem 9.5

Express the following in the form $x + iy$:

$$\text{(a) } \frac{1}{i}, \text{ (b) } \frac{1-i}{2-i} \text{ and (c) } \frac{i(2+i)}{(1-2i)(2-i)}.$$

9.4 The Argand Diagram

Since a complex number is defined in terms of two real numbers, it is convenient to use a graphical representation in which the real and imaginary parts define a point (x, y) in a plane. Such a representation is provided by an **Argand diagram**, as seen in Figure 9.2.

Problem 9.6

Plot your answers to Problem 9.3 as points in an Argand diagram.

The location of z in the Argand diagram can be specified by using either Cartesian coordinates (x, y), where $x = \text{Re } z$, $y = \text{Im } z$, *or* polar coordinates (r, θ) where, $r \geqslant 0$ and $-\pi < \theta \leqslant \pi$. The reason for choosing this range of θ values, rather than $0 < \theta \leqslant 2\pi$, derives from the convention that θ should be positive in the first two quadrants (above the x-axis), moving in an anticlockwise sense from the Re z axis, and negative in the third and fourth quadrants (below the x-axis), moving in a clockwise sense. The quadrant numbering runs from 1 to 4, in an anticlockwise direction, as indicated in Figure 9.2.

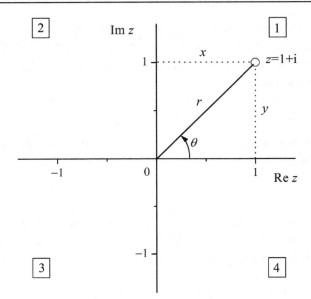

Figure 9.2 An Argand diagram displaying the complex number $z = 1 + i$, in terms of the Cartesian coordinates (1, 1) or, alternatively, in terms of the polar coordinates, $(r = \sqrt{2},\ \theta = \pi/4)$.

9.4.1 The Modulus and Argument of z

The polar coordinates r and θ, define the **modulus** (alternatively known as the **absolute value** and sometimes denoted by $|z|$) and **argument**, respectively, of z. From Pythagoras' Theorem, and simple trigonometry, the modulus and argument of z are defined as follows (see Figure 9.2):

$$r = \sqrt{x^2 + y^2},\ r \geqslant 0 \tag{9.11}$$

$$\tan \theta = \frac{y}{x} \Rightarrow \theta = \tan^{-1}(y/x) \tag{9.12}$$

Great care is required in determining θ, because it is easy to make a mistake in specifying the correct quadrant. For example, although the complex numbers $z = 1 - i$ and $z = -1 + i$ both have $\tan \theta = -1$, they lie in the fourth and second quadrants, respectively, as seen in Figure 9.3.

If we evaluate $\tan^{-1}(-1)$, using the $\tan^{-1}$ function on an electronic calculator, to determine θ, we obtain -0.7854 rad, which is equivalent to $-0.7854 \times 180\ /\ \pi = -45°$. This places z in the fourth quadrant, which is correct for $z = 1 - i$ but incorrect for $z = -1 + i$. In the latter case, we need to look at the values of $x = \mathrm{Re}\ z$ and $y = \mathrm{Im}\ z$, and choose the more appropriate value for θ, recognising that $\tan(\pi + \theta)$ and $\tan \theta$ have the same value. The use **radians** or **degrees** when evaluating the argument of a complex number is largely a matter of

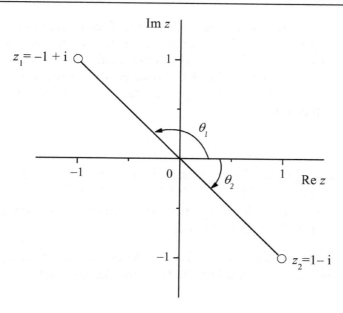

Im z

$z_1 = -1 + i$

1

θ_1

-1

0

θ_2

1 Re z

-1

$z_2 = 1 - i$

Figure 9.3 An Argand diagram showing the complex numbers $z_1 = -1 + i$ and $z_2 = 1 - i$ with modulus $\sqrt{2}$ and arguments $3\pi/4$ and $-\pi/4$, respectively.

taste or context. However, you may find it more convenient to work with degrees when referring to an Argand diagram because it is easier to associate a complex number with a given quadrant in this case.

Worked Problem 9.4

Q. (a) Given that $\tan\theta = \dfrac{\sin\theta}{\cos\theta}$, use the addition formulae for sine and cosine to show that $\tan(\pi + \theta) = \tan\theta$.

(b) Find the modulus and argument of the complex numbers $1 - 2i$ and $-1 + 2i$

A: (a) Since $\sin\theta = 0$ and $\cos\pi = -1$, the addition formulae for $\sin(\pi + \theta)$ and $\cos(\pi + \theta)$ give:

$\sin(\pi + \theta) = \sin\pi\cos\theta + \cos\pi\sin\theta = -\sin\theta$

$\cos(\pi + \theta) = \cos\pi\cos\theta - \sin\pi\sin\theta = -\cos\theta,$

and so:

$$\tan(\pi+\theta) = \frac{\sin(\pi+\theta)}{\cos(\pi+\theta)} = \frac{-\sin\theta}{-\cos\theta} = \tan\theta.$$

(b) For $z = 1 - 2i$, we can identify x and y with 1 and -2, respectively. Thus the modulus, r, is $\sqrt{5}$. The argument, expressed in degrees (for convenience), is found by solving $\tan^{-1}(-2) = \theta$: i.e. $\theta = -63.43°$ or $116.17°$ ($\theta = -1.107$ rad or 2.034 rad). In this case, an Argand diagram shows that θ is in the fourth quadrant, and so $\theta = -63.43°$ is the correct value for the argument. For $z = -1 + 2i$, it follows again that $r = \sqrt{5}$ and $\theta = -63.43°$ or $116.17°$; however, the correct argument this time is $116.17°$, because z lies in the second quadrant.

Problem 9.7

Find the modulus and argument (in degrees) of the complex numbers (a) $-1 - 2i$ and (b) $2i$.

Hint: You will need to exercise a little care in determining the argument for the second of the two complex numbers.

9.5 The Polar Form for Complex Numbers

So far we have assumed that a complex number takes the form $z = x + iy$, where x and y are the values of Re z and Im z, respectively. However, from trigonometry (see Figure 2.2) we see that:

$$x = r \cos \theta \text{ and } y = r \sin \theta.$$

Consequently, z may be expressed in terms of r, θ as:

$$z = r(\cos \theta + i \sin \theta). \tag{9.13}$$

Eqn (9.13) is not yet in a form that is fundamentally different from the Cartesian form expressed in eqn (9.5). However, we can obtain an alternative, more compact, and far more powerful way of writing the polar form of a complex number by re-visiting the Maclaurin series for the $\sin\theta$, $\cos\theta$ and exponential functions. The Maclaurin series for cosine and sine are:

$$\cos \theta = 1 - \frac{\theta^2}{2!} + \frac{\theta^4}{4!} + \cdots + \frac{(-1)^{n-1}}{(2n-2)!}\theta^{2n-2} + \cdots \tag{9.14}$$

$$\sin \theta = \theta - \frac{\theta^3}{3!} + \frac{\theta^5}{5!} + \cdots + \frac{(-1)^{n-1}}{(2n-1)!}\theta^{2n-1} + \cdots \tag{9.15}$$

If we substitute each of these into eqn (9.13) we obtain:

$$z = r(\cos\theta + i\sin\theta) = r\left\{1 + i\theta - \frac{\theta^2}{2!} - \frac{i\theta^3}{3!} + \frac{\theta^4}{4!} + \frac{i\theta^5}{5!} + \cdots\right\}, \qquad (9.16)$$

and on re-writing the right-hand side in terms of powers of $i\theta$ using $i^2 = -1$, $i^3 = -i$, $i^4 = 1$, *etc.*, we obtain:

$$z = r\left\{1 + i\theta + \frac{(i\theta)^2}{2!} + \frac{(i\theta)^3}{3!} + \frac{(i\theta)^4}{4!} + \frac{(i\theta)^5}{5!} + \cdots\right\}. \qquad (9.17)$$

The braces in eqn (9.17) contain the Maclaurin series for $e^{i\theta}$, and so we can rewrite the polar form for z more compactly as:

$$z = re^{i\theta}. \qquad (9.18)$$

Problem 9.8

Use the result that $i^2 = -1$, $i^3 = -i$, $i^4 = 1$ and $i^5 = i$ to evaluate:

(a) $(i\theta)^2$; (b) $(i\theta)^3$; (c) $(i\theta)^4$; (d) $(i\theta)^5$,

and hence show that eqn (9.17) is equivalent to eqn (9.16).

9.5.1 Euler's Formula

If we equate eqn (9.18) with eqn (9.13) we obtain:

$$z = re^{i\theta} = r(\cos\theta + i\sin\theta) \qquad (9.19)$$

which on cancelling r yields:

$$e^{i\theta} = \cos\theta + i\sin\theta. \qquad (9.20)$$

This important result is known as Euler's formula.

Worked Problem 9.5

Q. Express the complex conjugate of $z = r(\cos\theta + i\sin\theta)$ in polar form.

A. The complex conjugate of z can be written in terms of r, θ as above, using the Maclaurin series for $\cos\theta$ and $\sin\theta$, as:

$$z^* = r(\cos\theta - i\sin\theta)$$

$$= r\left\{1 - i\theta - \frac{\theta^2}{2!} + \frac{i\theta^3}{3!} + \frac{\theta^4}{4!} - \frac{i\theta^5}{5!} + \cdots\right\}$$

which can be rewritten as:

$$z^* = r\left\{1 - i\theta + \frac{(i\theta)^2}{2!} - \frac{(i\theta)^3}{3!} + \frac{(i\theta)^4}{4!} - \frac{(i\theta)^5}{5!} + \cdots\right\}.$$

The part in parentheses is the Maclaurin series for $e^{-i\theta}$ and so we can now express z^* as

$$z^* = re^{-i\theta}. \tag{9.21}$$

9.5.1.1 The Number $e^{i\pi}$

Using Euler's formula to evaluate $e^{i\pi}$ we see that:

$$e^{i\pi} = \cos\pi + i\sin\pi. \tag{9.22}$$

However, as $\cos\pi = -1$ and $\sin\pi = 0$, we obtain the extraordinary and elegant result that:

$$e^{i\pi} = -1, \tag{9.23}$$

which rearranges to a single relationship:

$$e^{i\pi} + 1 = 0, \tag{9.24}$$

containing the irrational numbers e and π, the imaginary number, i, as well as the numbers zero and unity.

9.5.2 Powers of Complex Numbers

The advantage of using the polar form for z is that it makes certain manipulations much easier. Thus, for example, we can obtain:

- The modulus of z directly from the product of z and z^*, using eqns (9.18) and (9.21):

$$zz^* = r^2 e^{i\theta} e^{-i\theta} = r^2 \Rightarrow r = \sqrt{zz^*} \tag{9.25}$$

- Positive and negative powers of z:

$$z^n = \left(re^{i\theta}\right)^n = r^n e^{in\theta} \, (n = \pm 1, \, \pm 2, \, \pm 3, \, \pm 4 \cdots) \qquad (9.26)$$

where, for a given value of n, z^n is seen to be a complex number, with modulus r^n and argument $n\theta$.

- Rational powers of z, where $n = p / q \, (q \neq 0)$:

$$z^{p/q} = r^{p/q} e^{i(p/q)\theta} \qquad (9.27)$$

Worked Problem 9.6

Q. If $z = \cos\theta + i\sin\theta$, show that $1 / z$ is the complex conjugate of z.

A. As z has a unit modulus $(r = 1)$, $z = e^{i\theta}$, and $1/z = \dfrac{1}{e^{i\theta}} = e^{-i\theta}$, which is the complex conjugate of z [see eqn (9.21)].

Problem 9.9

For the two complex numbers $z_1 = r_1 e^{i\theta_1}$ and $z_2 = r_2 e^{i\theta_2}$, give expressions for the modulus and argument of:

$$(a) \; z_1 z_2, \text{(b) } z_1/z_2 \text{ and (c) } z_1^2/z_2^4.$$

Problem 9.10

Express $z = -1 - i$ in polar form, and thus determine the modulus and argument (in radians) of z^2 and z^{-4}.

9.5.3 The De Moivre Theorem

We have seen from eqn (9.26) that the nth power of a complex number can be expressed as:

$$z^n = r^n e^{in\theta}, \qquad (9.28)$$

with the modulus and argument r^n and $n\theta$, respectively. Using Euler's formula, eqn (9.28) becomes:

$$z^n = r^n e^{in\theta} = r^n(\cos n\theta + i \sin n\theta) = r^n(\cos\theta + i\sin\theta)^n. \quad (9.29)$$

After cancelling the r^n factors in eqn (9.29), we obtain the **De Moivre theorem**:

$$(\cos\theta + i\sin\theta)^n = \cos n\theta + i\sin n\theta. \quad (9.30)$$

Problem 9.11

(a) Show that $\dfrac{1}{(\cos\theta + i\sin\theta)} = \cos\theta - i\sin\theta$.

(b) Give an expression for $(\cos\theta + i\sin\theta)^{1/2}$.

(c) Use eqn (9.29) to give expressions for the real and imaginary parts of z^n.

(d) Find the real and imaginary parts of z^3 and z^{-2}, where $z = -1 - i$.

9.5.4 Complex Functions

So far we have been concerned largely with the concept of the complex number, but we can see from our discussion of Euler's formula that the general form of a complex number actually represents a complex mathematical function, say $f(\theta)$, where:

$$f(\theta) = \cos\theta + i\sin\theta. \quad (9.31)$$

This function comprises a real part and an imaginary part, and so, in general, we can define a complex function in the form:

$$f(x) = g(x) + ih(x) \quad (9.32)$$

where the complex conjugate of the function is given by:

$$f(x)^* = g(x) - ih(x) \quad (9.33)$$

Thus $f(x)f(x)^*$ is a real function of the form:

$$f(x)f(x)^* = g(x)^2 + h(x)^2 \quad (9.34)$$

The property of complex functions given in eqn (9.34) plays a very important role in quantum mechanics, where the wave function of an electron, ψ, which may be complex in form, is related to the physically meaningful probability density through the product $\psi\psi^*$. If ψ is a complex function, then, from eqn (9.34), $\psi\psi^*$ is a real function.

9.5.4.1 The Periodicity of the Exponential Function

It may seem odd to think of the exponential function, e^x, as periodic because it is clearly not so when the exponent, x, is real. However, replacing x with the imaginary number $i\theta$ yields the complex number $z = e^{i\theta}$ whose modulus and argument are 1 and θ, respectively. If we represent the values of the function on an Argand diagram, we see that they lie on a circle of radius, $r = 1$, in the complex plane (see Figure 9.4). Different values of θ then define the location of complex numbers of modulus unity on the circumference of the circle. We can also see that the function is periodic, with period 2π:

$$e^{i(\theta+2m\pi)} = e^{i\theta}\cdot e^{i2m\pi} = e^{i\theta}. \tag{9.35}$$

Problem 9.12

Use Euler's formula to show that $e^{i2\pi} = 1$, and hence prove that $e^{i(\theta+2m\pi)} = e^{i\theta}$, for $m = 1, 2, \ldots$.

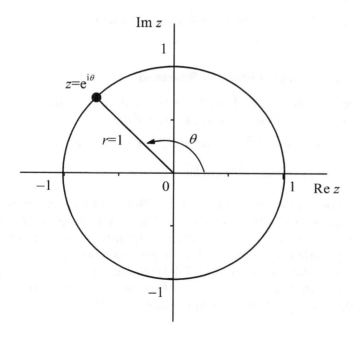

Figure 9.4 The function $z = e^{i\theta}$ is periodic in the complex plane, with period 2π.

Problem 9.13

(a) Use De Moivre's theorem to show that

$$e^{-i\theta} = \cos\theta - i\sin\theta.$$

(b) Use Euler's formula, and the result given in part (a), to show that:

(i) $\cos\theta = \dfrac{1}{2}\left(e^{i\theta} + e^{-i\theta}\right)$

(ii) $\sin\theta = \dfrac{1}{2i}\left(e^{i\theta} - e^{-i\theta}\right).$

Problem 9.14

The solution of the differential equation describing the simple harmonic oscillator problem [see Worked Problem 7.4(c)] is:

$$y = A\cos kt + B\sin kt.$$

Derive an alternative form for the solution, using the function e^{ikt}, and its complex conjugate, e^{-ikt}.

Hint: Use the results given in Problem 9.13(b).

9.5.4.2 The Eigenvalue Problem Revisited

The three 2p orbitals resulting from the solution of the Schrödinger equation for the hydrogen atom can be written as:

$$\psi_1 = N_1 e^{-r/2a_0} r \sin\theta e^{i\phi}; \quad \psi_0 = N_2 e^{-r/2a_0} r \cos\theta;$$

$$\psi_{-1} = N_1 e^{-r/2a_0} r \sin\theta e^{-i\phi},$$

where N_1 and N_2 are constants, a_0 is the Bohr radius, r is the distance of the electron from the nucleus, and the suffix attached to each ψ indicates the value of the orientation quantum number m_l. The Schrödinger equation, $\hat{H}\psi = E\psi$, is an example of an eigenvalue problem (see Sections 4.3.1 and 7.4.3) where, in this case, $\hat{H}$ is an operator known as the Hamiltonian, E is the eigenvalue (corresponding to the energy of the system), and ψ is the eigenfunction (or wavefunction). As we saw in the earlier chapters, if two functions are both solutions to an eigenvalue problem, associated with the same eigenvalue, then a linear combination of the functions will also be a solution. We can use this property to construct real orbital functions that we can visualise more easily. We explore this idea a little further in the next problem.

Problem 9.15

(a) Find the real and imaginary parts of each of the three 2p orbitals given above.

(b) Use the results given in Problem 9.13(b) to show that the following linear combinations yield real functions:

(i) $\dfrac{1}{\sqrt{2}}(\psi_1 + \psi_{-1}) = \sqrt{2}N_1 e^{-r/2a_0} r \sin\theta \cos\phi$

(ii) $-\dfrac{i}{\sqrt{2}}(\psi_1 - \psi_{-1}) = \sqrt{2}N_1 e^{-r/2a_0} r \sin\theta \sin\phi$

(c) Given that $x = r \sin\theta \cos\phi$, $y = r \sin\theta \sin\phi$ and $z = r \cos\theta$, rewrite the three *real* atomic orbital functions, ψ_0, $\dfrac{1}{\sqrt{2}}(\psi_1 + \psi_{-1})$ and $-\dfrac{i}{\sqrt{2}}(\psi_1 - \psi_{-1})$, in terms of the independent variables x, y and z, and hence propose new labels for the three wavefunctions.

9.5.4.3 Structure Factors in Crystallography

The intensity of the scattered beam of X-rays from the (*hkl*) plane of a crystal is proportional to FF^*, where F, the structure factor is given by:

$$F(hkl) = \sum_{j}^{cell} f_j e^{2\pi i [hx_j + ky_j + lz_j]}. \tag{9.36}$$

The summation runs over the appropriate number of atoms in the unit cell with (fractional) coordinates (x_j, y_j, z_j) and scattering factor, f_j.

Problem 9.16

Metallic sodium has a body-centred cubic structure with two atoms per unit cell located at $(0, 0, 0)$ and $(\frac{1}{2}, \frac{1}{2}, \frac{1}{2})$, respectively.

(a) Use eqn (9.36) to show that $F(hkl) = f_{Na} + f_{Na}e^{\pi i(h+k+l)}$ and, with the aid of Euler's formula, determine its real and imaginary parts.

(b) Show that reflections occur [*i.e.* $F(hkl) \neq 0$] only if $h + k + l$ is even.

9.5.5 Roots of Complex Numbers

The polar form of a complex number, z, raised to the power n, is given in eqn (9.28) as:

$$z^n = \left(re^{i\theta}\right)^n = r^n e^{in\theta} \tag{9.37}$$

De Moivre's theorem allows us to express z^n in the form:

$$z^n = r^n(\cos n\theta + i \sin n\theta). \tag{9.38}$$

It follows that one square root of a complex number (where $n = \frac{1}{2}$) is given by:

$$z^{\frac{1}{2}} = r^{\frac{1}{2}}\left(\cos\frac{\theta}{2} + i \sin\frac{\theta}{2}\right). \tag{9.39}$$

The method used to retrieve the second square root is now described in Worked Problem 9.7.

Worked Problem 9.7

Q. Use eqn (9.37) and the periodicity of $e^{i\theta}$ (see Problem 9.12) to find the two square roots of -1.

A. Substituting eqn (9.35) for $e^{i\theta}$ into eqn (9.37) yields:

$$z^n = \left(re^{i(\theta + 2m\pi)}\right)^n = r^n e^{i(\theta + 2m\pi)n}, \ m = 1, 2, 3, \ldots$$

Now, the number (-1) has $r = 1$ and $\theta = \pi$; hence:

$$(-1)^{1/2} = e^{i(\pi + 2m\pi) \times 1/2}, \; m = 1, 2, 3, \ldots$$

$$= e^{i(\pi/2 + m\pi)}, \; m = 1, 2, 3, \ldots$$

$$= \cos\left(\frac{\pi}{2} + m\pi\right) + i\sin\left[\frac{\pi}{2} + m\pi\right], \; m = 1, 2, 3, \ldots$$

Thus for $m = 1$:

$$z^{1/2} = \cos\frac{3\pi}{2} + i\sin\frac{3\pi}{2} = -i;$$

for $m = 2$:

$$z^{1/2} = \cos\frac{5\pi}{2} + i\sin\frac{5\pi}{2} = i;$$

for $m = 3$:

$$z^{1/2} = \cos\frac{7\pi}{2} + i\sin\frac{7\pi}{2} = -i;$$

for $m = 4$:

$$z^{1/2} = \cos\frac{9\pi}{2} + i\sin\frac{9\pi}{2} = i$$

and so on. We see that taking $m \geqslant 3$ merely replicates the roots already found, and so the two square roots of 1 are $\pm i$.

This method can be extended to find the n n^{th} roots of any number.

Problem 9.17

Show that the three cube roots of i (given by $i^{1/3}$) are $-\frac{\sqrt{3}}{2} + \frac{1}{2}i$, $-i$ and $\frac{\sqrt{3}}{2} + \frac{1}{2}i$.

Summary of Key Points

This chapter introduces imaginary and complex numbers as a legitimate extension of the number system. The key points discussed in this chapter include:

1. An introduction of the imaginary number $i = \sqrt{-1}$ as a means to finding all roots of polynomial equations.
2. A definition of the general form of a complex number, $z = x + iy$, comprising real and imaginary parts.
3. The algebra of complex numbers: addition, subtraction and multiplication.
4. The complex conjugate and division of complex numbers.
5. The graphical representation of the complex number through the Argand diagram.
6. The definition of modulus and argument of a complex number.
7. The polar form of complex numbers, $z = re^{i\theta}$.
8. Euler's formula, $e^{i\theta} = \cos\theta + i\sin\theta$.
9. Powers of complex numbers and de Moivre's theorem, $e^{in\theta} = \cos n\theta + i\sin n\theta$.
10. Complex functions.
11. The periodicity of the exponential function, $e^{i\theta}$, and the modelling of wave phenomena.
12. The real and complex forms of atomic orbitals.
13. Finding the roots of positive, negative and complex numbers.

10
Working with Arrays I: Determinants

In all areas of the physical sciences, we encounter problems that require the solution of sets of simultaneous linear equations. These range from seemingly mundane everyday problems to highly complex problems in quantum mechanics or spectroscopy requiring the solution of hundreds of simultaneous linear equations. For small numbers of such equations, the solutions may be most straightforwardly obtained using the methods of elementary algebra; but, as the number of equations increases, their alegbraic solution becomes cumbersome and ultimately intractable. In this chapter we introduce the concept of the determinant to provide one of the tools used to solve problems involving large numbers of simultaneous equations. The other tools required for solving systems of linear equations are provided by matrix algebra, which we discuss in detail in Chapter 11.

Aims:

This chapter introduces the determinant as a mathematical object which we can use to tackle problems involving large numbers of simultaneous linear equations. By the end of this chapter, you should be able to:

- Appreciate that a determinant expands to yield an expression or value.
- Recognise how determinants can be used to solve simultaneous linear equations.
- Expand determinants of second and third order about a given row or column.
- Use the properties of determinants to introduce as many zeros as possible to the right (or left) of the leading diagonal.
- Define and evaluate the first order cofactors of a determinant.

10.1 Origins – The Solution of Simultaneous Linear Equations

We begin our discussion of linear systems by introducing the determinant as a tool for solving sets of simultaneous linear equations in which the indices of the unknown variables are all unity. Consider the pair of equations:

$$a_{11}x + a_{12}y = b_1 \qquad (10.1)$$

$$a_{21}x + a_{22}y = b_2 \qquad (10.2)$$

where a_{11}, a_{12}, a_{21}, a_{22}, b_1 and b_2 are constant coefficients, and x and y are the 'unknowns'. We can determine the unknowns using elementary algebra as follows:

- Multiply eqn (10.1) by a_{22} and eqn (10.2) by a_{12} to give:

$$a_{11}a_{22}x + a_{12}a_{22}y = b_1a_{22} \qquad (10.3)$$

$$a_{12}a_{21}x + a_{12}a_{22}y = b_2a_{12} \qquad (10.4)$$

- Subtract eqn (10.4) from eqn (10.3) to yield:

$$(a_{11}a_{22} - a_{12}a_{21})x = b_1a_{22} - b_2a_{12},$$

which we can rearrange to give an expression for x in terms of the constant coefficients:

$$x = \frac{b_1a_{22} - b_2a_{12}}{a_{11}a_{22} - a_{12}a_{21}} \qquad (10.5)$$

- Now, multiply eqn (10.1) by a_{21} and eqn (10.2) by a_{11}, and subtract the resulting equations to give an analogous expression for y

$$y = \frac{b_2a_{11} - b_1a_{21}}{a_{11}a_{22} - a_{12}a_{21}} \qquad (10.6)$$

The denominators in eqns (10.5) and (10.6) are the same, and can be written alternatively as:

$$\begin{vmatrix} a_{11} & a_{12} \\ a_{21} & a_{22} \end{vmatrix} = a_{11}a_{22} - a_{12}a_{21} \qquad (10.7)$$

The symbol on the left side of eqn (10.7) defines a **determinant** of order 2, the value of which is given on the right. We can similarly express the numerators as determinants of order 2, and we see that the value of the two unknowns is then given by the ratio of two determinants:

$$x = \frac{\begin{vmatrix} b_1 & b_2 \\ a_{12} & a_{22} \end{vmatrix}}{\begin{vmatrix} a_{11} & a_{12} \\ a_{21} & a_{22} \end{vmatrix}} \text{ and } y = \frac{\begin{vmatrix} b_2 & b_1 \\ a_{21} & a_{11} \end{vmatrix}}{\begin{vmatrix} a_{11} & a_{12} \\ a_{21} & a_{22} \end{vmatrix}} \tag{10.8}$$

The purpose of introducing this notation is that it readily extends to n linear algebraic equations in n unknowns. The problem then reduces to one of evaluating the respective determinants of order n.

Problem 10.1

Solve the following simultaneous equations for x and y, by evaluating the appropriate determinants according to eqn (10.8).

$$2x + y = 5 \tag{10.9}$$

$$\tfrac{1}{2}x + 8y = 9 \tag{10.10}$$

Hint: You will need to associate each of the coefficients a_{11}, a_{12}, a_{21}, a_{22}, b_1 and b_2 in eqns (10.1) and (10.2) with those in eqns (10.9) and (10.10).

This type of problem arises in a chemical context quite frequently. For example, the activation energy of a chemical reaction can be determined by measuring the rate constant for a particular reaction at two different temperatures. The relationship between rate constant and temperature is given by the Arrhenius equation:

$$k = A\mathrm{e}^{-E_a/RT} \tag{10.11}$$

where E_a is the activation energy for the reaction, and A is the so-called pre-exponential factor. We can convert the Arrhenius equation to linear form by taking logs of both sides:

$$\ln k = \ln A - \frac{E_a}{RT}. \tag{10.12}$$

If we now measure the rate constant at two different temperatures, T_1 and T_2, we obtain a pair of simultaneous linear equations, which we can solve for the two unknowns, E_a and $\ln A$:

$$\ln k_1 = \ln A - \frac{E_a}{RT_1} \qquad (10.13)$$

$$\ln k_2 = \ln A - \frac{E_a}{RT_2} \qquad (10.14)$$

Problem 10.2

(a) By analogy with eqns (10.1) and (10.2), identify which terms in eqns (10.13) and (10.14) correlate with the constant coefficients a_{11}, a_{12}, a_{21}, a_{22}, b_1 and b_2, and which terms correlate with the unknowns, x and y.

(b) If the temperatures on the Farenheit and Centigrade scales are T and t, respectively, we can express the relationship between the two as $T = at + b$, where a and b are unknowns. By analogy with eqns (10.1) and (10.2), use the method of determinants to obtain the values of a and b by using the boiling and freezing points of water on the two temperature scales. Hence find the formula relating T to t.

Hint: The boiling and freezing points of water on the Farenheit scale are $T_b = 212\ ^\circ F$ and $T_f = 32\ ^\circ F$.

10.2 Expanding Determinants

In general, a determinant of order n is defined as a square array of n^2 elements arranged in n rows and n columns:

$$
\begin{vmatrix}
a_{11} & a_{12} & \cdots & a_{1n} \\
a_{21} & a_{22} & \cdots & a_{2n} \\
\vdots & \vdots & \ddots & \vdots \\
a_{n1} & a_{n2} & \cdots & a_{nn}
\end{vmatrix}
\qquad (10.15)
$$

The elements of this determinant are denoted by a_{ij} or b_{ij}, where i denotes the row and j the column number. Note that the letter used commonly derives from the label applied to a related square matrix – a consequence of the common definition of a determinant as an operation on a square matrix (see Section 11.1). We have seen above in eqn (10.7) that a determinant of order 2 is evaluated in terms of the elements a_{ij} which lie at the intersection of the ith row with the jth column of the determinant.

A determinant of order 3, which might result from a problem involving three simultaneous equations in three unknowns, is evaluated by expanding as follows:

$$\begin{vmatrix} a_{11} & a_{12} & a_{13} \\ a_{21} & a_{22} & a_{23} \\ a_{31} & a_{32} & a_{33} \end{vmatrix} = a_{11} \begin{vmatrix} a_{22} & a_{23} \\ a_{32} & a_{33} \end{vmatrix} - a_{12} \begin{vmatrix} a_{21} & a_{23} \\ a_{31} & a_{33} \end{vmatrix}$$
$$+ a_{13} \begin{vmatrix} a_{21} & a_{22} \\ a_{31} & a_{32} \end{vmatrix} \tag{10.16}$$

This expansion proceeds by taking the elements of the first row in turn, and multiplying each one by the determinant of what remains on crossing out the row and column containing the chosen element, and then attaching the sign $(-1)^{i+j}$. The signed determinants of order 2 in eqn (10.16) are known as the first order cofactors A_{11}, A_{12} and A_{13} of the three elements a_{11}, a_{12} and a_{13}, respectively.

In general, the n^2 cofactors of any determinant of order n are obtained by deleting *one* row and *one* column to form a determinant of order $n-1$, the value of which is multiplied by an odd or even power of -1, depending upon the choice of row index and column index. Thus, if the ith row and j^{th} column of a determinant of order n are both deleted, then the ij^{th} cofactor, A_{ij}, is formed from the value of the resulting determinant of order $n-1$, multiplied by $(-1)^{i+j}$. For example, the cofactor A_{12} of the determinant of order 3 in eqn (10.16) is obtained by deleting the *first* row and the *second* column of the determinant, and multiplying the resulting determinant of order 2 by $(-1)^{1+2}$:

$$A_{12} = (-1)^{1+2} \begin{vmatrix} a_{21} & a_{23} \\ a_{31} & a_{33} \end{vmatrix} = - \begin{vmatrix} a_{21} & a_{23} \\ a_{31} & a_{33} \end{vmatrix} \tag{10.17}$$

Rewriting eqn (10.16) in terms of the three cofactors:

$$A_{11} = \begin{vmatrix} a_{22} & a_{23} \\ a_{32} & a_{33} \end{vmatrix}, \ A_{12} = - \begin{vmatrix} a_{21} & a_{23} \\ a_{31} & a_{33} \end{vmatrix} \text{ and}$$
$$A_{13} = \begin{vmatrix} a_{21} & a_{22} \\ a_{31} & a_{32} \end{vmatrix} \tag{10.18}$$

yields:

$$\begin{vmatrix} a_{11} & a_{12} & a_{13} \\ a_{21} & a_{22} & a_{23} \\ a_{31} & a_{32} & a_{33} \end{vmatrix} = a_{11}A_{11} + a_{12}A_{12} + a_{13}A_{13}. \tag{10.19}$$

If we now expand each of the cofactors (all determinants of order 2) according to eqn (10.7), we obtain the full expansion of the determinant, given in eqn (10.20), expressed as a sum of three positive and three negative terms:

$$\begin{vmatrix} a_{11} & a_{12} & a_{13} \\ a_{21} & a_{22} & a_{23} \\ a_{31} & a_{32} & a_{33} \end{vmatrix} = a_{11}a_{22}a_{33} - a_{11}a_{23}a_{32} - a_{12}a_{21}a_{33} \tag{10.20}$$

$$+ a_{12}a_{23}a_{31} + a_{13}a_{21}a_{32} - a_{13}a_{22}a_{31}$$

In this example, the determinant is initially expanded from the first row but, in fact, we could just as easily expand from any row or column. Thus, for example, expanding from column two gives the alternative expansion:

$$\begin{vmatrix} a_{11} & a_{12} & a_{13} \\ a_{21} & a_{22} & a_{23} \\ a_{31} & a_{32} & a_{33} \end{vmatrix} = a_{12}A_{12} + a_{22}A_{22} + a_{32}A_{32} \tag{10.21}$$

which, upon expanding the cofactors, yields:

$$\begin{vmatrix} a_{11} & a_{12} & a_{13} \\ a_{21} & a_{22} & a_{23} \\ a_{31} & a_{32} & a_{33} \end{vmatrix} = -a_{12}\begin{vmatrix} a_{21} & a_{23} \\ a_{31} & a_{33} \end{vmatrix} + a_{22}\begin{vmatrix} a_{11} & a_{13} \\ a_{31} & a_{33} \end{vmatrix}$$

$$-a_{32}\begin{vmatrix} a_{11} & a_{13} \\ a_{21} & a_{23} \end{vmatrix}. \tag{10.22}$$

Expanding each of the determinants of order 2 in eqn (10.22) yields eqn (10.20), but with the six terms on the right in a different order.

A slightly quicker route to ensuring the correct signs in the sum of the cofactor values is obtained by remembering the general rule for expansion from any row or column in pictoral form:

$$\begin{vmatrix} + & - & + & - & + \\ - & + & - & + & - \\ + & - & + & - & \cdots \\ - & + & - & \cdots & \cdots \\ + & - & \cdots & \cdots & \cdots \end{vmatrix} \tag{10.23}$$

We shall discuss cofactors again when we meet matrix inverses in Chapter 11.

Worked Problem 10.1

Q. Expand the following determinants from the given row or column, as indicated:

(a) $\begin{vmatrix} 0 & 1 \\ 1 & 0 \end{vmatrix}$;

(b) $\begin{vmatrix} \cos\theta & -\sin\theta & 0 \\ \sin\theta & \cos\theta & 0 \\ 0 & 0 & 1 \end{vmatrix}$, from column 3.

A. From the definition given for the expansion of a determinant of order two, we have:

(a) $\begin{vmatrix} 0 & 1 \\ 1 & 0 \end{vmatrix} = 0 - 1 \times 1 = -1$;

(b) $\begin{vmatrix} \cos\theta & -\sin\theta & 0 \\ \sin\theta & \cos\theta & 0 \\ 0 & 0 & 1 \end{vmatrix} = 0 + 0 + 1 \times \begin{vmatrix} \cos\theta & -\sin\theta \\ \sin\theta & \cos\theta \end{vmatrix}$

$$= \cos^2\theta + \sin^2\theta = 1.$$

Problem 10.3

Expand $\begin{vmatrix} 1 & -1 & 2 \\ 0 & 3 & 0 \\ 2 & -2 & -2 \end{vmatrix}$ from (a) column 2 and (b) row 2.

Problem 10.4

(a) Evaluate the cofactors A_{33}, A_{22}, A_{32} and A_{23} of
$\begin{vmatrix} 1 & 0 & -2 \\ 2 & 8 & 4 \\ 3 & 2 & 2 \end{vmatrix}$.

(b) Evaluate the cofactors A_{12} and A_{21} of $\begin{vmatrix} \cos\theta & -\sin\theta & 0 \\ \sin\theta & \cos\theta & 0 \\ 0 & 0 & 1 \end{vmatrix}$.

10.3 Properties of Determinants

(1)   A determinant is unaltered in value if *all* rows and columns are interchanged:

$$e.g. \quad \begin{vmatrix} 1 & 2 \\ 3 & 4 \end{vmatrix} = \begin{vmatrix} 1 & 3 \\ 2 & 4 \end{vmatrix} = -2 \qquad (10.24)$$

(2)   A determinant changes sign if two rows or columns are interchanged:

$$\begin{vmatrix} 1 & 2 \\ 3 & 4 \end{vmatrix} = - \begin{vmatrix} 2 & 1 \\ 4 & 3 \end{vmatrix} = -2 \qquad (10.25)$$

(3)   A factor can be removed from each element of *one* row (or column) to give a new determinant, the value of which when multiplied by the factor gives the original value of the determinant. For example:

$$\begin{vmatrix} 1 & 2 \\ 3 & 4 \end{vmatrix} = 2 \begin{vmatrix} 1 & 1 \\ 3 & 2 \end{vmatrix} = -2 \qquad (10.26)$$

Here, the factor 2 has been removed from column 2. Conversely, when a determinant is multiplied by a constant, the constant can be absorbed into the determinant by multiplying the elements of *one* row (or column) by that constant.

(4)   The value of a determinant is unaltered if a constant multiple of one row or column is added to or subtracted from another row or column, respectively: for example, if we subtract twice column 1 from column 2, we obtain:

$$\begin{vmatrix} 1 & 2 \\ 3 & 4 \end{vmatrix} = \begin{vmatrix} 1 & 0 \\ 3 & -2 \end{vmatrix} = -2. \qquad (10.27)$$

10.4 Strategies for Expanding Determinants Where $n > 3$

We can take a number of different approaches for evaluating determinants of higher order:

(a)   For determinants of order 3 or lower, it is easiest to expand in full from the row or column containing the greatest number of zeros (for example, see Problem 10.3).

(b) For determinants of order 4 to about 6, it is best to introduce as many zeros as possible to the right (or left) of the leading diagonal using properties (1) – (4) (Section 10.3). If all the elements to the right or left of the leading diagonal are zero, then:

$$\begin{vmatrix} a_{11} & 0 & 0 & \cdots & 0 \\ a_{21} & a_{22} & 0 & \cdots & 0 \\ a_{31} & a_{32} & a_{33} & \cdots & 0 \\ \vdots & \vdots & \vdots & \ddots & \vdots \\ a_{n1} & a_{n2} & a_{n3} & \cdots & a_{nn} \end{vmatrix} = a_{11}a_{22}a_{33}\cdots a_{nn}, \qquad (10.28)$$

and the expansion of the determinant is given by the product of the elements lying on the leading diagonal.

(c) When expanding determinants of high order ($n > 5$), it is much more convenient to use one of the widely available computer algebra systems (Maple, Mathematica, *etc.*) or a numerical computer algorithm. There are many chemical situations in which we have to expand determinants of large order. For example, in computing the vibrational frequencies of ethene, it is necessary to expand a determinant of order 12 (for a non-linear molecule containing N nuclei, the order will be $3N - 6$).

Problem 10.5

Expand each of the following determinants:

$$\text{(a)} \begin{vmatrix} 1 & 2 & 3 \\ 0 & 8 & 2 \\ -2 & 4 & 2 \end{vmatrix}; \text{ (b)} \begin{vmatrix} 1 & 0 & -2 \\ 2 & 8 & 4 \\ 3 & 2 & 2 \end{vmatrix}$$

(i) from row 2;
(ii) using row/column operations to transform the first row to 1 0 0 before expansion from row 1;
(iii) using row/column operations to transform all the elements to the right of the leading diagonal to zero, before expanding from row 1.

Problem 10.6

In using the Hückel model for calculating the molecular orbital energies, ε, for electrons in the π shell of the allyl system, it is necessary to solve the following equation:

$$\begin{vmatrix} \alpha - \varepsilon & \beta & 0 \\ \beta & \alpha - \varepsilon & \beta \\ 0 & \beta & \alpha - \varepsilon \end{vmatrix} = 0, \tag{10.29}$$

where the symbols α and β are parameters (both negative in value) of the model.

(a) Use property (3) of determinants to remove a factor of β from each row (or column) of the determinant shown in eqn (10.29).

Hint: Division of each element in *one* row or column by β results in a new determinant, the value of which is multiplied by β. Thus division of every element in the determinant results in a new determinant, the value of which is multiplied by β^3.

(b) Show that, on making the substitution $x = (\alpha - \varepsilon) / \beta$, the expansion of the determinant yields $x^3 - 2x = 0$.
(c) Find the three roots of this equation.
(d) Deduce the three orbital energies.

Summary of Key Points

This chapter develops the concept of the determinant as a precursor to a more complete treatment of matrix algebra in Chapter 4. The key points discussed include:

1. The use of determinants to solve sets of simultaneous linear equations.
2. The expansion of determinants of low order in full.
3. Cofactors of determinants.
4. Properties of determinants and their use to simplify the evaluation of determinants of high order.

11
Working with Arrays II: Matrices and Matrix Algebra

In the previous chapter, we saw how determinants are used to tackle problems involving the solution of systems of linear equations. In general, the branch of mathematics which deals with linear systems is known as **linear algebra**, in which matrices and vectors play a dominant role. In this chapter we shall explore how matrices and matrix algebra are used to address problems involving coordinate transformations, as well as revisiting the solution of sets of simultaneous linear equations. Vectors are explored in Chapter 12.

Matrices are two-dimensional arrays (or tables) with specific shapes and properties:

$$\begin{pmatrix} 2 & -1 \\ 0 & 3 \end{pmatrix}, \begin{pmatrix} x_1 \\ x_2 \\ x_3 \end{pmatrix}, (1 \quad 2 \quad 3).$$

Their key property is that they give us a formalism for systematically handling sets of objects – called **elements** – which, for example, can be numbers, chemical property values, algebraic quantities or integrals. Superficially, matrices resemble determinants, insofar as they are constructed from arrays of elements. However, as we shall see, they are really quite distinct from one another. The most important difference is that while a determinant expands to yield an expression (and a value, when its elements are numbers), a matrix does not.

Aims:

In this chapter we develop matrix algebra from two key perspectives: one makes use of matrices to facilitate the handling of coordinate transformations, in preparation for a development of symmetry theory; the other revisits determinants, and through the definition of the matrix inverse, provides a means for solving sets of linear equations.

By the end of this chapter, you should:

- Recognise the difference between a matrix and a determinant.
- Recognise how matrices can be used to handle large linear systems in a compact way.
- Be comfortable working with basic operations of matrix algebra (addition, subtraction, multiplication).
- Recognise specific kinds of matrix.
- Use the special properties of a square matrix to evaluate its determinant and inverse.
- Understand the basic principles of group theory.

11.1 Introduction: Some Definitions

A **matrix** is an array of **elements**, comprising n rows and m columns, enclosed in parentheses (round brackets). By convention, matrices are named using bold typeface letters of upper or lower case, such as $\mathbf{A}$ or $\mathbf{b}$, so we could, for example, label the matrices above as:

$$\mathbf{B} = \begin{pmatrix} 2 & -1 \\ 0 & 3 \end{pmatrix}, \ \mathbf{c} = \begin{pmatrix} x_1 \\ x_2 \\ x_3 \end{pmatrix}, \ \mathbf{d} = \begin{pmatrix} 1 & 2 & 3 \end{pmatrix}.$$

The elements of the matrix are usually denoted a_{ij} or b_{ij} (depending on the letter used to label the matrix itself), where i denotes the row and j the column number. Thus, for example, the matrix $\mathbf{B}$ above has two rows and two columns, and is said to be a 2×2 matrix; however, as the matrix is square, it is commonly named a **square matrix** of **order** two, with elements assigned as follows:

$$b_{11} = 2, \ b_{12} = -1, \ b_{21} = 0, \ b_{22} = 3.$$

Sometimes, it is more convenient to use the notation $(\mathbf{B})_{ij}$ to indicate the ij^{th} element of matrix $\mathbf{B}$. Similarly, the 3×1 matrix $\mathbf{c}$ is called a **column matrix**, and the 1×3 matrix $\mathbf{d}$ is called a **row matrix**. The general matrix, $\mathbf{A}$, having order $(n \times m)$ is called a **rectangular matrix** with elements:

$$\mathbf{A} = \begin{pmatrix} a_{11} & a_{12} & \cdots & a_{1m} \\ a_{21} & a_{22} & \cdots & a_{2m} \\ \vdots & \vdots & \vdots & \vdots \\ a_{n1} & a_{n2} & \cdots & a_{nm} \end{pmatrix}. \tag{11.1}$$

Two matrices $\mathbf{A}$ and $\mathbf{B}$ are equal if, and only if, $a_{ij} = b_{ij}$ for all i,j. This also implies that the two matrices have the same order.

Problem 11.1

For each of the matrices $\mathbf{b} = \begin{pmatrix} 1 & 1 & 1 \\ 2 & -2 & 2 \end{pmatrix}$, $\mathbf{c} = \begin{pmatrix} 3 & -1 \\ 1 & -3 \end{pmatrix}$, $\mathbf{d} = \begin{pmatrix} 1 \\ 0 \end{pmatrix}$ and $\mathbf{e} = (0 - i \quad 1 \quad i)$:

(a) Name their shapes.
(b) List their elements.
(c) Give their order.

11.2 Rules for Combining Matrices

In this section, we explore the matrix analogues of addition, subtraction and multiplication of numbers. The analogue for division (the inverse operation of multiplication) has no direct counterpart for matrices.

11.2.1 Multiplication of a Matrix by a Constant

The multiplication of a matrix, $\mathbf{A}$, by a constant c (a real, imaginary or complex number), is achieved by simply multiplying each element by the constant, resulting in the elements changing from a_{ij} to $c\,a_{ij}$, for all i, j.

Problem 11.2

Multiply the following matrices by 2:

(a) $\mathbf{B} = \begin{pmatrix} 4 & 5 \\ 1 & 6 \\ -4 & 3 \end{pmatrix}$; (b) $\mathbf{C} = \begin{pmatrix} 2 & \frac{5}{2} \\ \frac{1}{2} & 3 \\ -2 & \frac{3}{2} \end{pmatrix}$.

11.2.2 Addition and Subtraction of Matrices

If two matrices have the same order, then addition and subtraction are defined as:

$$\mathbf{C} = \mathbf{A} \pm \mathbf{B}, \text{ with } a_{ij} \pm b_{ij}, \text{ for all } i, j. \qquad (11.2)$$

Neither addition nor subtraction is defined for combining matrices of different orders.

Worked Problem 11.1

Q. Given the following matrices (notice that **A** has real, complex and imaginary elements):

$$\mathbf{A} = \begin{pmatrix} 2 & 2i & -3 \\ 1 & 2 & 8 \\ 5 & -4+i & 1 \end{pmatrix}, \mathbf{B} = \begin{pmatrix} 4 & 5 \\ 1 & 6 \\ -4 & 3 \end{pmatrix},$$

$$\mathbf{C} = \begin{pmatrix} 3 \\ 1 \\ 2 \end{pmatrix}, \mathbf{D} = \begin{pmatrix} 1 & 1 \\ 4 & 1 \\ 5 & 1 \end{pmatrix}, \mathbf{F} = \begin{pmatrix} 2 \\ 1 \\ 4 \end{pmatrix},$$

evaluate:

(a) $\mathbf{G} = 2\mathbf{B} + \mathbf{D}$,
(b) $\mathbf{M} = \mathbf{C} - 2\mathbf{F}$,
(c) $\mathbf{H} = i\mathbf{A}$

A. (a) $\mathbf{G} = \begin{pmatrix} 8 & 10 \\ 2 & 12 \\ -8 & 6 \end{pmatrix} + \begin{pmatrix} 1 & 1 \\ 4 & 1 \\ 5 & 1 \end{pmatrix} = \begin{pmatrix} 9 & 11 \\ 6 & 13 \\ -3 & 7 \end{pmatrix}$

(b) $\mathbf{M} = \begin{pmatrix} 3 \\ 1 \\ 2 \end{pmatrix} - 2 \begin{pmatrix} 2 \\ 1 \\ 4 \end{pmatrix} = \begin{pmatrix} 3-4 \\ 1-2 \\ 2-8 \end{pmatrix} = \begin{pmatrix} -1 \\ -1 \\ -6 \end{pmatrix}$

(c) $\mathbf{H} = i \begin{pmatrix} 2 & 2i & -3 \\ 1 & 2 & 8 \\ 5 & -4+i & 1 \end{pmatrix} = \begin{pmatrix} 2i & -2 & -3i \\ i & 2i & 8i \\ 5i & -4i-1 & i \end{pmatrix}$

Problem 11.3

If $\mathbf{A} = \begin{pmatrix} 1 & i \\ -i & 1 \end{pmatrix}$, $\mathbf{B} = \begin{pmatrix} 1 & -i \\ i & 1 \end{pmatrix}$,

$\mathbf{R} = \begin{pmatrix} \cos\theta & \sin\theta \\ -\sin\theta & \cos\theta \end{pmatrix}$, $\mathbf{S} = \begin{pmatrix} \cos\theta & -\sin\theta \\ \sin\theta & \cos\theta \end{pmatrix}$,

express (a) $\mathbf{A} + \mathbf{B}$, (b) $\mathbf{A} - \mathbf{B}$, (c) $\mathbf{R} + \mathbf{S}$ and (d) $\mathbf{R} - \mathbf{S}$ in terms of $\mathbf{C} = \begin{pmatrix} 0 & 1 \\ -1 & 0 \end{pmatrix}$ and $\mathbf{D} = \begin{pmatrix} 1 & 0 \\ 0 & 1 \end{pmatrix}$.

11.2.3 Matrix Multiplication

Given an $n \times m$ matrix, $\mathbf{A}$, and an $m \times p$ matrix, $\mathbf{B}$, then the ij^{th} element, c_{ij}, of the resulting $n \times p$ product matrix $\mathbf{C} = \mathbf{AB}$, is found by selecting the i, j values and then for each choice, summing the products of the elements in *row i* of $\mathbf{A}$ with those in *column j* of $\mathbf{B}$ (Figure 11.1):

$$\begin{array}{ccccc} \mathbf{C} & = & \mathbf{A} & \mathbf{B} \\ (n \times p) & & (n \times m) & (m \times p) \end{array} \tag{11.3}$$

A number of features relating to matrix multiplication are worthy of note.

- If the number of columns in $\mathbf{A}$ is not equal to the number of rows in $\mathbf{B}$, then multiplication is undefined.
- In general, even if $\mathbf{AB}$ is defined, then $\mathbf{BA}$ may not be defined.
- If $\mathbf{AB}$ and $\mathbf{BA}$ are both defined, their orders may differ.
- Even if $\mathbf{AB}$ and $\mathbf{BA}$ have the same order, the two product matrices may not be equal. In these circumstances, matrix multiplication is non-commutative, *i.e.* $\mathbf{AB} \neq \mathbf{BA}$.

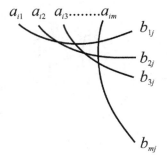

Figure 11.1 The product of an $n \times m$ matrix, **A**, and an $m \times p$ matrix, **B** is an $n \times p$ matrix **C**, whose ij^{th} element, c_{ij}, is found by summing the products of the elements in *row i* of **A** with those in *column j* of **B**.

Worked Problem 11.2

Q. Where defined, determine the products (a) **AB** and (b) **BA** of the following matrices:

$$A = \begin{pmatrix} 1 & 3 \\ 3 & 1 \end{pmatrix} \quad B = \begin{pmatrix} 1 & 1 & 2 \\ 1 & 2 & 1 \end{pmatrix}$$

A. (a) Matrix **A** is 2×2 and **B** is 2×3; thus the product **AB** is defined, as the number of columns in **A** is same as the number of rows in **B**. The product matrix will have order (2×3) (the number of rows in **A** and the number of columns in **B**):

$$AB = \begin{pmatrix} 1 & 3 \\ 3 & 1 \end{pmatrix} \begin{pmatrix} 1 & 1 & 2 \\ 1 & 2 & 1 \end{pmatrix}$$

$$= \begin{pmatrix} (1 \times 1) + (3 \times 1) & (1 \times 1) + (3 \times 2) & (1 \times 2) + (3 \times 1) \\ (3 \times 1) + (1 \times 1) & (3 \times 1) + (1 \times 2) & (3 \times 2) + (1 \times 1) \end{pmatrix}$$

$$= \begin{pmatrix} 4 & 7 & 5 \\ 4 & 5 & 7 \end{pmatrix}.$$

(b) **BA** is undefined because the number of columns in **B** is not same as the number of rows in **A**.

Problem 11.4

For the matrices:

$$A = \begin{pmatrix} 1 & 2 \\ 2 & 1 \end{pmatrix}, \quad B = \begin{pmatrix} 1 & -1 \\ -1 & 2 \end{pmatrix},$$

$$C = \begin{pmatrix} -1 & 1 \\ -1 & 1 \end{pmatrix}, \quad D = (1 \quad 2), \quad E = \begin{pmatrix} 3 \\ -1 \end{pmatrix},$$

find each product matrix specified below, where defined, and give its order, as appropriate:

$$AB, \, BA, \, AC, \, BC, \, DE, \, ED, \, DA, \, AD, \, EA, \, AE,$$

$$AB - BA, \, (AB)C, \, A(BC), \, A(B+C), \, AB+AC$$

11.2.3.1 Properties of Matrix Multiplication

You may have observed from your answers to Problem 11.4 that multiplication of matrices follows similar rules to that of numbers, insofar as it is:

- **Associative** :

$$(AB)C = A(BC) \tag{11.4}$$

- **Distributive** :

$$A(B+C) = AB + AC. \tag{11.5}$$

One exception is the commutative law. In general, matrix multiplication is:

- **Non-Commutative** :

$$AB \neq BA \text{ (except in certain special situations).} \tag{11.6}$$

As we suggested earlier, there is no general way of defining matrix division. However, for some square matrices, we can define an operation that looks superficially like division, but is really only multiplication (see Section 11.6).

11.3 Origins of Matrices

11.3.1 Coordinate Transformations

Matrices have their origin in **coordinate transformations**, where, in two dimensions, for example, a chosen point, with coordinates (x, y), is transformed to a new location with coordinates (x', y'). For example, consider an anticlockwise rotation of the point P in the xy-plane, about the z-axis, through an angle θ, as shown in Figure 11.2.

We can use simple trigonometry to relate the coordinates of Q to those of P by expressing the Cartesian coordinates in terms of polar coordinates. Thus the (x, y) coordinates of point P become:

$$x = r \cos \alpha \text{ and } y = r \sin \alpha, \tag{11.7}$$

and those of point Q become:

$$x' = r \cos(\theta + \alpha) \text{ and } y' = r \sin(\theta + \alpha). \tag{11.8}$$

If we now use the addition theorems for cosine and sine (see Section 2.3.3), we obtain the expansions:

$$x' = r \cos(\theta + \alpha)$$
$$= r \cos \theta \cos \alpha - r \sin \theta \sin \alpha \tag{11.9}$$
$$= x \cos \theta - y \sin \theta;$$

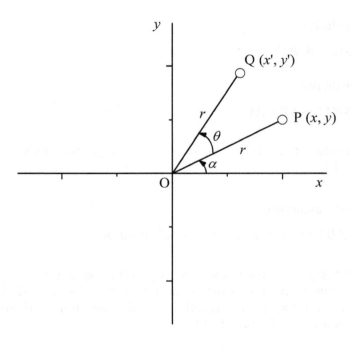

Figure 11.2. Rotation about the z-axis of the point P(x, y) through an angle θ to Q (x', y'). α is the angle between OP and the x-axis.

$$y' = r \sin(\theta + \alpha)$$

$$= r \sin \theta \cos \alpha + r \cos \theta \sin \alpha \qquad (11.10)$$

$$= x \sin \theta + y \cos \theta.$$

which allows us to express the coordinates of Q (x', y') in terms of those of P (x, y).

$$\left. \begin{array}{l} x' = x \cos \theta - y \sin \theta \\ y' = x \sin \theta + y \cos \theta \end{array} \right\} \qquad (11.11)$$

Eqn (11.11) describes the transformation of coordinates under an anticlockwise rotation by an angle, θ. This coordinate transformation is completely characterised by a square matrix, $\mathbf{A}$, with elements $\cos \theta$ and $\pm \sin \theta$, and the column matrices, $\mathbf{r}$, and $\mathbf{r}'$, involving the initial and final coordinates, respectively:

$$\mathbf{A} = \begin{pmatrix} \cos \theta & -\sin \theta \\ \sin \theta & \cos \theta \end{pmatrix}, \mathbf{r}' = \begin{pmatrix} x' \\ y' \end{pmatrix}, \mathbf{r} = \begin{pmatrix} x \\ y \end{pmatrix}. \qquad (11.12)$$

We can now use matrix notation to replace the two eqns (11.11) by the single matrix eqn (11.13):

$$\begin{array}{ccc} \begin{pmatrix} x' \\ y' \end{pmatrix} & = & \begin{pmatrix} \cos \theta & -\sin \theta \\ \sin \theta & \cos \theta \end{pmatrix} & \begin{pmatrix} x \\ y \end{pmatrix} \\ \mathbf{r}' & = & \mathbf{A} & \mathbf{r} \end{array} \qquad (11.13)$$

We can confirm that eqn (11.13) correctly represents the coordinate transformation by evaluating the product matrix $\mathbf{Ar}$ on the right side:

$$\begin{array}{ccc} \begin{pmatrix} x' \\ y' \end{pmatrix} & = & \begin{pmatrix} x \cos \theta - y \sin \theta \\ x \sin \theta + y \cos \theta \end{pmatrix}. \\ \mathbf{r}' & = & \mathbf{A} \, \mathbf{r} \end{array} \qquad (11.14)$$

Since $\mathbf{r}'$ and $\mathbf{Ar}$ are both 2×1 matrices, we can equate the elements in $\mathbf{r}'$ with those in $\mathbf{Ar}$, to restore the original equations, which confirms eqn (11.13) as the correct matrix representation of eqns (11.11).

Worked Problem 11.3

Consider the coordinate transformation involving reflection in the y-axis (Figure 11.3). We can see that this transformation simply involves a change in sign of x, with the value of y remaining unchanged. Thus the transformed point, Q, will have coordinates $(x', y') = (-x, y)$.

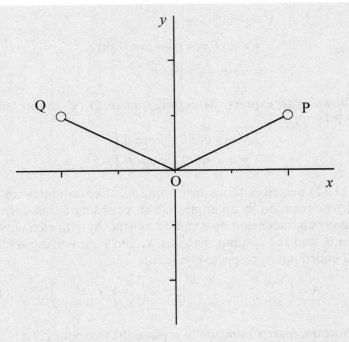

Figure 11.3 Reflection of the point P(x, y) in the y-axis to obtain the point Q($-x$, y).

We can represent this transformation in terms of a matrix equation $\mathbf{r}' = \mathbf{Cr}$, where $\mathbf{C}$ is a 2×2 matrix characterising reflection in the y-axis, and $\mathbf{r}$ and $\mathbf{r}'$ are the column matrices containing the initial and final coordinates, respectively.

Q. Show that the matrix $\mathbf{C}$ characterising reflection in the y-axis is $\mathbf{C} = \begin{pmatrix} -1 & 0 \\ 0 & 1 \end{pmatrix}$.

A. The coordinate transformation written as a matrix equation is:

$$\underset{\mathbf{r}'}{\begin{pmatrix} x' \\ y' \end{pmatrix}} = \begin{pmatrix} -x \\ y \end{pmatrix} = \underset{\mathbf{C}}{\begin{pmatrix} c_{11} & c_{12} \\ c_{21} & c_{22} \end{pmatrix}} \underset{\mathbf{r}}{\begin{pmatrix} x \\ y \end{pmatrix}}$$

where c_{11}, c_{12}, c_{21} and c_{22} are the elements of the matrix $\mathbf{C}$ which characterise reflection in the y-axis. Multiplying out the right side, we have:

$$c_{11}x + c_{12}y = -x \text{ and } c_{21}x + c_{22}y = y$$

Thus, if we compare the x and y coefficients on each side of these equations, we obtain:

$$c_{11} = -1, c_{12} = 0, c_{21} = 0 \text{ and } c_{22} = 1 \text{ and so } \mathbf{C} = \begin{pmatrix} -1 & 0 \\ 0 & 1 \end{pmatrix}.$$

11.3.1.1 Sequential Coordinate Transformations

The effect of applying two sequential coordinate transformations on a point, $\mathbf{r}$, can be represented by the product of the two matrices, each one of which represents the respective transformation. We need to take care, however, that the matrices are multiplied in the correct order because, as we saw above, matrix multiplication is often non-commutative. For example, in order to find the matrix representing an anticlockwise rotation by θ, followed by a reflection in the y-axis, we need to find the product $\mathbf{CA}$ (and not $\mathbf{AC}$ as we might initially assume).

Problem 11.5

(a) Find the matrix, $\mathbf{D}$, describing the coordinate transformation resulting from reflection in the line $y = x$.

(b) (i) Find the matrix, $\mathbf{E}$, describing the coordinate transformation resulting from a reflection in the line $y = x$, followed by a reflection in the y-axis (see Worked Problem 11.3).

(ii) Find the matrix, $\mathbf{F}$, describing the coordinate transformation that results from a reflection in the y-axis followed by a reflection in the line $y = x$.

In each case check your answer graphically, by using the matrices $\mathbf{E}$ and $\mathbf{F}$ to transform the coordinates (1, 2) to their new location (x'', y'').

11.3.2 Coordinate Transformations in Three Dimensions: A Chemical Example

In preparation for the discussion of group theory in Section 11.9, let us consider how we might use matrix representations of coordinate transformations to characterise the shape of a molecule – something of vital importance, for example, in describing the vibrational motions in a molecule. In order to accomplish this objective, we need to consider only those linear transformations in three dimensions that interchange equivalent points in a molecule. One example of such a transformation involves the interchange of coordinates defining the positions of two fluorine nuclei in the planar molecule BF_3. We can achieve this result by extending the rotation and reflection coordinate transformations in Figures 11.2 and 11.3 to three dimensions. For BF_3, there are four mirror planes but, for the moment, let us focus only on the yz mirror plane, which is perpendicular to the plane of the molecule, and contains the boron and the fluorine nucleus F_1 (Figure 11.4).

The matrix $\mathbf{C}$, defined in Worked Problem 11.3, describes reflection in the y-axis of a point defined by the two coordinates (x, y). We can rewrite matrix $\mathbf{C}$ in terms of all three coordinates as follows:

$$
\begin{pmatrix} x' \\ y' \\ z' \end{pmatrix} = \begin{pmatrix} -1 & 0 & 0 \\ 0 & 1 & 0 \\ 0 & 0 & 1 \end{pmatrix} \begin{pmatrix} x \\ y \\ z \end{pmatrix}, \tag{11.15}
$$

$$\mathbf{r'} = \mathbf{C} \quad \mathbf{r}$$

where we note that the z-coordinate is unchanged by the coordinate transformation. Thus, a reflection in the yz-plane interchanges points located at the nuclei F_2 and F_3.

If we now rotate an arbitrary point (x, y, z) about the z-axis, the x and y coordinates are transformed according to matrix $\mathbf{A}$, defined in eqn (11.12), but the z coordinate is unchanged: thus:

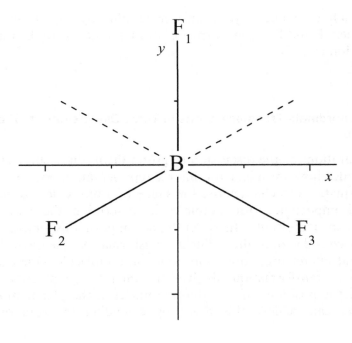

Figure 11.4   The nuclear configuration for BF_3 in the xy-plane. The z-axis is perpendicular to the paper, and passes through B. Three of the mirror planes are perpendicular to the paper, and contain the boron and one of the fluorine nuclei, respectively. The fourth mirror plane lies in the plane of the molecule and contains all of the nuclei.

$$\begin{pmatrix} x' \\ y' \\ z' \end{pmatrix} = \begin{pmatrix} \cos\theta & -\sin\theta & 0 \\ \sin\theta & \cos\theta & 0 \\ 0 & 0 & 1 \end{pmatrix} \begin{pmatrix} x \\ y \\ z \end{pmatrix} \qquad (11.16)$$

$$\mathbf{r}' \quad = \qquad\qquad \mathbf{B} \qquad\qquad \mathbf{r}$$

An anticlockwise rotation of $\theta = 2\pi/3$ (equivalent to $120°$) about the z-axis described in eqn (11.16) transforms a point located at either F_1, F_2 or F_3 to an equivalent point located at F_2, F_3 or F_1, respectively.

The important point here is that if the coordinates of points are represented in matrix form, then the geometrical actions involved in carrying out a rotation or reflection may also be represented by matrices, which enables us to mimic problems in geometry using matrix algebra; that is, geometrical operations on points can be replaced by matrix representations acting on column matrices containing the coordinates of points. We shall re-visit these ideas in Section 11.9, where we develop a brief introduction to the principles of symmetry theory.

11.4 Operations on Matrices

11.4.1 The Transpose of a Matrix

Given an $n \times m$ matrix $\mathbf{B}$, we can construct its transpose, $\mathbf{B}^T$, by interchanging the rows and columns. Thus the ij^{th} element of $\mathbf{B}$ becomes the ji^{th} element of $\mathbf{B}^T$ according to:

$$(\mathbf{B})_{ij} = (\mathbf{B}^T)_{ji} \qquad (11.17)$$

Worked Problem 11.4

Q. Find the transpose of:

$$\text{(a) } \mathbf{B} = \begin{pmatrix} 1 & 1 \\ 3 & 2 \\ 4 & 1 \end{pmatrix}, \text{ (b) } \mathbf{C} = (0 \ -1 \ 1).$$

A.

$$\text{(a) } \mathbf{B}^T = \begin{pmatrix} 1 & 3 & 4 \\ 1 & 2 & 1 \end{pmatrix}, \text{ (b) } \mathbf{C}^T = \begin{pmatrix} 0 \\ -1 \\ 1 \end{pmatrix}.$$

Problem 11.6

Find the transpose of each of the following matrices:

$$\text{(a) } \mathbf{A} = \begin{pmatrix} \cos \theta & -\sin \theta \\ \sin \theta & \cos \theta \end{pmatrix}, \text{(b) } \mathbf{C} = \begin{pmatrix} -1 & 1 \\ -1 & 1 \end{pmatrix},$$

$$\text{(c) } \mathbf{D} = \begin{pmatrix} 1 & 3 & 4 \\ 1 & 2 & 1 \end{pmatrix}.$$

Problem 11.7

If $\mathbf{X}$ is an $n \times m$ matrix:

(a) Give the order of $\mathbf{XX^T}$ and $\mathbf{X^TX}$;
(b) Use the matrix $\mathbf{B}$ from Worked Problem 11.2 to find $\mathbf{BB^T}$ and $\mathbf{B^TB}$.

11.4.2 The Complex Conjugate Matrix

Taking the complex conjugate of every element of a matrix, $\mathbf{A}$, yields the **complex conjugate matrix**, $\mathbf{A}^*$: that is, $(\mathbf{A}^*)_{ij} = (\mathbf{A})_{ij}^*$. If all the elements of $\mathbf{A}$ are real, then $\mathbf{A}^* = \mathbf{A}$.

11.4.3 The Complex Conjugate Transposed Matrix

The transpose of the complex conjugate matrix (sometimes termed the **adjoint matrix**), is written as $\mathbf{A}^\dagger$ and defined such that:

$$\mathbf{A}^\dagger = (\mathbf{A}^*)^T \equiv (\mathbf{A}^T)^* \Rightarrow (\mathbf{A}^\dagger)_{ij} = (\mathbf{A}^*)_{ji.} \qquad (11.18)$$

If $\mathbf{A}^* = \mathbf{A}$ (a real matrix) then $\mathbf{A}^\dagger = \mathbf{A}^T$.

Problem 11.8

If $\mathbf{A} = \begin{pmatrix} 1+i & i \\ -i & 1 \end{pmatrix}$, give the forms of $\mathbf{A}^*$ and $\mathbf{A}^\dagger$.

Problem 11.9

Show that for the matrices $\mathbf{A} = \begin{pmatrix} 1 & 1-i \\ 1+i & -1 \end{pmatrix}, \mathbf{B} = \begin{pmatrix} 1 & 1+i \\ 1+i & 0 \end{pmatrix}$:

(a) $(\mathbf{AB})^* = \mathbf{A}^*\mathbf{B}^*$,   (b) $(\mathbf{AB})^\dagger = \mathbf{B}^\dagger\mathbf{A}^\dagger$

Note: These results are valid for any matrices $\mathbf{A}$ and $\mathbf{B}$, for which multiplication is defined.

11.4.4 The Trace of a Square Matrix

The trace of a square matrix, $\mathbf{A}$, of order n, denoted by $\mathrm{tr}\mathbf{A}$, is defined as the sum of its diagonal elements:

$$\mathrm{tr}\,\mathbf{A} = \sum_{i=1}^{n}(\mathbf{A})_{ii} \tag{11.19}$$

For example, the matrix $\mathbf{A} = \begin{pmatrix} 1 & -1 & 0 \\ 2 & -3 & 1 \\ 1 & -2 & 0 \end{pmatrix}$, has $\mathrm{tr}\mathbf{A} = 1 - 3 + 0 = -2$.

Since the transpose of a square matrix leaves the diagonal unchanged, we see that $\mathrm{tr}\mathbf{A} = \mathrm{tr}\mathbf{A}^\mathrm{T}$.

Problem 11.10

For the matrices:

$$\mathbf{A} = \begin{pmatrix} 1 & -1 \\ 0 & 3 \end{pmatrix}, \mathbf{B} = \begin{pmatrix} 0 & 1 \\ 1 & -2 \end{pmatrix},$$

$$\mathbf{C} = \begin{pmatrix} -1 & 1 \\ 1 & 0 \end{pmatrix}, \mathbf{D} = \begin{pmatrix} 1 & -1 & 0 \\ 1 & -2 & 0 \end{pmatrix},$$

show that:

(a) $\mathrm{tr}(\mathbf{AB}) = \mathrm{tr}(\mathbf{BA})$, (b) $\mathrm{tr}(\mathbf{ABC}) = \mathrm{tr}(\mathbf{CAB}) = \mathrm{tr}(\mathbf{BCA})$,
(c) $\mathrm{tr}(\mathbf{D}^\mathrm{T}\mathbf{D}) = \mathrm{tr}(\mathbf{DD}^\mathrm{T})$.

11.4.5 The Matrix of Cofactors

The cofactor of a determinant, which we first defined in Section 10.2, is characterised by a row and column index, in much the same way as we characterise the elements in a matrix. Thus, we can form the matrix of cofactors by accommodating each cofactor in its appropriate position. For example, the determinant:

$$\det \mathbf{A} = \begin{vmatrix} a_{11} & a_{12} \\ a_{21} & a_{22} \end{vmatrix} \tag{11.20}$$

gives rise to the four cofactors A_{11}, A_{12}, A_{21} and A_{22}, which may be collected together in the matrix of cofactors, $\mathbf{B}$:

$$\mathbf{B} = \begin{pmatrix} A_{11} & A_{12} \\ A_{21} & A_{22} \end{pmatrix} = \begin{pmatrix} a_{22} & -a_{21} \\ -a_{12} & a_{11} \end{pmatrix} \tag{11.21}$$

11.5 The Determinant of a Product of Square Matrices

For two square matrices, $\mathbf{A}$ and $\mathbf{B}$ of order n, the determinant of the product matrix $\mathbf{AB}$ is given by the product of the two determinants:

$$\det(\mathbf{AB}) = \det(\mathbf{BA}) = \det\mathbf{A} \; \det\mathbf{B} \tag{11.22}$$

We now return to a further discussion of some special matrices that arise in a chemical context.

11.6 Special Matrices

So far, we have met matrices of different orders, but we have not been concerned with the properties of their constituent elements. In this section, we introduce the null and unit matrices, and then present a catalogue of important kinds of matrix that are common in developing mathematical models used, for example, in the calculation of vibrational frequencies of molecules, distributions of electron density and other observable properties of molecules.

11.6.1 The Null Matrix

The general null matrix is an n by m matrix, all of whose elements are zero. If the matrix is:

- Rectangular, it is named as $\mathbf{O}_{nm}$.
- Square ($n = m$) it is named as $\mathbf{O}_n$

- A column matrix, it is named O_{n1} or more commonly and simply as O.

Given an $m \times n$ matrix X,

$$O_{nm}X_{mn} = O_n, \quad X_{mn}O_{nm} = O_m. \tag{11.23}$$

11.6.2 The Unit Matrix

The unit matrix is a square matrix of order n, denoted here by E_n whose leading diagonal elements are all unity (*i.e.* have value 1), and whose off-diagonal elements are zero. Thus, for example:

$$E_2 = \begin{pmatrix} 1 & 0 \\ 0 & 1 \end{pmatrix}, \, E_3 = \begin{pmatrix} 1 & 0 & 0 \\ 0 & 1 & 0 \\ 0 & 0 & 1 \end{pmatrix}. \tag{11.24}$$

The elements of E_n may be denoted e_{ij} but, in practice, they are usually specified using the Kronecker delta, which is written as:

$$e_{ij} = \delta_{ij} = \begin{cases} 1, & \text{for } i=j \\ 0, & \text{for } i \neq j \end{cases} \tag{11.25}$$

where $i = j$ designates a diagonal position and $i \neq j$ a non-diagonal position.

As E_n is an $n \times n$ matrix, E_nA (pre-multiplication of A by E_n) is equal to A if A has order $(n \times p)$; likewise AE_n (post-multiplication of A by E_n) yields A if A has order $(p \times n)$.

Problem 11.11

For each of the following matrix products:

(a) $O_{23} \begin{pmatrix} 1 & 3 \\ 2 & 2 \\ 0 & 1 \end{pmatrix}$; (b) $\begin{pmatrix} 1 & 2 & 3 \\ 4 & 5 & 6 \end{pmatrix} O_{23}$;

(c) $E_3 \begin{pmatrix} 1 & 3 \\ 2 & 2 \\ 0 & 1 \end{pmatrix}$; (d) $\begin{pmatrix} 1 & 2 & 3 \\ 4 & 5 & 6 \end{pmatrix} E_3$; (e) $\begin{pmatrix} 1 & 3 \\ 2 & 2 \\ 0 & 1 \end{pmatrix} E_3$,

give the resultant matrix, where the product is defined.

Problem 11.12

$$\text{If } \mathbf{A} = \begin{pmatrix} \cos\theta & -\sin\theta \\ \sin\theta & \cos\theta \end{pmatrix}, \text{ and } \mathbf{G} = \begin{pmatrix} 0 & -1 & 1 \\ 2 & -1 & 2 \\ 1 & 1 & 1 \end{pmatrix}$$

(a)

    (i) Find $\det\mathbf{A}$, and the matrix of cofactors, $\mathbf{B}$;

    (ii) Show that $\mathbf{B}^{\mathrm{T}}\mathbf{A} = \mathbf{E}_2\det\mathbf{A}$.

(b) Show that $\mathbf{H}^{\mathrm{T}}\mathbf{G} = \mathbf{E}_3\det\mathbf{G}$, where $\mathbf{H}$ is the matrix of cofactors of $\det\mathbf{G}$.

11.7 Matrices with Special Properties

11.7.1 Symmetric Matrices

A square matrix, $\mathbf{A}$, is said to be symmetric if it has the property $(\mathbf{A})_{ij} = (\mathbf{A})_{ji}$: that is,

$$\mathbf{A} = \mathbf{A}^{\mathrm{T}}. \tag{11.26}$$

For example, the following matrix is symmetric:

$$\mathbf{A} = \begin{pmatrix} 2 & 1 & 3 \\ 1 & 4 & 3 \\ 3 & 3 & 0 \end{pmatrix}$$

as reflection in the leading diagonal leaves the array of elements unchanged in appearance.

For any n by m matrix $\mathbf{X}$, both $\mathbf{X}^{\mathrm{T}}\mathbf{X}$ and $\mathbf{X}\mathbf{X}^{\mathrm{T}}$ are symmetric matrices.

11.7.2 Orthogonal Matrices

An orthogonal matrix, $\mathbf{A}$, is a square matrix of order n with the property

$$\mathbf{A}^{\mathrm{T}}\mathbf{A} = \mathbf{A}\mathbf{A}^{\mathrm{T}} = \mathbf{E}_n. \tag{11.27}$$

It follows from Property (1) of determinants (see Section 10.3), that $\det\mathbf{A} = \det\mathbf{A}^{\mathrm{T}}$, since the value of a determinant is unchanged if all columns and rows are interchanged. It also follows from eqn (11.22) that $\det(\mathbf{A}\mathbf{A}^T) = \det \mathbf{A} \times \det \mathbf{A} = (\det \mathbf{A})^2$ and from the property of an orthogonal matrix given in eqn (11.27) that $(\det \mathbf{A})^2 = \det\mathbf{E}_n = 1$. Consequently, since $(\det\mathbf{A})^2 = 1$, it follows that, for an orthogonal matrix, $\det\mathbf{A} = \pm 1$. However, it does not necessarily follow that an arbitrary matrix satisfying this criterion is orthogonal, since it must also satisfy eqn (11.27).

Problem 11.13

(a) Find the value(s) of k for which the matrix

$$\mathbf{A} = \begin{pmatrix} \frac{1}{\sqrt{2}} & k \\ \frac{1}{\sqrt{2}} & \frac{-1}{\sqrt{2}} \end{pmatrix}$$

   is orthogonal. Check your answer by verifying that $\mathbf{A}^T\mathbf{A} = \mathbf{A}\mathbf{A}^T = \mathbf{E}_n$.

(b) Find the value(s) of θ for which $\mathbf{R} = \begin{pmatrix} \cos\theta & \sin\theta & 0 \\ \sin\theta & \cos\theta & 0 \\ 0 & 0 & 1 \end{pmatrix}$ is orthogonal.
   Hint: $\cos 2\theta = \cos^2\theta - \sin^2\theta$.

(c) Find the value(s) of k for which the matrix $\mathbf{A} = \begin{pmatrix} 1 & k \\ -1 & 1 \end{pmatrix}$, satisfies the condition $\det\mathbf{A} = \pm 1$. Check your answer against eqn (11.27) and comment on whether the matrix $\mathbf{A}$ is orthogonal or not.

As we shall see in later Sections, orthogonal matrices play an important role in defining the coordinate transformations that are used in characterising the symmetry properties of molecules.

11.7.3 Singular Matrices

A square matrix, $\mathbf{A}$, for which $\det\mathbf{A} = 0$, is said to be singular. Such matrices usually arise when the number of variables (or degrees of freedom) is over-specified for the chosen model, as would occur, for example, in:

- Using the same atomic orbital twice in constructing molecular orbitals in the linear combination of atomic orbitals (LCAO) model,
- Solving an inconsistent set of equations.

Superficially, using the same atomic orbital twice in constructing molecular orbitals using the LCAO method may seem misguided. However, there are cases when the second occurrence of the atomic orbital is disguised.

11.7.4 Hermitian Matrices

A complex square matrix, that is equal to the transpose of its complex conjugate, is called an Hermitian matrix, which is:

$$\mathbf{A} = \mathbf{A}^\dagger. \qquad (11.28)$$

Problem 11.14

(a) Verify that the matrix $\mathbf{A} = \begin{pmatrix} 0 & 3+i \\ 3-i & 1 \end{pmatrix}$ is Hermitian.

(b) If $\mathbf{x}$ is the 2 by 1 column matrix $\begin{pmatrix} 1 \\ i \end{pmatrix}$, and $\mathbf{A}$ is the Hermitian matrix in part (a), show that $\mathbf{x}^\dagger \mathbf{A} \mathbf{x} = -1$.

11.7.5 Unitary Matrices

A square matrix $\mathbf{U}$, of order n, is said to be unitary if :

$$\mathbf{U}^\dagger \mathbf{U} = \mathbf{U}\mathbf{U}^\dagger = \mathbf{E}_n \qquad (11.29)$$

It follows from the definition of a unitary matrix that $\det\mathbf{U} = \pm 1$. As for orthogonal matrices, an arbitrary matrix having the property $\det\mathbf{U} = \pm 1$, is unitary only if the requirements of eqn (11.29) are satisfied.

Hermitian and unitary matrices play the same role for matrices with complex elements that symmetric and orthogonal matrices do for matrices with real elements. These features are summarised in Table 11.1:

Table 11.1 A summary of special square matrices.

Matrices with real elements	*Matrices with complex elements*
Transpose, $\mathbf{A}^T$	Complex conjugate transpose, $\mathbf{A}^\dagger$
$\begin{pmatrix} 2 & 5 \\ 3 & 4 \end{pmatrix} \Rightarrow \begin{pmatrix} 2 & 3 \\ 5 & 4 \end{pmatrix}$	$\begin{pmatrix} 2 & 3+i \\ -i & 4 \end{pmatrix} \Rightarrow \begin{pmatrix} 2 & i \\ 3-i & 4 \end{pmatrix}$
Symmetric $\mathbf{A} = \mathbf{A}^T$	Hermitian $\mathbf{A} = \mathbf{A}^\dagger$
$\begin{pmatrix} 2 & 5 \\ 5 & 3 \end{pmatrix}$	$\begin{pmatrix} 2 & 3+i \\ 3-i & 1 \end{pmatrix}$
Orthogonal $\mathbf{A}^T\mathbf{A} = \mathbf{A}\mathbf{A}^T = \mathbf{E}_n$	Unitary $\mathbf{U}^\dagger\mathbf{U} = \mathbf{U}\mathbf{U}^\dagger = \mathbf{E}_n$
$\begin{pmatrix} \cos\theta & -\sin\theta \\ \sin\theta & \cos\theta \end{pmatrix}\begin{pmatrix} \cos\theta & \sin\theta \\ -\sin\theta & \cos\theta \end{pmatrix} = \mathbf{E}_2$	$\begin{pmatrix} \frac{1}{\sqrt{2}} & \frac{-i}{\sqrt{2}} \\ \frac{i}{\sqrt{2}} & \frac{-1}{\sqrt{2}} \end{pmatrix}\begin{pmatrix} \frac{1}{\sqrt{2}} & \frac{-i}{\sqrt{2}} \\ \frac{i}{\sqrt{2}} & \frac{-1}{\sqrt{2}} \end{pmatrix} = \mathbf{E}_2$
A consequence of the above is that $\det\mathbf{A} = \pm 1$	A consequence of the above is that $\det\mathbf{U} = \pm 1$

Symmetric, Hermitian, orthogonal and unitary matrices all arise in the quantum mechanical models used to probe aspects of molecular structure.

Problem 11.15

Classify each of the following matrices according to whether they are symmetric, Hermitian, orthogonal or unitary:

$$\text{(a) } \mathbf{A} = \begin{pmatrix} 1 & i \\ -i & 0 \end{pmatrix}; \quad \text{(b) } \mathbf{B} = \frac{1}{\sqrt{2}} \begin{pmatrix} 1 & i \\ -i & -1 \end{pmatrix};$$

$$\text{(c) } \mathbf{C} = \begin{pmatrix} 0 & -1 \\ 1 & 0 \end{pmatrix}; \quad \text{(d) } \mathbf{D} = \begin{pmatrix} 1 & -1 \\ -1 & 0 \end{pmatrix}.$$

We now proceed to identify the last of the special matrices that are important to us.

11.7.6 The Inverse Matrix

The **inverse** of a square matrix $\mathbf{A}$, of order n is written $\mathbf{A}^{-1}$, has the property:

$$\mathbf{A}\mathbf{A}^{-1} = \mathbf{A}^{-1}\mathbf{A} = \mathbf{E}_n. \tag{11.30}$$

and exists only if $\det\mathbf{A} \neq 0$. If $\det\mathbf{A} = 0$ then $\mathbf{A}$ is singular and $\mathbf{A}^{-1}$ does not exist. We saw in Problem 11.12(a) that the transposed matrix of cofactors, $\mathbf{B}^{\mathrm{T}}$ is related to $\mathbf{A}$ and $\det\mathbf{A}$ – irrespective of the order of $\mathbf{A}$ – according to the formula:

$$\mathbf{B}^{\mathrm{T}}\mathbf{A} = \mathbf{E}_n \det\mathbf{A}. \tag{11.31}$$

Rearranging eqn (11.31) gives:

$$\frac{1}{\det\mathbf{A}}\mathbf{B}^{\mathrm{T}} = \frac{\mathbf{E}_n}{\mathbf{A}} \tag{11.32}$$

but we know from eqn (11.30) that $\mathbf{A}^{-1} = \dfrac{\mathbf{E}_n}{\mathbf{A}}$, and so:

$$\mathbf{A}^{-1} = \frac{1}{\det\mathbf{A}}\mathbf{B}^{\mathrm{T}} \tag{11.33}$$

which provides us with a formula for obtaining A^{-1} from B^T and $\det A$. It should be remembered, however, that A^{-1} exists only if A is non-singular.

Worked Problem 11.5

Q. Find the inverse of the matrix $A = \begin{pmatrix} 1 & -1 \\ 1 & 1 \end{pmatrix}$.

A. First, $\det A = 1 + 1 = 2$;

The matrix of cofactors of A is

$$B = \begin{pmatrix} A_{11} & A_{12} \\ A_{21} & A_{22} \end{pmatrix} = \begin{pmatrix} 1 & -1 \\ 1 & 1 \end{pmatrix} \Rightarrow B^T = \begin{pmatrix} 1 & 1 \\ -1 & 1 \end{pmatrix}.$$

Thus the inverse of matrix A is given by:

$$A^{-1} = \tfrac{1}{2} \begin{pmatrix} 1 & 1 \\ -1 & 1 \end{pmatrix} = \begin{pmatrix} \tfrac{1}{2} & \tfrac{1}{2} \\ -\tfrac{1}{2} & \tfrac{1}{2} \end{pmatrix}$$

Check: Use the definition of the matrix inverse to confirm that $AA^{-1} = A^{-1}A = E_n$

$$\begin{pmatrix} 1 & -1 \\ 1 & 1 \end{pmatrix} \begin{pmatrix} \tfrac{1}{2} & \tfrac{1}{2} \\ -\tfrac{1}{2} & \tfrac{1}{2} \end{pmatrix} = \begin{pmatrix} 1 & 0 \\ 0 & 1 \end{pmatrix} = E_2 \ \checkmark$$

Problem 11.16

Find the inverse of the matrix $A = \begin{pmatrix} 1 & -1 & 1 \\ -1 & -1 & 1 \\ 1 & 1 & 1 \end{pmatrix}$, and demonstrate that $AA^{-1} = A^{-1}A = E_3$.

The inverse matrix has many uses, but of particular relevance to us as chemists is the role they play in:

- The solution of sets of simultaneous linear equations.
- Developing the concept of a group which, in turn, underpins the basis of symmetry theory.

11.8 Solving Sets of Linear Equations

Suppose we have a set of three equations, each of which is linear in the unknowns x_1, x_2, x_3:

$$a_{11}x_1 + a_{12}x_2 + a_{13}x_3 = b_1$$

$$a_{21}x_1 + a_{22}x_2 + a_{23}x_3 = b_2 \qquad (11.34)$$

$$a_{31}x_1 + a_{32}x_2 + a_{33}x_3 = b_3$$

where the a_{ij} and b_i ($i, j = 1, 2, 3$) are constant coefficients. If all the b_i are zero, then the equations are called homogeneous but, if one or more of the b_i are non-zero, then the equations are called inhomogeneous.

We can write the three linear eqns (11.34) as a single matrix equation:

$$\begin{pmatrix} a_{11} & a_{12} & a_{13} \\ a_{21} & a_{22} & a_{23} \\ a_{31} & a_{32} & a_{33} \end{pmatrix} \begin{pmatrix} x_1 \\ x_2 \\ x_3 \end{pmatrix} = \begin{pmatrix} b_1 \\ b_2 \\ b_3 \end{pmatrix} \qquad (11.35)$$

and then check that eqns (11.35) and (11.34) are equivalent, by evaluating the matrix product in the left side of eqn (11.35) to give:

$$\begin{pmatrix} a_{11}x_1 + a_{12}x_2 + a_{13}x_3 \\ a_{21}x_1 + a_{22}x_2 + a_{23}x_3 \\ a_{31}x_1 + a_{32}x_2 + a_{33}x_3 \end{pmatrix} = \begin{pmatrix} b_1 \\ b_2 \\ b_3 \end{pmatrix}. \qquad (11.36)$$

We now have two 3×1 matrices, which are equal to one another and, because this implies equality of the elements, we regenerate the original linear equations given in eqn (11.34). If we now rewrite eqn (11.35) in a more compact form as:

$$\mathbf{Ax} = \mathbf{b} \qquad (11.37)$$

and pre-multiply by $\mathbf{A}^{-1}$, the matrix inverse of $\mathbf{A}$, we obtain:

$$\mathbf{A}^{-1}\mathbf{Ax} = \mathbf{A}^{-1}\mathbf{b}. \qquad (11.38)$$

Since $\mathbf{A}^{-1}\mathbf{A} = \mathbf{E}_n$ and $\mathbf{E}_n\mathbf{x} = \mathbf{x}$, it follows that the unique solution is given by:

$$\mathbf{x} = \mathbf{A}^{-1}\mathbf{b} \qquad (11.39)$$

However, this solution is meaningful only if det$\mathbf{A}$ is non-singular. If $\mathbf{A}$ is singular, the equations are inconsistent, in which case, no solution is forthcoming.

Worked Problem 11.6

Q. (a) Confirm that the following equations have a single, unique solution.

$$x_1 - x_2 + x_3 = 1$$
$$-x_1 - x_2 + x_3 = 2 \qquad (11.40)$$
$$x_1 + x_2 + x_3 = -1$$

(b) Find the solution.

A. (a) Rewriting eqn (11.40) in matrix form gives:

$$\begin{pmatrix} 1 & -1 & 1 \\ -1 & -1 & 1 \\ 1 & 1 & 1 \end{pmatrix} \quad \begin{pmatrix} x_1 \\ x_2 \\ x_3 \end{pmatrix} \quad = \quad \begin{pmatrix} 1 \\ 2 \\ -1 \end{pmatrix}.$$
$$\quad\quad\quad \mathbf{A} \quad\quad\quad\quad\quad \mathbf{x} \quad\quad\quad\quad \mathbf{b}$$

The set of equations has a unique solution as $\det\mathbf{A} = -4$ (see Problem 11.16) indicating that the equations are consistent.

(b) Following the procedure in Worked Problem 11.5, and with reference to the answer to Problem 11.16, we find

$\mathbf{A}^{-1} = \begin{pmatrix} \frac{1}{2} & -\frac{1}{2} & 0 \\ -\frac{1}{2} & 0 & \frac{1}{2} \\ 0 & \frac{1}{2} & \frac{1}{2} \end{pmatrix}$. Thus, the solution, according to eqn

(11.39), is given by:

$$\begin{pmatrix} x_1 \\ x_2 \\ x_3 \end{pmatrix} = \begin{pmatrix} \frac{1}{2} & -\frac{1}{2} & 0 \\ -\frac{1}{2} & 0 & \frac{1}{2} \\ 0 & \frac{1}{2} & \frac{1}{2} \end{pmatrix} \begin{pmatrix} 1 \\ 2 \\ -1 \end{pmatrix} = \begin{pmatrix} -\frac{1}{2} \\ -1 \\ \frac{1}{2} \end{pmatrix},$$

from which we see that $x_1 = -\frac{1}{2}$, $x_2 = -1$, $x_3 = \frac{1}{2}$.

Problem 11.17

Find the values of x, y, z that satisfy the equations:

$$x+2y+3z=1$$
$$8y+2z=1$$
$$-2x+4y+2z=2$$

So far we have considered only the solutions to sets of inhomogeneous linear equations where at least one of the b_i is non-zero. If, however, we have a set of homogeneous equations, where all of the b_i are zero, then we may define two further possible limiting cases:

- If $\det \mathbf{A} \neq 0$, and $\mathbf{b} = \mathbf{0}$ (all b_i are zero), then this approach will only ever yield the solution, $\mathbf{x} = \mathbf{0}$, *i.e.* $x_1 = x_2 = x_3 = 0$ since $\mathbf{x} = \mathbf{A}^{-1}\mathbf{0} = \mathbf{0}$.
- If $\det \mathbf{A} = 0$, and $\mathbf{b} = 0$, then, again, $\mathbf{A}^{-1}$ will be undefined. However, although the solution may yield the so-called trivial result $\mathbf{x} = \mathbf{0}$, other solutions may also exist.

11.8.1 Solution of Linear Equations: A Chemical Example

In Problem 10.6, we saw how the molecular orbital energies for the allyl system are determined from the solution of a determinantal equation. At this point, we are now in the position to understand the origin of this equation.

In the Hückel model, the result of minimising the energy of the appropriately occupied π molecular orbitals results in the following set of linear equations in the unknown atomic orbital coefficients, c_r, together with the molecular orbital energy, ε:

$$c_1(\alpha-\varepsilon)+c_2\beta=0$$
$$c_1\beta+c_2(\alpha-\varepsilon)+c_3\beta=0 \qquad (11.41)$$
$$c_2\beta+c_3(\alpha-\varepsilon)=0$$

Eqns (11.41) may be more succinctly expressed as a single matrix equation:

$$\begin{pmatrix} (\alpha-\varepsilon) & \beta & 0 \\ \beta & (\alpha-\varepsilon) & \beta \\ 0 & \beta & (\alpha-\varepsilon) \end{pmatrix} \begin{pmatrix} c_1 \\ c_2 \\ c_3 \end{pmatrix} = 0, \qquad (11.42)$$

or in more compact form as $\mathbf{Ac} = \mathbf{0}$, where $\mathbf{0}$ is a null column matrix. Eqns (11.41) provide an example of a set of homogeneous equations, because the right side constant coefficients, equivalent to b_i in eqn (11.34), are all zero [and hence the appearance of the null column matrix $\mathbf{0}$ in the equivalent matrix eqn (11.42)]. The trivial solution to eqn (11.42), where $c_1 = c_2 = c_3 = 0$ ($\mathbf{c} = \mathbf{0}$) is of no physical significance, as the molecular orbitals do not then exist – another example of how important it is to use physical intuition to interpret the significance of a mathematical result! A more detailed study of the mathematics indicates that eqn (11.42) has a non-trivial solution if $\det\mathbf{A} = 0$, the solution of which yields the orbital energies, $\varepsilon = \alpha, \mp\sqrt{2}\beta$, as seen in Problem 10.6.

Problem 11.18

The three molecular orbitals for the allyl system are obtained by solving the set of simultaneous eqns (11.41) for each value of ε, in turn, to obtain the atomic orbital coefficients, c_i.

(a) For $\varepsilon = \alpha$, show that $c_3 = -c_1$ and that $c_2 = 0$.
(b) For $\varepsilon = \alpha + \sqrt{2}\beta$, show that $c_2 = \sqrt{2}c_1$ and $c_3 = c_1$.
(c) For $\varepsilon = \alpha - \sqrt{2}\beta$, show that $c_2 = -\sqrt{2}c_1$ and $c_3 = c_1$.
(d) For each of the three orbital energies, construct the column matrix, $\mathbf{c_i}$, where each element is expressed in terms of, c_1.

11.9 Molecular Symmetry and Group Theory

One of the key applications of matrices in chemistry is in the characterisation of molecular symmetry. In Section 11.3, we saw how it was possible to represent the coordinate transformations associated with rotation and reflection in terms of matrices. These notions are now explored in the next Section, where we develop some of the basic ideas of group theory.

11.9.1 An Introduction to Group Theory

A group consists of a set of elements (*e.g.* numbers or square matrices), for which there is a specified mode of combination (for example, addition, subtraction or multiplication), subject to the four following requirements:

(a) For any **R, S** in the set, the combination **RS** is a member of the set (closure);

(b) For any **R, S** and **T** in the set, the mode of combination must be associative: that is:

$$R(ST) = (RS)T;$$

(c) There is an identity element **E** such that, for any element, **R**, in the set, **RE = ER = R;**

(d) for each **R,** there is an inverse element **R**$^{-1}$ such that:

$$RR^{-1} = R^{-1}R = E.$$

The number of elements in the group is termed its order, which may be finite or infinite.

The elements of a group should not to be confused with the elements of a matrix or determinant.

Worked Problem 11.7

Q. Investigate whether the set of integers forms a group under each of the following modes of combination: (a) addition, (b) subtraction and (c) multiplication.

A.

(a) *Addition*: the sum of any two integers is an integer (closure satisfied); addition of integers is associative; the identity element is zero (*e.g.* $2 + 0 = 0 + 2 = 2$); the inverse of any integer n is $-n$ [*e.g.* $2 + (-2) = 0$, the identity element], and $-n$ is an integer which is in the set. Since all four criteria are satisfied, the set of integers forms a group of infinite order under addition.

(b) *Subtraction*: the difference of any two integers is an integer (closure satisfied); subtraction of integers is not associative, *e.g.* $(3 - 4) - 2 = -3$, while $3 - (4 - 2) = 1$, and so the set of integers does not form a group under subtraction.

(c) *Multiplication*: the product of any two integers is an integer (closure satisfied); multiplication is associative; the inverse of any non-zero integer n is the rational number $1 / n$ which is not an integer and so the set of integers does not form a group under multiplication.

Problem 11.19

Demonstrate that the set of numbers $G_1 = \{1, -1, i, -i\}$ forms a group of order 4 under multiplication.

11.9.2 Groups of Matrices

Groups of non-singular (square) matrices are of special interest in chemistry, because they are used to characterise molecular and solid state structures according to their symmetry properties. This is vital when determining:

- Whether spectroscopic transitions of all kinds are forbidden or allowed.
- The most likely mechanisms of some classes of organic reaction, where symmetry controls the outcome.
- The arrangement of species in the unit cell of solid state structures.

When deciding whether a given set of matrices of order n forms a group under multiplication, we can disregard associativity as one of the criteria because multiplication of matrices is *always* necessarily associative – and so we only need to check for closure, the presence of E_n and identify all inverses.

Worked Problem 11.8

Let us consider, as an example, the set, G_3, of the following three matrices of order 2 to see whether they form a group under multiplication:

$$G_3 = \left\{ A = \begin{pmatrix} 1 & 0 \\ 0 & 1 \end{pmatrix}, B = \begin{pmatrix} -\frac{1}{2} & -\frac{\sqrt{3}}{2} \\ \frac{\sqrt{3}}{2} & -\frac{1}{2} \end{pmatrix}, C = \begin{pmatrix} -\frac{1}{2} & \frac{\sqrt{3}}{2} \\ -\frac{\sqrt{3}}{2} & -\frac{1}{2} \end{pmatrix} \right\},$$

The best way of checking the group requirements is to construct the multiplication table as shown in Table 11.2:

Table 11.2   Multiplication table for the set of matrices G_3

		→ First Operation		
		A	B	C
Second Operation ↓	A	A	B	C
	B	B	C	A
	C	C	A	B

We can see that this set of matrices forms a group, as:

1. The set is closed under multiplication;
2. **B** is the inverse of **C**, and *vice versa*;
3. **A** is the identity element, which is its own inverse.

Problem 11.20

Construct a multiplication table for each of the following sets of matrices to confirm that the sets constitute groups of order 4 under matrix multiplication.

(a) $G_4 = \left\{ A = \begin{pmatrix} 1 & 0 \\ 0 & 1 \end{pmatrix}, B = \begin{pmatrix} -1 & 0 \\ 0 & -1 \end{pmatrix}, \right.$

$\left. C = \begin{pmatrix} 0 & 1 \\ -1 & 0 \end{pmatrix}, D = \begin{pmatrix} 0 & -1 \\ 1 & 0 \end{pmatrix} \right\}$

(b) $S_4 = \left\{ A = \begin{pmatrix} 1 & 0 \\ 0 & 1 \end{pmatrix}, B = \begin{pmatrix} -1 & 0 \\ 0 & -1 \end{pmatrix}, \right.$

$\left. C = \begin{pmatrix} 0 & i \\ -i & 0 \end{pmatrix}, D = \begin{pmatrix} 0 & -i \\ i & 0 \end{pmatrix} \right\}$

List the inverse of each element in groups G_4 and S_4.

11.9.3 Group Theory in Chemistry

Molecules are classified in terms of their symmetry properties by constructing groups of matrices that describe coordinate transformations, resulting in the interchange of chemically equivalent points. For any given molecule, these coordinate transformations form the elements of a particular **point group** that describes its symmetry properties.

For example, the water molecule is bent in its ground state, with a bond angle of approximately 105° (see Figure 11.5). If we rotate an initial point lying above the plane of the molecule, directly over *one* of the hydrogen nuclei through 180° about the principal axis passing through the oxygen nucleus in the molecular plane (the *z*-axis in Figure 11.5), the transformed point will lie *below* the *other* hydrogen nucleus. Likewise, reflection in the plane perpendicular to the molecular plane, and containing the principal axis of rotation (the *xz*-plane), transforms the initial point to an equivalent point lying *above* the other hydrogen nucleus. However, reflection in the plane of the molecule (the *yz*-plane) transforms the initial point to one lying directly below the original hydrogen nucleus. Clearly, the identity

Point groups are so called because there is always one point that is unmoved under every symmetry operation. This is not the case in **space groups** that are used to characterise the symmetry properties of crystals.

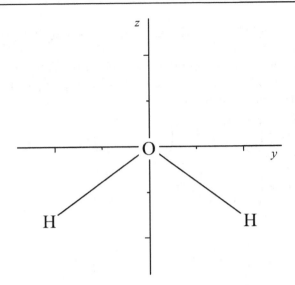

Figure 11.5 The nuclear configuration for H_2O in the yz-plane. The x-axis is perpendicular to the paper, and passes through O. One of the two mirror planes lies in the plane of the paper, whilst the second is perpendicular to the paper and contains the O atom. The principal axis of rotation is the z-axis.

operation leaves the initial point unmoved. These three symmetry operations are known as the C_2, σ_v and σ_v', respectively, and together with the identity element, E, constitute the four elements of the C_{2v} point group to which water belongs.

If the location of one of the nuclei is taken as the initial point, then we see that each symmetry operation just exchanges (or leaves unmoved) the coordinates of chemically equivalent nuclei. It transpires that if we follow the procedure of physically moving the nuclei, rather than a representative point in space, then each matrix generated is the inverse of the one associated with the appropriate coordinate transformation although the traces of the respective matrices are the same. This is helpful because in most applications of group theory, we work with the traces rather than the elements of the transformation matrices.

In the context of group theory the trace of a matrix is usually referred to as the **character**.

Problem 11.21

How many symmetry operations can you list that interchange chemically equivalent nuclei in the planar molecule, BF_3 (see Section 11.3.2)?
Hint: How many mirror planes and axes of rotation are there?

Summary of Key Points

In this chapter we have introduced the matrix as a means of handling sets of objects and discussed the key aspects of matrix algebra. A great deal of the chapter has involved a cataloguing of the properties and types of matrices but we have also tried to emphasise the chemical importance of matrices, in particular in the vital role they play in the classification of molecular symmetry and the development of group theory. The key points discussed include:

1. An introduction to matrix notation.
2. Rules for combining matrices through addition, subtraction and multiplication.
3. How matrices are used to represent coordinate transformations, and hence to characterise the symmetry properties of molecules.
4. Operations on matrices containing real and complex elements.
5. Special matrices, including the unit matrix and the null matrix.
6. Matrices with special properties.
7. The matrix of cofactors and the definition of the inverse matrix.
8. The application of matrix algebra for solving sets of simultaneous linear equations: homogeneous and inhomogeneous equations.
9. An introduction to molecular symmetry and group theory.
10. Elucidating the characteristic electronic structures associated with molecules.
11. Introducing some of the concepts necessary in the study and use of vectors.

12
Vectors

Many of the physical quantities which we deal with from day to day, such as mass, temperature or concentration, require only a single number (with appropriate units) to specify their value. Such quantities are called scalar quantities, specified exclusively in terms of their value. However, we frequently encounter other quantities, called vectors, which require us to specify a magnitude (a positive value), and a direction. Velocity is an example of a vector quantity, whereas speed is a scalar quantity (in fact speed defines the magnitude of velocity). This is why we say that an object travelling on a circular path, with constant speed (such as an electron orbiting a nucleus in the Bohr model of the atom) is accelerating: its velocity changes with time because its direction is constantly changing, in spite of the fact that the speed is constant (see Figure 12.1).

In the example shown in Figure 12.1, we define both the position and velocity in terms of vectors: The position of the electron at any given time is given by a position vector, referenced to an origin O. So, when the electron is at point P, its location is defined by the vector a, whereas when it has moved to point Q, the position is defined by a different position vector, b. We can also represent the velocity at points P and Q by the two vectors v_a and v_b, both of which have the same magnitude (length) but whose direction is different.

In chemistry, we meet many physical quantities and properties that require us to specify both magnitude and direction. These include:

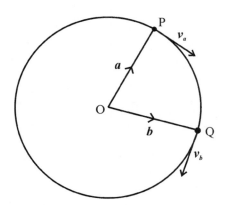

Figure 12.1 The velocity and position vectors of an electron at two points P and Q in a circular Bohr orbit.

- The magnetic and electrical properties of atoms, molecules and solids.
- Forces between molecules.
- Velocity of a molecule in the gas phase.
- Angular momentum associated with molecular rotational motion and with the orbital and spin motions of electrons.

In order to qualify properly as a vector, a quantity must obey the rules of vector algebra (scalar quantities obey the rules of arithmetic). Consequently, we need to describe and define these rules before we can solve problems in chemistry involving vector quantities. Linear algebra is the field of mathematics that provides us with the notation and rules required to work with directional quantities.

Aims:

In this chapter, we discuss the concept of the vector from a number of perspectives, ranging from the graphical description, to a presentation of vector algebra and on to examples of how we can apply vector algebra to specific chemical problems involving directional properties. By the end of the chapter, you should understand how:

- Vectors are defined geometrically in terms of direction and magnitude.
- Vectors are defined algebraically, using base vectors.
- Vectors are combined using addition or subtraction.
- The scalar and vector products are defined and used.
- The triple scalar product is defined and used for calculating the volume of a parallelepiped.
- Matrix representatives of vectors are formulated and used.

12.1 The Geometric Construction of Vectors

A vector is represented mathematically by a directed line segment, the length of which corresponds to the magnitude of the vector, whilst its orientation, taken together with an attached arrow, indicates its direction. To simplify matters, we first consider vectors in two dimensions (2-D space), and then extend the concepts to dealing with vectors in three or more dimensions.

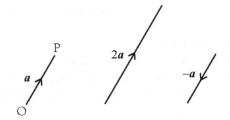

Figure 12.2 Representations of the vectors **a**, 2**a** and −**a** as directed line segments.

12.1.1 Vectors in Two Dimensions

The arrow indicating direction is placed at any appropriate point on the line segment. We have chosen for the most part to place the arrow at the mid point.

Consider the three directed line segments representing the vectors shown in Figure 12.2: all three have initial and final points, which may or may not be labelled, whereas the arrow indicates the direction. In each case the position of the initial point is of no significance. The magnitudes (lengths) of the left-most and right-most vectors are the same, but their directions are opposite; the middle vector has the same direction as the left vector, but twice its magnitude.

12.1.2 Conventions

1. Vectors are represented by symbols such as **a**, **b**, ... and their respective magnitudes are given by $|a|$, $|b|$, ..., or just a, b, ... An alternative notation, $\overrightarrow{OP}$, is sometimes used when we wish to describe a displacement in space between two points (in this case, points O and P).

2. The vectors **a** and **b** are said to be equal if their magnitudes and directions are the same – irrespective of the locations of their initial points. Hence, any directed line segment with the same length and direction as **a** is represented by **a**.

3. A **unit vector** is a vector having unit magnitude (or length). Unit vectors are symbolised by $\hat{a}$, $\hat{b}$,..., and correspond to the vectors **a**, **b**,... divided by their own magnitude. For example:

$$\hat{a} = \frac{a}{|a|}. \tag{12.1}$$

4. A null vector, **0**, has zero magnitude and consequently no direction is defined.

12.2 Addition and Subtraction of Vectors

12.2.1 Vector Addition

Consider the two vectors **a** and **b** shown in Figure 12.3.

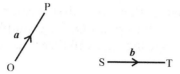

Figure 12.3   Vectors *a* and *b* with initial points O and S, respectively.

The sum of *a* and *b* is given by the vector *c*, which is found in the following series of steps:

(a) Translate the vector *b* until its initial point coincides with that of *a*:

(b) Construct a parallelogram as indicated in Figure 12.4.

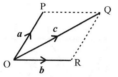

Figure 12.4   The parallelogram formed by the addition of the two vectors *a* and *b*.

The directed line segment $\overrightarrow{OQ}$ represents the vector *c*, defined as the sum of *a* and *b*. Furthermore, as $\overrightarrow{OQ} = \overrightarrow{OP} + \overrightarrow{PQ} = \overrightarrow{OR} + \overrightarrow{RQ}$, it follows that *c* = *a* + *b* = *b* + *a*, from which we see that addition is commutative; in other words, a displacement $\overrightarrow{OR}$ followed by $\overrightarrow{RQ}$ clearly leads to the same final point as a displacement $\overrightarrow{OP}$ followed by $\overrightarrow{PQ}$.

Since $\overrightarrow{OR}$ and $\overrightarrow{PQ}$ are equivalent, and represent the same vector *b*, we can use a triangle to summarise the relationship between *a*, *b* and the resultant vector *c* (see Figure 12.5). For this reason, the equality *c* = *a* + *b* is often known as the triangle rule.

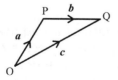

Figure 12.5   The triangle rule, in which *c* = *a* + *b*.

12.2.2 Vector Subtraction

The subtraction of two vectors can be thought of as the addition of two vectors that differ in their sign. If we think of this in terms of displacements in space, then the first vector corresponds to a displacement from point P to point Q, for example, whereas a second identical vector with opposite sign will direct us back to point P from point Q.

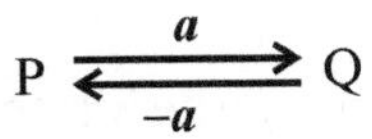

$$P \underset{-a}{\overset{a}{\rightleftarrows}} Q$$

The net result is the null vector – we end up where we started.

$$a + -a = a - a = 0 \tag{12.2}$$

It follows that subtraction of two vectors, a and b, is equivalent to adding the vectors a and $-b$, and so we can define vector subtraction in a general sense as:

$$d = a + (-b) = a - b, \tag{12.3}$$

which can be expressed in terms of a variant of the triangle rule, as seen in Figure 12.6.

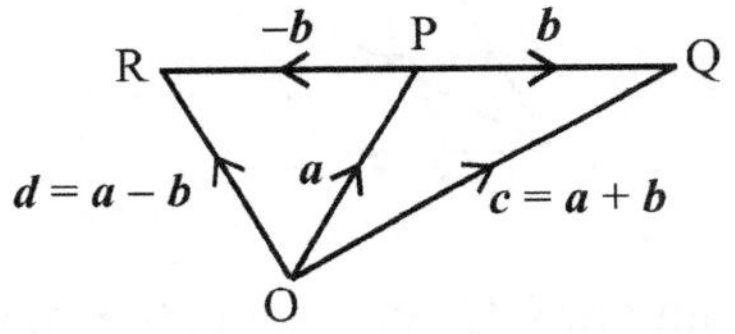

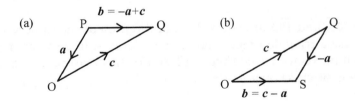

Figure 12.6 Vector subtraction represented in terms of the triangle rule.

It also follows from Figure 12.6, that if $a + b = c$ then $c - a = b$, which we represent graphically in two ways in Figure 12.7:

(a) ... (b) ...

Figure 12.7 Two alternative representations of the subtraction of two vectors.

Note that both representations are equivalent, in spite of the fact that the initial and final points of vector b are located at different points in space in the two representations. Thus, since the vector is fully defined simply by its direction and magnitude, the locations of the initial and final points are unimportant – unless we define them to act in specific locations.

Problem 12.1

Use the vectors a and b in Figure 12.5 to construct parallelograms, defined by the vectors: (a) $c = a + 2b$ and (b) $d = 2a - b$.

12.3 Base Vectors

Any kind of operation on a vector, including addition and subtraction, can be somewhat laborious when working with its graphical representation. However, by referring the vectors to a common set of unit vectors, termed base vectors, we can reduce the manipulations of vectors to algebraic operations.

In three dimensional space, a convenient set of three unit vectors is provided by $\hat{i}, \hat{j}, \hat{k}$, which are directed along the x-, y-, and z-Cartesian axes, respectively (Figure 12.8).

In this system of coordinates, if a point P has the coordinates (x, y, z), then the directed line segment $\overrightarrow{OP}$, extending from the origin O to point P, corresponds to the vector r. If we apply the triangle rule twice, we obtain:

$$r = \overrightarrow{OQ} + \overrightarrow{QP} \tag{12.4}$$

$$= \overrightarrow{OR} + \overrightarrow{RQ} + \overrightarrow{QP} \tag{12.5}$$

$$\Rightarrow \quad r = x\hat{i} + y\hat{j} + {}_1z\hat{k} \tag{12.6}$$

Eqn (12.6) expresses r as a sum of the vectors $x\hat{i}$, $y\hat{j}$ and $z\hat{k}$, which are called the projections of r in the direction of the x-, y- and z-axes. The magnitudes of each projection are given by the x-, y- and z-values, respectively, defining the location of P; however, in the context

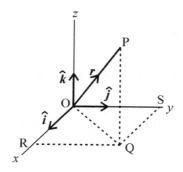

Figure 12.8 Base vectors in three dimensions for the Cartesian coordinate system.

of vectors, these values (coordinates) are known as the components of r; if the components are all zero, then this defines the null vector. Note that for problems in two dimensions, only two base vectors are required, such as, for example $\hat{i}$ and $\hat{j}$.

12.3.1 The Magnitude of a Vector in Three-Dimensional Space

If we apply the Pythagoras' Theorem, first to triangle ORQ in Figure 12.8, and then to triangle OQP, we obtain an expression for the magnitude of r in terms of its components:

$$|r| = \left(x^2 + y^2 + z^2\right)^{1/2} \qquad (12.7)$$

12.3.2 Vector Addition, Subtraction and Scalar Multiplication using Algebra

The algebraic approach to vector addition and subtraction simply involves adding or subtracting the respective projections, $x\hat{i}$, $y\hat{j}$ and $z\hat{k}$, of the two (or more) vectors. Scalar multiplication requires each projection to be multiplied by the scalar quantity.

Worked Problem 12.1

Q. If $u = \hat{i} + \hat{j} + 2\hat{k}$ and $v = -2\hat{i} - \hat{j} + \hat{k}$, find:
(a) $2v$, (b) $u - 2v$ and (c) $|u - 2v|$.

A. (a) $2v = -4\hat{i} - 2\hat{j} + 2\hat{k}$
(b) $u - 2v = \left(\hat{i} + \hat{j} + 2\hat{k}\right) - \left(-4\hat{i} - 2\hat{j} + 2\hat{k}\right) = \left(5\hat{i} + 3\hat{j}\right)$
(c) $|u - 2v| = \sqrt{5^2 + 3^2} = \sqrt{34}$.

Problem 12.2

If $a = \hat{i} + \hat{j} - 2\hat{k}$, $b = \hat{i} + \hat{k}$, $c = \hat{i} + \hat{j} + \hat{k}$ and $d = \hat{i} - 2\hat{k}$, find :

(a) $3a - 2b$, (b) $-2a - b$, (c) $a + b - c - d$, (d) $|a - d|$,
(e) $(a + c) / |a + c|$, (f) the magnitude of the vector in (e),
(g) $|a| - |c|$.

Problem 12.3

Consider the planar complex ion $Co(CN)_4^{2-}$, shown schematically in Figure 12.9. The central Co lies at the origin, and the four CN^- ligands lie on either the x- or y-axis; R is the Co–C interatomic distance.

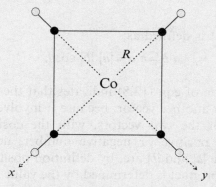

Figure 12.9   The planar complex ion $Co(CN)_4^{2-}$, where the carbon atoms are represented by ● and the nitrogen atoms by ◐.

(a) Identify the unit vectors directed toward each of the four CN^- ligands.
(b) Give the forms of the four vectors, directed from Co to each C atom.
(c) Find the vectors specifying one of the shortest and one of the longest C–C distances, and hence determine these distances in terms of R.

Hint: The representation of vector subtraction shown in Figure 12.7(a) may be helpful.

12.4 Multiplication of Vectors

In algebra, as we saw in Chapter 2, the act of multiplication is an unambiguous and well-defined operation indicated by the sign ×. In the algebra of vectors, however, multiplication and division have no obvious conventional meaning. Despite this drawback, the two kinds of multiplication operation on pairs of vectors in widespread use are defined in the following subsections.

12.4.1 Scalar Product of Two Vectors

Consider the vectors a and b in Figure 12.10, in which the angle between the two vectors is θ:

The × symbol used in the multiplication of numbers and symbols is commonly suppressed: thus, 6xy is shorthand for the product

Objects formed by placing vectors in juxtaposition, such as **ab**, are called dyadics and have a role in theoretical aspects of Raman spectroscopy, for example.

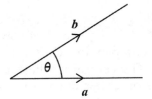

Figure 12.10 Two vectors *a* and *b*, inclined with respect to one another at an angle, θ.

The scalar product is defined as:

$$a\cdot b=b\cdot a=|a|\ |b|\ \cos\theta. \tag{12.8}$$

The right-hand side of eqn (12.8) indicates that the result is a scalar (number), and **not** another vector, because it involves the product of the magnitudes of the two vectors, with the cosine of the angle between them (a positive or negative number, depending on the angle). Thus, since $|a|$ and $|b|$ are, by definition positive numbers, the sign of the scalar product is determined by the value of the angle θ. In particular, the scalar product is:

- Positive, for an acute angle ($\theta < 90°$).
- Zero for $\theta = 90°$.
- Negative for an obtuse angle ($90° < \theta \leqslant 180°$).

By convention, the θ angle is restricted to the range $0 \leqslant \theta \leqslant 180°$. If $\theta = 90°$, then $a\cdot b = 0$, and a and b are said to be orthogonal. On the other hand, the scalar product of a vector with itself ($\theta = 0°$; $\cos \theta = 1$), yields the square of its magnitude:

$$a\cdot a=|a|^{2}, \text{ implying that } |a|=\sqrt{a\cdot a}. \tag{12.9}$$

Furthermore, if a is of unit magnitude, then $a\cdot a = 1$, and a is then said to be normalised.

12.4.1.1 Specifying the Angle θ

In some situations, it is important to be aware of how the sense of direction of the two vectors a and b affects the choice for the value of the angle between them. For example, the angle θ between the vectors in Figure 12.10 constitutes the correct choice, because the two vectors are directed away from the common point of origin. However, if vector a (or vector b) is directed in the opposite sense (the dashed directed line segment in Figure 12.11), then we determine the angle between a and b by realigning the two vectors to ensure once again that they are directed away from the common origin point. The angle is then defined as $180° - \theta$ (Figure 12.11).

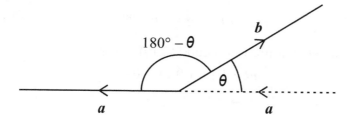

Figure 12.11 The relationship between vector direction and angle.

12.4.1.2 The Scalar Product in the Chemical Context

Scalar products arise in a number of important areas in chemistry. For example, they are involved in:

- Determining the energy, W, of a molecular electric or magnetic dipole interacting with an electric or magnetic field , $W = -\mu_e \cdot E$, or $W = -\mu_m \cdot H$, respectively.
- Evaluating the consequences of the intermolecular dipole–dipole interactions in crystalline substances.
- Crystallography where the **scalar triple product** (see Section 12.5.3) is used to evaluate the volume of a crystallographic unit cell.

12.4.1.3 Scalar Products of Vectors Expressed in Terms of Base Vectors

The scalar product of two vectors a and b, expressed in terms of base vectors, is obtained by taking the sum of the scalar products of each base vector pair, together with the appropriate product of components.

Worked Problem 12.2

Q. Find the scalar product of the vectors $a = \hat{i} + \hat{j} - 2\hat{k}$ and $b = \hat{i} + \hat{k}$.

A. We find the scalar product of a and b using the respective components $(1,1,-2)$ and $(1,0,1)$. Thus, expanding the brackets yields:

$$a \cdot b = \left(\hat{i} + \hat{j} - 2\hat{k}\right) \cdot \left(\hat{i} + \hat{k}\right)$$
$$= \hat{i} \cdot \hat{i} + \hat{i} \cdot \hat{k} + \hat{j} \cdot \hat{i} + \hat{j} \cdot \hat{k} - 2\hat{k} \cdot \hat{i} - 2\hat{k} \cdot \hat{k} \qquad (12.10)$$

We now use eqn (12.8) to evaluate each scalar product of base vectors, to obtain:

$$\hat{i} \cdot \hat{i} = \hat{j} \cdot \hat{j} = \hat{k} \cdot \hat{k} = 1 (\theta = 0) \qquad (12.11)$$

$$\hat{\pmb{i}}\cdot\hat{\pmb{j}}=\hat{\pmb{i}}\cdot\hat{\pmb{k}}=\hat{\pmb{k}}\cdot\hat{\pmb{j}}=0(\theta=90^{\circ}) \qquad (12.12)$$

which, on substitution into eqn (12.10), gives:

$$\pmb{a}\cdot\pmb{b}=1+0+0+0-0-2$$
$$=-1.$$

Problem 12.4

If $\pmb{a}=2\hat{\pmb{i}}+3\hat{\pmb{k}}$, $\pmb{b}=\hat{\pmb{i}}+\hat{\pmb{j}}+\hat{\pmb{k}}$ and $\pmb{c}=\hat{\pmb{i}}-2\hat{\pmb{j}}+\hat{\pmb{k}}$, find:
(a) $\pmb{a}\cdot\pmb{c}$, (b) $\pmb{a}\cdot(\pmb{b}-2\pmb{c})$, (c) $\pmb{a}\cdot(\pmb{b}+\pmb{a})$ and (d) $\pmb{b}\cdot\pmb{c}$

12.4.1.4 Finding the Angle Between Two Vectors

In the previous section we saw that, in spite of appearances, we do not need to know the angle between two vectors in order to evaluate the scalar product according to eqn (12.8): we simply exploit the properties of the orthonormal base vectors to evaluate the result algebraically. However, we can approach from a different perspective, and use the right-hand side of eqn (12.8) to find the angle between two vectors, having evaluated the scalar product using the approach detailed above. The next Worked Problem details how this is accomplished.

Vectors which are orthogonal to one another, as well as being normalised, are said to be orthonormal.

Worked Problem 12.3

Q. Use eqn (12.8) to find the angle between the vectors $\pmb{a}=\hat{\pmb{i}}+\hat{\pmb{j}}-2\hat{\pmb{k}}$ and $\pmb{b}=\hat{\pmb{i}}+\hat{\pmb{k}}$.

A. As the scalar product of these two vectors is negative, and has the value -1 (Worked Problem 12.2), we know that the angle θ is obtuse. The next step involves substituting the vector magnitudes $|\pmb{a}|=\sqrt{6}$ and $|\pmb{b}|=\sqrt{2}$ into eqn (12.8), in order to determine the value of $\cos\theta$:

$$-1=\sqrt{6}\times\sqrt{2}\times\cos\theta \Rightarrow \cos\theta=-\frac{1}{\sqrt{2}\sqrt{6}}=-\frac{1}{2\sqrt{3}}.$$

Since $\cos\theta$ is negative (an obtuse angle), and θ is restricted to $0 \leqslant \theta \leqslant 180°$, we obtain the result $\theta = 106.7°$.

Problem 12.5

(a) Using the definition of the vectors a, b and c in Problem 12.4, find the angle between (i) a and $(b-2c)$ and (ii) b and c.

(b) Find a value of λ for which the two vectors $d = 3\hat{i} - 2\hat{j} - \hat{k}$ and $e = \hat{i} + \lambda\hat{j} + 2\hat{k}$ are orthogonal.

12.4.1.5 Simple Application of the Scalar Product: the Cosine Law for a Triangle

If the sides of the triangle OPQ in Figure 12.5, formed from the vectors a, b and c have magnitudes a, b and c, respectively, and B is the angle opposite b, then we can use eqns (12.8) and (12.9) to find a useful relationship between a, b, c and B.

The triangle rule $c = a + b$ may be rewritten as $b = c - a$, from which we form the scalar product $b \cdot b$:

$$b \cdot b = (c - a) \cdot (c - a) \cdot = a \cdot a + c \cdot c - 2a \cdot c \qquad (12.13)$$

However, since we know that the scalar product of a vector with itself yields the square of its magnitude [eqn (12.9)], and that the angle between a and c is B (and not $180° - B$), it follows that:

$$b \cdot b = b^2 = a^2 + c^2 - 2ac \cos B. \qquad (12.14)$$

This can be extended to construct analogous expressions involving the angles A and C, opposite vectors a and c, respectively.

Problem 12.6

Use the triangle rule in the form $c = a + b$ to derive the form of the cosine formula, involving the angle C.

Hint: Find an expression for $c \cdot c$ and decide, using Figure 12.11, whether the angle θ (in degrees) between the vectors a and b is the same as angle C or the angle $180° - C$. Remember: the cosine formula given above always has the same form, regardless of our choice of a, b and c or angles A, B and C.

Problem 12.7

The complex ion $CoCl_4^{2-}$ adopts a tetrahedral shape, in which the Co lies at the centre of a cube of side $2a$, and the Cl^- species are located on alternate cube vertices; the Co–Cl interatomic distance is taken as R. The coordinate axes are chosen to pass through the centres of opposite pairs of cube faces, with the Co lying at the origin, as shown in Figure 12.12:

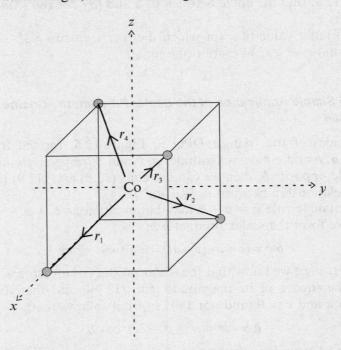

Figure 12.12 The complex ion $CoCl_4^{2-}$, where ● represents a Cl^- species.

(a) Given that the coordinates of the four Cl^- ligands are $(a, -a, -a)$, $(-a, a, -a)$, (a, a, a) and $(-a, -a, a)$, write down the algebraic form of the four vectors, r_1, r_2, r_3 and r_4, directed from the central Co $(0, 0, 0)$ to the four ligands.

(b) Find the magnitude of any one of the Co–Cl vectors, and hence express a in terms of R.

(c) Use the triangle rule shown in Figure 12.7(a) to find a vector associated with the inter-ligand distance, and hence find its magnitude in terms of R.

12.4.2 Vector Product of Two Vectors

In the previous section, we defined the scalar product as a vector operation resembling the act of multiplication, which results in a scalar (or number, with or without units). We can now define a second type of vector multiplication known as the **vector product**, which results in another vector rather than a number. The vector product is defined as:

$$\boldsymbol{a} \times \boldsymbol{b} = -\boldsymbol{b} \times \boldsymbol{a} = |\boldsymbol{a}|\,|\boldsymbol{b}|\, \sin\theta \cdot \hat{\boldsymbol{n}}, \qquad (12.15)$$

where $\hat{\boldsymbol{n}}$ is the unit vector orthogonal (perpendicular in 2-D or 3-D space) to the plane containing $\boldsymbol{a}$ and $\boldsymbol{b}$. Since there are two possible choices for $\hat{\boldsymbol{n}}$ (up or down), the convention for selecting the appropriate direction for $\hat{\boldsymbol{n}}$ requires the vectors $\boldsymbol{a}$, $\boldsymbol{b}$ and $\hat{\boldsymbol{n}}$ to form a right-handed system of axes as shown in Figure 12.13:

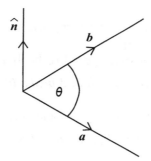

Figure 12.13 The axis convention for determining the sign of the unit vector $\hat{n}$ directed perpendicular to the plane containing the vectors $\boldsymbol{a}$ and $\boldsymbol{b}$.

If we imagine the action of a right-hand corkscrew, in which a is rotated towards $\boldsymbol{b}$, in an anti-clockwise sense when viewed from above, the corkscrew moves in the direction of $\hat{\boldsymbol{n}}$; it follows that the analogous corkscrew motion taking $\boldsymbol{b}$ to $\boldsymbol{a}$ (clockwise) yields a movement in the direction of $-\hat{\boldsymbol{n}}$. Consequently, the vector products involving the base vectors $\hat{\boldsymbol{j}}$ and $\hat{\boldsymbol{k}}$, or, by suitable changes, any other pair of base vectors, are determined by forming a right-hand clockwise system as seen in Figure 12.14.

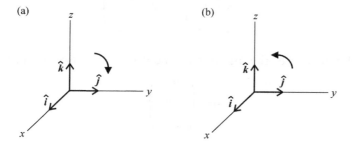

Figure 12.14 Formation of the vector product (a) $\hat{\boldsymbol{k}} \times \hat{\boldsymbol{j}} = -\hat{\boldsymbol{i}}$, and (b) $\hat{\boldsymbol{j}} \times \hat{\boldsymbol{k}} = \hat{\boldsymbol{i}}$ determined by looking down the x-axis and imagining the action of a right-handed corkscrew motion (see the text for details).

Although analogous results can be derived for other pairs of base vectors, the simplest aid for obtaining the appropriate result is to use the diagram shown in Figure 12.15. The vector product $\hat{i} \times \hat{j}$ for example, is verified by moving in a clockwise manner from $\hat{i}$ to $\hat{j}$ to the next base vector $\hat{k}$. However, $\hat{j} \times \hat{i}$ yields $-\hat{k}$ because anticlockwise circulation introduces a negative sign.

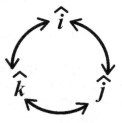

Figure 12.15 The vector product of any two base vectors, moving in a clockwise or anticlockwise direction, yields resultant vectors of positive or negative signs, respectively.

In forming the vector product of two vectors a and b, we should remember that:

- The order of operation is very important – the operation is not commutative [eqn (5.15)].
- The resulting vector is orthogonal to both a and b, implying that:

$$a \cdot (a \times b) = 0 \text{ and } b \cdot (a \times b) = 0$$

Although the vector product is not generally associative, there are examples where this rule is violated – one being the three vectors given in Problem 12.11.

- The operation is not generally associative:

$$(a \times b) \times c \neq a \times (b \times c)$$

12.4.2.1 Vector Products in a Chemical Context

Vector products arise when:

- Working with the angular momentum, ℓ (a vector property), associated with the circular motion of a particle of mass, m, moving under a constant potential about a fixed point with velocity and position described by the vectors,

$$v = \begin{pmatrix} v_x \\ v_y \\ v_z \end{pmatrix} ; \quad x = \begin{pmatrix} x \\ y \\ z \end{pmatrix}$$

In this instance, the angular momentum $\ell = r \times mv = r \times p$, where

$$p = \begin{pmatrix} mv_x \\ mv_y \\ mv_z \end{pmatrix}$$

is the linear momentum. Such model systems have particular relevance when considering the orbital motion of an electron about a nucleus in an atom or about the internuclear axis in a linear molecule.

- Evaluating the volume of a crystallographic unit cell through the scalar triple product (see Section 12.5.3).

Worked Problem 12.4

Q. (a) Use eqn (12.15) to express the vector products $\hat{i} \times \hat{i}, \hat{j} \times \hat{j}$ and $\hat{k} \times \hat{k}$, in terms of the base vectors $\hat{i}, \hat{j}$ and $\hat{k}$;
(b) With the aid of Figure 12.15, find the vector products $\hat{i} \times \hat{j}, \hat{i} \times \hat{k}$ and $\hat{j} \times \hat{k}$.

A. (a) $\hat{i} \times \hat{i} = |\hat{i}||\hat{i}| \sin \theta \cdot \hat{n} = 1 \times 1 \cdot \sin 0 \cdot \hat{n} = 0$, since $\sin 0 = 0$.
The outcome is the same for $\hat{j} \times \hat{j}$ and $\hat{k} \times \hat{k}$, and so:

$$\hat{i} \times \hat{i} = \hat{j} \times \hat{j} = \hat{k} \times \hat{k} = 0$$

It is important not to confuse this result with the analogous scalar products.

(b) $\hat{i} \times \hat{j} = \hat{k}, \ \hat{i} \times \hat{k} = -\hat{j}$ and $\hat{j} \times \hat{k} = \hat{i}$.

Problem 12.8

Use the definitions of a and c in Problem 12.4, and the results of Worked Problem 12.4, to find:

(a) $a \times c$, (b) $c \times a$, (c) $|c \times a|$, (d) $\left(\hat{i} \times \hat{j}\right) \times \hat{j}$ and (e) $\hat{i} \times \left(\hat{j} \times \hat{j}\right)$.

12.4.3 Area of a Parallelogram

The vector product of a and b provides a route for calculating the area of a parallelogram. We explore this method in Worked Problem 12.5.

Worked Problem 12.5

Consider the parallelogram OPQR in shown in Figure 12.16. If we extend OR to point T and drop perpendicular lines from P to S and from Q to T, we construct a rectangle with the same area as the original parallelogram: a result achieved by chopping off the triangle OPS from the left side of the parallelogram and reattaching it at the right side.

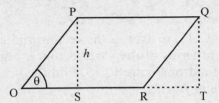

Q. (a) Explain why the areas of the triangles OPS and RQT are equal.
   (b) If the directed line segments $\overrightarrow{OR}$ and $\overrightarrow{OP}$, are represented by the vectors a and b, show that the area of the parallelogram is given by $|a|h$.
   (c) Deduce that the area, A, of the parallelogram is given by $|a \times b|$.

A. (a) The areas of the triangles OPS and RQT are equal because the lengths of the sides OP and RQ are the same, as are the angles $P\hat{O}S$ and $Q\hat{R}T$.
   (b) Given the equivalency of the two triangles OPS and RQT, the area of the rectangle SPQT must be the same as that of the parallelogram OPQR. Consequently the area of the parallelogram must be equal to the magnitude of $\overrightarrow{ST}$ multiplied by the height of the rectangle, h. Since $\overrightarrow{ST}$ has the same magnitude as $\overrightarrow{OR}$, and $\overrightarrow{OR}$ is represented by the vector a, it follows that the magnitude of $\overrightarrow{ST}$ is equal to the magnitude of a. Thus, the area $A = |a|h$.
   (c) As triangle OPS is right-angled, it follows that $h = |b|\sin\theta$, and
   $$A = |a||b|\sin\theta = |a \times b|, \text{ where } a \times b = |a||b|\sin\theta \cdot \hat{n}.$$

Problem 12.9

For $a = a_1\hat{i} + a_2\hat{j} + a_3\hat{k}$, $b = b_1\hat{i} + b_2\hat{j} + b_3\hat{k}$, $c = c_1\hat{i} + c_2\hat{j} + c_3\hat{k}$,

show that :

(a) $a \cdot b = a_1b_1 + a_2b_2 + a_3b_3$;

(b) $b \times c = (b_2c_3 - b_3c_2)\hat{i} - (b_1c_3 - b_3c_1)\hat{j} + (b_1c_2 - b_2c_1)\hat{k}$.

12.5 Matrices and Determinants Revisited: Alternative Routes to Evaluating Scalar and Vector Products

12.5.1 The Scalar Product

If the components of the vectors a and b in Problem 12.9 form the elements of the column matrices $v_1 = \begin{pmatrix} a_1 \\ a_2 \\ a_3 \end{pmatrix}$ and $v_2 = \begin{pmatrix} b_1 \\ b_2 \\ b_3 \end{pmatrix}$, then the scalar product $a \cdot b$ takes the form:

$$v_1^T v_2 = (a_1 \quad a_2 \quad a_3) \begin{pmatrix} b_1 \\ b_2 \\ b_3 \end{pmatrix} = a_1b_1 + a_2b_2 + a_3b_3,$$

giving the same result as in Problem 12.9(a).

12.5.2 The Vector Product

If we compare the form of the vector product, given in the answer to Problem 12.9(b), with the expansion of a determinant of order three, given in eqn (10.20), we see that, if the correspondences:

$$a_{11} = \hat{i}, \ a_{12} = \hat{j}, \ a_{13} = \hat{k};$$

$$a_{21} = b_1, \ a_{22} = b_2, \ a_{23} = b_3;$$

$$a_{31} = c_1, \ a_{32} = c_2, \ a_{33} = c_3;$$

are made, then:

$$\boldsymbol{b} \times \boldsymbol{c} = \begin{vmatrix} \hat{\boldsymbol{i}} & \hat{\boldsymbol{j}} & \hat{\boldsymbol{k}} \\ b_1 & b_2 & b_3 \\ c_1 & c_2 & c_3 \end{vmatrix}. \tag{12.16}$$

Using the properties of determinants, we see that exchanging rows two and three results in a change of sign of the determinant. Such a change corresponds to the vector product $\boldsymbol{c} \times \boldsymbol{b}$ and is consistent with eqn (12.15) and $\boldsymbol{b} \times \boldsymbol{c} = -\boldsymbol{c} \times \boldsymbol{b}$.

Problem 12.10

If $\boldsymbol{a} = \hat{\boldsymbol{i}} + \hat{\boldsymbol{j}} + \hat{\boldsymbol{k}}$ and $\boldsymbol{b} = \hat{\boldsymbol{i}} - \hat{\boldsymbol{j}} + \hat{\boldsymbol{k}}$:

(a) Use eqn (12.16) to find the vector $\boldsymbol{a} \times \boldsymbol{b}$;
(b) Find $|\boldsymbol{a} \times \boldsymbol{b}|$, and specify a unit vector in the direction of $\boldsymbol{a} \times \boldsymbol{b}$.

12.5.3 The Scalar Triple Product

If we define three vectors $\boldsymbol{a}$, $\boldsymbol{b}$ and $\boldsymbol{c}$, as in Problem 12.9, the expression $\boldsymbol{a} \cdot (\boldsymbol{b} \times \boldsymbol{c})$, known as the scalar triple product, yields a scalar quantity, the magnitude of which provides the formula for the volume, V, of a parallelepiped with adjacent edges defined by vectors $\boldsymbol{a}$, $\boldsymbol{b}$ and $\boldsymbol{c}$ (an example in chemistry being a crystalline unit cell). If the determinantal representation of $\boldsymbol{b} \times \boldsymbol{c}$ is used, then, on expanding the determinant from the first row, and evaluating the three scalar products, we obtain:

$$\boldsymbol{a} \cdot (\boldsymbol{b} \times \boldsymbol{c}) = \boldsymbol{a} \cdot \begin{vmatrix} \hat{\boldsymbol{i}} & \hat{\boldsymbol{j}} & \hat{\boldsymbol{k}} \\ b_1 & b_2 & b_3 \\ c_1 & c_2 & c_3 \end{vmatrix} \tag{12.17}$$

$$= \boldsymbol{a} \cdot \left\{ \hat{\boldsymbol{i}}(b_2 c_3 - b_3 c_2) - \hat{\boldsymbol{j}}(b_1 c_3 - b_3 c_1) + \hat{\boldsymbol{k}}(b_1 c_2 - b_2 c_1) \right\}$$

$$= a_1(b_2 c_3 - b_3 c_2) - a_2(b_1 c_3 - b_3 c_1) + a_3(b_1 c_2 - b_2 c_1) \tag{12.18}$$

which, in turn, may be converted back into the determinantal form:

$$a \cdot (b \times c) = \begin{vmatrix} a_1 & a_2 & a_3 \\ b_1 & b_2 & b_3 \\ c_1 & c_2 & c_3 \end{vmatrix}. \tag{12.19}$$

Thus, the volume (a positive quantity), V, of the parallelepiped formed from a, b and c has the formula:

$$V = |a \cdot (b \times c)|. \tag{12.20}$$

Since $a \cdot (b \times c)$ may be negative, we take the modulus to ensure a positive result.

We explore the application of the scalar triple product for evaluating the volume of a crystallographic unit cell in the final two problems of this chapter.

Problem 12.11

If $a = a_2\hat{j} + a_3\hat{k}$, $b = b_1\hat{i}$, $c = c_2\hat{j}$, use eqn (12.19) to find an expression for $a \cdot (b \times c)$.

Problem 12.12

Crystalline naphthalene has a monoclinic unit cell, defined by the vectors in Problem 12.11, where $|a| = 0.824$ nm, $|b| = 0.600$ nm and $|c| = 0.866$ nm, and the angles between a and b, a and c, and b and c are $\alpha = 90°$, $\beta = 122.9°$ and $\gamma = 90°$, respectively.

(a) Give the values of b_1 and c_2.
(b) Use eqn (12.8) for the scalar product $a \cdot c$ and your answer to Problem 12.11 to show that $a_2 \times 0.866$ nm $= 0.824 \times 0.866 \times \cos \beta$ nm^2, and hence find the value of a_2.
(c) Use eqn (12.9) to show that $|a| = \sqrt{a_2^2 + a_3^2}$, and hence find the two possible values for a_3.
(d) Use eqn (12.19) to calculate the volume of the unit cell for naphthalene, using the positive value for a_3 obtained in (c).

Note: Repeating the calculation of the volume of the unit cell using the negative value for a_3 yields an identical result for the volume of the unit cell. The negative value for a_3 arises as a legitimate mathematical solution, but has little physical relevance other than to reflect the unit cell in the xy-plane.

Summary of Key Points

This chapter provides a description of some of the mathematical tools required to understand the properties of chemical and physical quantities that are defined not only by magnitude but also by direction. The key points discussed include:

1. The graphical definition of a vector.
2. A geometrical method for the addition and subtraction of vectors.
3. The properties of Cartesian base vectors.
4. An algebraic method for the addition and subtraction of vectors.
5. The scalar and vector products of two vectors.
6. The scalar triple products, involving three vectors.
7. Working with vectors using matrix and determinantal notation.
8. A selection of mathematical and chemically based examples, to illustrate practical applications of vectors.

13
Simple Statistics and Error Analysis

In this 13th and final chapter we start by addressing a question that might interest the more superstitious reader in particular, but which also makes an important point about the significance of sample size in experimental measurement. The number 13 is noteworthy in mathematics for a number of reasons but most obviously because it is a prime number. It also has many associations in folk law, and in many cultures is considered to be an unlucky number. So just how unlucky is it? How might we determine, for example, whether or not Friday 13th is a day to stay home or to venture out? A study published in the British Medical Journal concluded that 'The risk of hospital admission as a result of a transport accident [on a Friday 13th] may be increased by as much as 52%'.[1] This conclusion resulted from a statistical analysis of the number of hospital admissions due to traffic accidents on six different Friday 13th's over a four year period compared to the previous Friday the 6th in each of the months sampled. The sampling over six sets of data is of course an important component of the analysis but, at the outset of our discussion, it may not be obvious whether the sample is large enough to eliminate doubt about the significance of the findings. However, what should be clear is that comparison of a single Friday 6th with a single Friday 13th would be meaningless: in order to draw any meaningful conclusion about such an observation, we have to demonstrate reproducibility. In the case of their study, the authors will also have had to normalise the data set against other factors such as the volume of traffic on the road on each pair of days and take account of whether or not a particular Friday the 13th or Friday 6th falls on a bank holiday. They would also need to be sure that each hospital admission entry is definitely associated with an accident occurring on the day in question, with any resulting uncertainties accounted for in their conclusions. Such principles of sampling, statistical significance, the presentation and analysis of data and the assessment of its reliability lie at heart of scientific observation and of course of experimental chemistry.

The conclusions we draw about the significance of our experimental measurements depend critically on how carefully we design our experiment, our choice of method and how well we assess errors associated with our experiments. Sometimes we are making a simple measurement of some physical quantity, the eventual recorded value of which derives from repeated measurements using a tried and tested experimental approach. However, sometimes we might measure a quantity under a range of different experimental conditions in order to determine some other quantity that is related to the measured one through a mathematical relationship.

In this final chapter, we will look first of all at the types of errors associated with a measurement before going on to look in closer detail at statistical distributions of measurements and how we can extract a best estimate of a true value from a distribution as well as a measure of the uncertainty associated with that value. We then take a look at linear regression analysis, a topic that is immensely important when measurements take the form of a determination of a value of a dependent variable, such as a rate of reaction, for each of several values of an independent variable, such as concentration of a reactant. We close the chapter by looking at how errors propagate (discussed briefly in the context of differentials in Section 5.3) when the value of a physical quantity is not measured directly from an experiment but derives from a functional relationship with another measured quantity.

Aims:

The aim of this chapter is to develop an appreciation of how simple statistics can be used to understand the nature of repeated experimental measurement, how it can help us process and interpret experimental results and, importantly, how it enables us to make meaningful and useful assessments of errors arising from common experimental situations. By the end of the chapter, you should be able to:

- Recognise the sources and nature of different types of error in an experiment.
- Understand the difference between systematic and random errors.
- Understand the key features of the Gaussian or normal distribution and to be able to define the terms: mean, variance, standard deviation and standard error.

- Calculate a sample mean, sample standard deviation and standard error of the sample mean of a set of data and deduce the 95% confidence interval.
- Perform a least-squares analysis of a data set, deducing the best values of the parameters appearing in the linear relationship and the uncertainties associated with them.
- Determine how errors propagate through functions of single and multiple variables
- Understand the distinction between absolute and relative errors.

13.1 Errors

The uncertainty in a measurement is commonly also called the error but we have to exercise a little care in how we use this term. Errors are often associated with mistakes or with something that is definitely wrong but scientists tend to think of them as either:

1) The discrepancy between a measured value and some generally accepted true value – generally referred to as the accuracy of the measurement, or
2) The degree to which repeated measurement of a particular quantity gives the same result – generally referred to as the precision of the measurement.

An experimental method may be accurate but not especially precise if it suffers from random errors, but equally it may yield inaccurate results with high precision if it is subject to systematic errors (see Figure 13.1). In an experiment free from serious systematic errors, we can improve the precision considerably by increasing the sample size.

Low accuracy
High precision

High accuracy
Low precision

High accuracy
High precision

Figure 13.1 Arrows striking a target close to the bull's eye achieve high accuracy but do so not necessarily with high precision (centre). Conversely, a closely clustered group of shots off to one side demonstrate high precision but low accuracy (left). Both high precision and high accuracy is only achieved through a tight clustering of shots close to the bull's eye (right).

13.1.1 Systematic Errors

A laboratory measurement of the boiling point (b.p.) of isoamyl acetate, produced in an undergraduate experiment to synthesize banana oil, yields a temperature of 123 °C, a value the student notes to be 19 °C below the known reference value of 142 °C. What can the student conclude about this observation? Perhaps they made the wrong product, or perhaps their technique was somehow at fault. Repeating the measurement with the same thermometer but on a sample produced by a different student yields a boiling point of 124 °C, a result close enough to the first measurement to suggest that the method or apparatus may be at fault. A demonstrator suggests trying a different type of alcohol thermometer and this time the measured boiling point is 130 °C. A third measurement with a mercury thermometer that had been calibrated against a known standard yields a value of 141 °C. We now have four data points and are significantly closer towards being able to draw a meaningful conclusion about the reason for the low initial value of the measured boiling point. Repetition of this process using twenty different thermometers then builds up a data set describing a statistical distribution of thermometer performance at temperatures above 115 °C (shown in Figure 13.2), but perhaps is not terribly helpful in providing us with the definitive boiling point measurement of the ester. This type of error, resulting from some deficiency in the apparatus, is known as a systematic error and is not revealed by repeated measurement using the same apparatus and technique. We

Figure 13.2 The frequency distribution of twenty boiling point measurements of an ester made with twenty different thermometers is shown in a frequency histogram. In this case, the data reveal a systematic error between the values recorded and the reference value of 142 °C. The systematic error derives in this case from deficiencies in the types of thermometer used to make the measurements.

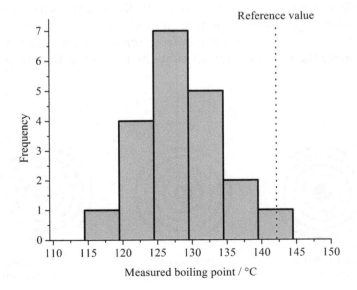

only discover its existence having considered the apparatus used and perhaps the process too. The true value of the boiling point of the ester will be that which has been determined by the most reliable and accurate method, with that having been established by benchmarking or calibration against some sort of standard.

Systematic errors always affect the accuracy of the result and will tend to do so in the same direction – *i.e.* the answer will always tend to be too high or too low rather than fluctuating about some central value following successive measurements. As a consequence of the fact that they derive from deficiencies of one sort or another in the experimental apparatus, they cannot be quantified by a statistical analysis of repeated observations.

13.1.2 Random Errors

The somewhat uncomfortable reality of experimental measurement is that two measurements of the same physical property with identical apparatus, using a nominally identical procedure, will almost invariably yield slightly different results. Repeating the measurement further will continue to yield slightly different values each time. Such variations result from random fluctuations in the experimental conditions from one measurement to the next and from limitations associated with the precision of the apparatus or the technique of whoever is conducting the experiment. The statistical nature of these fluctuations means that the discrepancies with respect to the 'true' value are equally likely to be positive or negative. The precision of the final result can always be improved by performing more and more measurements until the distribution of the scattered results becomes better defined, at which point the average value and an uncertainty associated with it can be established. It is this type of uncertainty or error that can be best treated using statistical methods.

Problem 13.1

Write down some examples of sources of random errors that might affect experiments you have performed at school or college or as part of your degree course. In each case, try to think of ways in which the impact of each source of error might be minimised.

13.2 The Statistics of Repeated Measurement

13.2.1 The Binomial Distribution

In the example given in the introduction to this chapter, the analysis of hospital admissions following road traffic accidents was based on six sets of data, five of which showed larger numbers of hospital admissions on a given Friday 13th compared with the previous Friday 6th, and one of which showed the opposite result. We can get some idea of the extent to which this data set is large enough by considering the probabilities associated with the possible outcomes of the tossing of a coin. For each toss of the coin, there are two equally likely outcomes with the probability of getting heads, $P(\text{H})$, $\frac{1}{2}$, and the probability of getting tails, $P(\text{T})$, which is also $\frac{1}{2}$. For two tosses of the coin, there are 2^2 possible outcomes with the probability of each outcome, HH, HT, TH or TT, being the product of the probability associated with each toss, $i.e.$ $\frac{1}{2} \times \frac{1}{2} = \left(\frac{1}{2}\right)^2 = \frac{1}{4}$. In other words, there is a one in four chance that any prediction of the outcome of two tosses will be correct. For three tosses, there are 2^3 possible outcomes, each with equal probability, some of which will be associated with the same nominal outcome but distinguished by the order in which the heads and tails fall. For example, the probability that we get HHT is the same as HTH or THH but each outcome counts as a distinguishable result, contributing as a result to the total probability that we get two heads and one tails. If we define $P(n/N)$ as the total probability that n heads results from N tosses of the coin, accounting for the fact that each result may be achievable in more than one distinguishable way (with the number of tails being $N - n$), then the total probabilities associated with 3, 2, 1 and 0 heads following three tosses are:

Three heads

$$P(3/3) = P(\text{HHH}) = \frac{1}{2} \times \frac{1}{2} \times \frac{1}{2} = \left(\frac{1}{2}\right)^3 = \frac{1}{8}$$

Two heads, one tail

$$P(2/3) = P(\text{HHT}) + P(\text{HTH}) + P(\text{THH}) = 3 \times \left(\frac{1}{2}\right)^3 = \frac{3}{8}$$

One head, two tails

$$P(1/3) = P(\text{TTH}) + P(\text{THT}) + P(\text{HTT}) = 3 \times \left(\frac{1}{2}\right)^3 = \frac{3}{8}$$

No heads, three tails

$$P(0/3) = P(\text{TTT}) = \left(\frac{1}{2}\right)^3 = \frac{1}{8}$$

In this example, the weightings associated with each type of result (1 3 3 1) are the **binomial coefficients** associated with the **binomial distribution** which applies to any experiment yielding one of two possible outcomes. Thus the probability of n heads resulting from N tosses, $P(n/N)$, is

$$P(n/N) = \binom{N}{n} p^n q^{N-n}$$

$$= \frac{N!}{n!(N-n)!} p^n (1-p)^{N-n} \tag{13.1}$$

where p denotes the probability of heads in a single toss (*i.e.* $\frac{1}{2}$) and $q = 1 - p$, the probability of tails in a single toss (also $\frac{1}{2}$).

Worked Problem 13.1

Q. Use the binomial distribution in eqn (13.1) to verify that the probability of getting two heads and one tail from three tosses of a coin is $\frac{3}{8}$.

A. For $N = 3$ and $n = 2$:

$$P(2/3) = \frac{3!}{2!(3-2)!} \left(\frac{1}{2}\right)^2 \left(\frac{1}{2}\right)^1 = 3 \times \left(\frac{1}{2}\right)^3 = \frac{3}{8}$$

where $p = \frac{1}{2}$ for a heads outcome and $q = \left(1 - \frac{1}{2}\right) = \frac{1}{2}$ for tails.

Problem 13.2

Use the binomial distribution in eqn (13.1) to calculate the probability of getting five heads and one tail from six tosses of a coin.

It is instructive at this point to extend Problem 13.2 to encompass all possible outcomes of six tosses of a coin, and then to compare the theoretical result with results obtained from two sets of experiments. Each set of experiments records the number of heads obtained from six tosses of a coin: in the first experiment the data derive from ten consecutive sets of six tosses and in the second from twenty sets. The result of this comparison is shown in Figure 13.3.

We can make two important observations from this comparison:

1) The theoretical probability of tossing five heads is about 1 in 10 (or to be precise, 3 in 32), a value large enough to suggest that some care is required in drawing conclusions from any one experiment in which one particular result is obtained five times out of six (and to wander at what point one might speculate about whether or not the coin is weighted!). The road traffic accident data in the example discussed in the introduction is binary in the sense that we might equate 'heads' with 'more accidents on Friday 13th' and 'tails' with 'more accidents on the preceding Friday 6th'. Thus a batch of data taken from one 4-year period is just as likely to throw up an odd result as one particular set of six tosses of a coin.

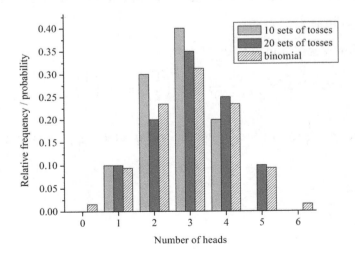

Figure 13.3 A comparison of two sets of experimental data, reporting the results of sets of six consecutive coin tosses with the theoretical binomial probability distribution. It is worth noting that the absence of experimental bars against the 0 and the 6 simply reflects the very low probability of tossing six heads or six tails from six consecutive tosses. This absence of data should not come as a surprise given the odds (1 in 64) and the sample sizes (10 and 20). It is also worth noting the absence of a five heads sequence in the smaller 10 sets sample for similar reasons.

2) As the sample size increases from ten to twenty, the differences between the experimental relative frequencies and the theoretical probability distribution get smaller – with our faith in the experimental results increasing as a result.

This last observation serves as a good point to consider in more detail how we use a distribution deriving from a large sample of data to derive a reliable estimate of the quantity being measured and an estimate of the uncertainties associated with that measurement.

13.2.2 The Gaussian Distribution

As we have seen above, the results of an experiment subject to random uncertainties will tend to cluster about a central most probable value. In such a distribution, the probability of obtaining a value different from the most probable value decreases as that difference increases. For a very large number of measurements, the distribution will adopt a relatively smooth probability density function, $p(x)$, typically having the mathematical form:

$$p(x) = \frac{1}{\sigma\sqrt{2\pi}} e^{-\frac{(x-\mu)^2}{2\sigma^2}}$$ (13.2)

The physical significance of the symmetrical function $p(x)$ is best understood from its definite integral $\int_a^b p(x)dx$ which gives the fraction of measurements that fall between $x = a$ and $x = b$ with $p(x)dx$ being the probability that any one measurement falls between x and $x + dx$. The function $p(x)$ in eqn (13.2) is known as the Gaussian distribution, or sometimes the normal distribution, and is characterised by a central mean (average) value, μ, and a width parameter, σ. A plot of the function is shown in Figure 13.4 for two different values of σ.

13.2.2.1 The Mean, Variance and Standard Deviation

For an experiment in which we take N measurements, x_i ($i = 1, 2, \ldots N$), the mean value $\bar{x}$ is given by:

$$\bar{x} = \frac{1}{N}\sum_{i=1}^{N} x_i$$ (13.3)

This experimental mean is our best estimate of the 'true' mean, μ, something we could only find by making an infinite number of measurements. The spread of the distribution, parameterised by σ, provides us with a measure of the uncertainty associated with the

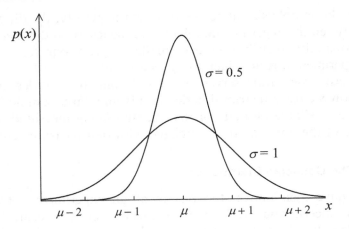

Figure 13.4 The normal or Gaussian distribution is a continuous probability distribution used to describe random measured values clustered symmetrically about a single mean value, μ. The spread or width of the distribution is parameterised by σ. Regardless of the width of the distribution, the total area will always be unity.

mean. A first step to quantifying this scatter about the mean might be to calculate the average value of the individual deviations from the mean, $x_i - \bar{x}$, but, for a symmetrical distribution, any one deviation is more or less equally likely to be either positive or negative and so an average deviation would then be close to zero. Consequently, it is conventional to calculate the square root of the average of the squares of all the deviations. This is known as the **standard deviation** and given the symbol, σ. Its square, σ^2, is called the **variance** and is defined by:

$$\sigma^2 = \frac{1}{N} \sum_{i=1}^{N} (x_i - \bar{x})^2 \tag{13.4}$$

13.2.2.2 The Sample Variance and Sample Standard Deviation

In practice, we do not know the 'true' value of the measured quantity, but only have an estimate of it provided from our sample mean. We must then reflect this additional uncertainty in our estimate of the standard deviation from our limited sample by defining our best estimate of the true variance, σ^2, as the **sample variance**, s^2, given by:

$$s^2 = \frac{1}{N-1} \sum_{i=1}^{N} (x_i - \bar{x})^2 \tag{13.5}$$

The **sample standard deviation** is then given by:

$$s = \sqrt{s^2} \tag{13.6}$$

In going from σ^2 to s^2, we say that the number of degrees of freedom has been reduced by 1 from N to $N - 1$ because we have already used the sample once to calculate the value of $\bar{x}$ required in the calculation of s.

13.2.2.3 Confidence Intervals

The probability that a particular value of x lies within a certain interval, $\pm a$, of the mean can be determined from the area enclosed by the Gaussian distribution function between $\mu - a$ and $\mu + a$. We can use the calculation of the integral and hence area to arrive at a range within which we can expect the values to lie within a particular probability. For the Gaussian distribution, the probability that x lies within σ of the mean is 0.683; that it lies within 2σ is 0.954; and that it lies within 3σ is 0.997. In other words, approximately 68% of the measured values will lie within one standard deviation of the mean; approximately 95% will lie within two standard deviations and over 99% will lie within three standard deviations. We can then define the '68% confidence interval' as $\mu \pm \sigma$; the '95% confidence interval' as $\mu \pm 2\sigma$ and the '99% confidence interval' as $\mu \pm 3\sigma$ (Figure 13.5).

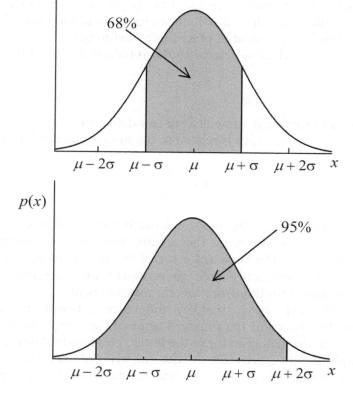

Figure 13.5 In a Gaussian distribution, approximately 68% of the measured values will lie within one standard deviation of the mean (upper panel) and 95% will lie within two standard deviations (lower panel).

13.2.2.4 Standard Error of the Mean and Standard Error of the Sample Mean

One defining feature of the discussion so far is that, if in doubt, make more measurements! A natural intuitive expectation that follows from this is that as the number of measurements, N, increases, so should the precision with which we determine the mean. This is particularly important because the mean is generally what we are most interested in finding and what the experiment was designed to measure in the first place. We can visualise this as the experimental distribution smoothing out as more measurements are made, with the centre of the distribution becoming better defined. This happens in spite of the fact that the spread or width of the distribution remains more or less constant. A good illustration of this process is shown in Figure 13.3 in which experimental frequency distributions in the numbers of heads in samples of six successive coin tosses are compared with the theoretical binomial distribution. What is clear from the plots is that the '10 sets of tosses' distribution deviates the most from the binomial distribution and is skewed to lower values. The average number of heads from the 10 sets of tosses is 2.7; from 20 sets, 3.05; and, of course from the perfectly symmetrical binomial distribution, 3. This difference in the evaluation of the mean happens in spite of the fact that the widths of all three distributions look about the same. We can quantify this improvement in precision by defining the standard error of the mean, σ_μ, which is related to the standard deviation by:

$$\sigma_\mu = \frac{\sigma}{\sqrt{N}} \tag{13.7}$$

In terms of sample parameters, we can define the standard error of the sample mean, $s_{\bar{x}}$, which is related to the sample standard deviation, s, by:

$$s_{\bar{x}} = \frac{s}{\sqrt{N}}. \tag{13.8}$$

Eqn (13.8) provides the important result that the precision of the mean value improves with the square root of the number of measurements, N. The standard error of the sample mean, $s_{\bar{x}}$, is the quantity most commonly used to represent the uncertainty in a measured mean resulting from several measurements.

As highlighted above, the key difference between the sample standard deviation and the standard error of the sample mean is the factor $\sqrt{N}$ in the denominator of the latter. The standard deviation of the sample mean, s, provides us with the average uncertainty in the individual measurement of the x_i and consequently will not change noticeably as we make more measurements. On the other hand, the standard error of the sample mean, $s_{\bar{x}}$, *will* decrease in value as we increase the number of measurements. This is in keeping with our expectation that the reliability of our measured mean improves as the number of measurements increases. Unfortunately, the square root means that in order to improve our precision by a factor of ten, we would need to make a hundred times as many measurements!

In reporting our results, it is conventional, if our best estimate of our quantity x is the mean value of several measurements to write:

$$\text{measured value of } x = \bar{x} \pm s_{\bar{x}}$$

where $s_{\bar{x}}$ provides a good measure of its uncertainty. This expression implies that we expect approximately 68% of any measurements of x, made using the same method, to fall within the range $\bar{x} \pm s_{\bar{x}}$.

How we choose to express the uncertainty associated with our measurement depends on how certain we are of our reported value. The convention described above assigns a confidence level that most scientists would view as reasonable and useful. However, sometimes we might want to broaden the limits of our confidence and you will often see uncertainties reported in terms of 95% confidence limits.

13.2.2.5 Student t Factors and the 95% Confidence Limit

For a Gaussian distribution, there is a 95% probability that the range $\bar{x} \pm 1.96 s/\sqrt{N}$ includes the true mean but this is only true for very large values of N. For smaller sample sizes, the range will be larger than this and is defined as $\bar{x} \pm t_v \left(s/\sqrt{N} \right)$ (or more concisely as $\bar{x} \pm t_v s_{\bar{x}}$) where the t_v are a set of multiplication factors known as the Student t-factors that extend from as large as 12.71 for a sample size of 1 to 2.01 for a sample size of 50. An infinitely large sample size would carry a Student t-factor of 1.96. We can thus define the **95% Confidence Limit** as the range $\bar{x} \pm 1.96 s/\sqrt{N}$ for an infinitely large sample size or, for smaller samples, $\bar{x} \pm t_v \left(s/\sqrt{N} \right)$.

The 'Student' in the Student t-factor was the pseudonym of William Sealy Gosset who published a derivation of the t-distribution in 1908 while working for the Guinness Brewery in Dublin.

Problem 13.3

The thermodynamic data in Table 13.1 derive from the same experiment conducted by 14 separate pairs of students in an undergraduate practical laboratory. The aim of the experiment was to find the enthalpy and Gibbs free energy changes associated with the decomposition of ammonium carbamate.

$$NH_4CO_2NH_2(s) \rightarrow 2NH_3(g) + CO_2(g)$$

Each data point in the table was determined by measuring the equilibrium vapour pressure as a function of temperature.

Table 13.1 Enthalpy and Gibbs free energy changes associated with the decomposition of ammonium carbamate, deriving from 14 separate sets of experiments.

i	$\Delta H_i^\ominus$/kJ mol^{-1}	$\Delta G_i^\ominus$/kJ mol^{-1}	i	$\Delta H_i^\ominus$/kJ mol^{-1}	$\Delta G_i^\ominus$/kJ mol^{-1}
1	126.0	21.3	8	155.0	20.7
2	146.0	20.3	9	141.0	20.0
3	142.0	20.1	10	131.0	19.0
4	140.0	20.1	11	145.0	19.9
5	114.0	21.7	12	149.0	20.6
6	150.0	22.1	13	146.0	21.0
7	152.0	20.0	14	153.0	19.6

For each set of $\Delta H^\ominus$ or $\Delta G^\ominus$ values, find the following:

(a) The mean.
(b) The sample standard deviation.
(c) The standard error of the sample mean, reporting your answer in the form $\bar{x} \pm s_{\bar{x}}$.
(d) The 95% confidence limit, assuming a Student t-factor of 2.14.

Hint: Once you have calculated the mean values for the enthalpy change, $\overline{\Delta H}^\ominus$, and free energy change, $\overline{\Delta G}^\ominus$, you may find it helpful to tabulate $\Delta H_i^\ominus - \overline{\Delta H}^\ominus$ and $\Delta G_i^\ominus - \overline{\Delta G}^\ominus$.

13.3 Linear Regression Analysis

In many experiments, the measurements involve the determination of a value of some dependent variable (*i.e.* the one being measured) for each of several values of an independent variable (*i.e.* one that the experimenter controls or chooses). For example, in the experiment covered in Problem 13.3, the equilibrium vapour pressure of ammonium carbamate is measured at a number of different temperatures. It will often be the case that a linear relationship exists between the dependent and independent variables, with the measured values of the dependent variable being distributed about an imagined straight line describing the perfect linear relationship between the two variables.

Of course many relationships are not linear functions but we can still often use linear regression analyses in these examples by linearising the relationship into a form which yields a straight line plot. In the example of ammonium carbamate, the equilibrium constant, K_p, varies with temperature according to the relationship:

$$K_p = e^{-\frac{\Delta H^{\ominus}}{RT} + \frac{\Delta S^{\ominus}}{R}} \qquad (13.9)$$

This function is clearly not linear in form but can easily be converted into linear form by taking natural logs each side:

$$\ln K_p = -\frac{\Delta H^{\ominus}}{RT} + \frac{\Delta S^{\ominus}}{R} \qquad (13.10)$$

A plot of $\ln K_p$ vs. $1/T$ will then yield a straight line, with gradient $-\Delta H^{\ominus}/R$ and y-intercept at $\Delta S^{\ominus}/R$.

The advantage of linear regression analysis is the simplicity of the concept and of the process (although this latter observation may be regarded as irrelevant given that most regressions will be performed on a computer), but the disadvantage is that uncertainties are often propagated through the linearisation process in a distorted way. This can result in misleading evaluation of errors if, for example, an equation is linearised by taking logs, such as in the transformation of eqn (13.9) into eqn (13.10).

13.3.1 The Least-Squares Method

13.3.1.1 Finding the Best Gradient and y-Intercept

For any experimental data set describing a linear relationship between two variables, we will have an intuitive sense of what the 'best' straight line through a scatter plot of the points should look like, but it is also possible to define the best fit line *quantitatively*. Given a data set of N measurements of a dependent variable, y, labelled y_1, y_2, y_3, ..., y_i, ..., y_N for each corresponding independent variable x_1, x_2, x_3, ..., x_i, ..., x_N with a theoretical linear relationship existing between the two sets:

$$y = mx + c \qquad (13.11)$$

a plot of the y_i values against the x_i values should yield a set of data similar to that shown in Figure 13.6. Once a line has been drawn through the data points, we can measure the distance of each point from the line, with this distance being defined as the vertical distance

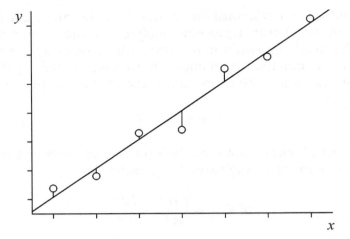

Figure 13.6 In a linear relationship between dependent (y) and independent (x) variables, the measured values of the dependent variable will be distributed about an imagined straight line.

between the data point and a point on the line directly above or below it. These distances, commonly referred to as residuals, $y_i - y_{line}$, will either be positive or negative, depending on whether the point lies above or below the line. However the *squares* of residuals $(y_i - y_{line})^2$ will all be positive. The method of least squares works by choosing the line with gradient and intercept that makes the sum of these squares over all the points as small as possible. In other words, we aim to minimise the sum:

$$\chi^2 = \sum_{i=1}^{N} (y_i - (mx_i + c))^2 \qquad (13.12)$$

where $mx_i + c = y_{line}$, with respect to the parameters m and c given in eqn (13.11). We achieve this by finding the partial derivatives of χ^2 (pronounced chi squared) with respect to m and c, and setting the result of each derivative to zero (see Section 4.4.1 for discussion on the finding and defining of stationary points):

$$\frac{\partial \chi^2}{\partial c} = -2 \sum_{i=1}^{N} (y_i - mx_i - c) = 0 \qquad (13.13)$$

$$\frac{\partial \chi^2}{\partial m} = -2 \sum_{i=1}^{N} x_i(y_i - mx_i - c) = 0 \qquad (13.14)$$

These two equations can be rewritten as a pair of simultaneous equations for m and c:

$$\sum_{i=1}^{N} y_i = m \sum_{i=1}^{N} x_i + cN \qquad (13.15)$$

$$\sum_{i=1}^{N} x_i y_i = m \sum_{i=1}^{N} x_i^2 + c \sum_{i=1}^{N} x_i, \qquad (13.16)$$

commonly referred to as normal equations and are straightforwardly soluble for m and c to yield:

$$m = \frac{N \sum_{i=1}^{N} x_i y_i - \sum_{i=1}^{N} x_i \sum_{i=1}^{N} y_i}{N \sum_{i=1}^{N} x_i^2 - \left(\sum_{i=1}^{N} x_i\right)^2} \qquad (13.17)$$

$$c = \frac{\sum_{i=1}^{N} y_i \sum_{i=1}^{N} x_i^2 - \sum_{i=1}^{N} x_i \sum_{i=1}^{N} x_i y_i}{N \sum_{i=1}^{N} x_i^2 - \left(\sum_{i=1}^{N} x_i\right)^2}. \qquad (13.18)$$

Consequently, m and c can be calculated by gathering together the appropriate sums from the experimental data and substituting them into eqns (13.17) and (13.18), respectively. Of course, we also need to find the uncertainties associated with the parameters m and c but in order to do that we need first to establish the uncertainty in our measurements of the y_i.

13.3.1.2 Finding the Uncertainties Associated with the Gradient and y-Intercept

If we assume that our measurement of each y_i is distributed normally about its true value, $y_{\text{line}} = mx_i + c$, with a width parameter σ_y, then the residuals $y_i - y_{\text{line}}$ will also be normally distributed about the same central value and with the same width, σ_y. A sensible approach at this point might be to suggest that an estimate for σ_y may be obtained by using a sum of squares approach:

$$\sigma_y = \sqrt{\frac{1}{N} \sum_{i=1}^{N} (y_i - y_{\text{line}})^2} \qquad (13.19)$$

which when substituting $mx_i + c$ for y_{line} gives:

$$\sigma_y = \sqrt{\frac{1}{N} \sum_{i=1}^{N} (y_i - mx_i - c)^2} \qquad (13.20)$$

However, the constants m and c in eqn (13.20) are the true values of the constants, rather than the best estimates that we arrived at from eqns (13.17) and (13.18), and so we need to reflect this in our estimate of σ_y by reducing the degrees of freedom by 2 (having used up 2 in computing m and c). Thus eqn (13.20) becomes modified to:

$$\sigma_y = \sqrt{\frac{1}{N-2} \sum_{i=1}^{N} (y_i - mx_i - c)^2} \qquad (13.21)$$

We are now in a position to calculate the uncertainties in the constants m and c through a simple propagation of errors (see Section 13.4) from the y_i to the gradient, m, and intercept, c. These are given below:

$$\sigma_m = \sigma_y \sqrt{\frac{N}{N \sum_{i=1}^{N} x_i^2 - \left(\sum_{i=1}^{N} x_i\right)^2}} \qquad (13.22)$$

and:

$$\sigma_c = \sigma_y \sqrt{\frac{\sum_{i=1}^{N} x_i^2}{N \sum_{i=1}^{N} x_i^2 - \left(\sum_{i=1}^{N} x_i\right)^2}}. \qquad (13.23)$$

Worked Problem 13.2

Q. Perform a linear regression analysis on the data in Table 13.2.

Table 13.2 The data set used to generate the plot in Figure 13.6

x_i	1	2	3	4	5	6	7
y_i	1.375	1.800	3.290	3.400	5.500	5.900	7.200

A. We start by evaluating each of the terms that appear in eqns (13.17) and (13.18) by expanding the Table 13.2 as follows:

x_i	1	2	3	4	5	6	7	$\sum x_i = 28$
y_i	1.375	1.800	3.290	3.400	5.500	5.900	7.200	$\sum y_i = 28.465$
x_i^2	1	4	9	16	25	36	49	$\sum x_i^2 = 140$
$x_i y_i$	1.375	3.600	9.870	13.600	27.500	35.400	50.400	$\sum x_i y_i = 141.745$

Thus:

$$N = 7, \quad \sum_{i=1}^{N} x_i = 28, \quad \sum_{i=1}^{N} y_i = 28.465, \quad \sum_{i=1}^{N} x_i^2 = 140,$$

$$\sum_{i=1}^{N} x_i y_i = 141.745, \quad \left(\sum_{i=1}^{N} x_i \right)^2 = 784$$

Inserting these values into eqn (13.17) yields for the gradient:

$$m = \frac{(7 \times 141.745) - (28 \times 28.465)}{(7 \times 140) - 784} = \frac{195.195}{196} = 0.9959$$

and into eqn (13.18) for the y-intercept

$$c = \frac{(28.465 \times 140) - (28 \times 141.745)}{(7 \times 140) - 784} = \frac{16.24}{196} = 0.08286$$

The uncertainty in the y_i is given by eqn (13.21) for which in this case the term:

$$\sum_{i=1}^{N} (y_i - mx_i - c)^2 = 0.8934$$

and so:

$$\sigma_y = \sqrt{\frac{1}{5} \times 0.8934} = 0.4227.$$

We can now use this to find the uncertainties associated with m and c:

$$\sigma_m = 0.4227 \sqrt{\frac{7}{196}} = 0.0799$$

and:

$$\sigma_c = 0.4227 \sqrt{\frac{140}{196}} = 0.3572.$$

Problem 13.4

In a study of the pyrolysis of 1-butene, methane was produced as a major product, with the following values determined for the first order rate constant between 766 and 828 K.

T / K	766	782	787	795	814	819	828
$k / 10^{-5}\ s^{-1}$	8.4	24.1	24.2	38.1	90.2	140.0	172.0

Use linear regression to evaluate the Arrhenius parameters, E_a and A for this reaction.

Hint: The Arrhenius equation describes how the rate constant depends upon temperature according to $k = Ae^{-E_a/RT}$. In order to perform a linear regression, you will first need to convert this equation into linear form.

Eqns (13.17) and (13.18) assumed that the measurements of the y_i values all have the same uncertainties. However, if they have different, *known* uncertainties σ_i, then we can introduce a weighting to reflect these differences. Eqns (13.17) and (13.18) thus become modified to:

$$m = \frac{\sum_{i=1}^{N} \frac{1}{\sigma_i^2} \sum_{i=1}^{N} \frac{x_i y_i}{\sigma_i^2} - \sum_{i=1}^{N} \frac{x_i}{\sigma_i^2} \sum_{i=1}^{N} \frac{y_i}{\sigma_i^2}}{\sum_{i=1}^{N} \frac{1}{\sigma_i^2} \sum_{i=1}^{N} \frac{x_i^2}{\sigma_i^2} - \left(\sum_{i=1}^{N} \frac{x_i}{\sigma_i^2} \right)^2} \tag{13.24}$$

$$c = \frac{\sum_{i=1}^{N} \frac{y_i}{\sigma_i^2} \sum_{i=1}^{N} \frac{x_i^2}{\sigma_i^2} - \sum_{i=1}^{N} \frac{x_i}{\sigma_i^2} \sum_{i=1}^{N} \frac{x_i y_i}{\sigma_i^2}}{\sum_{i=1}^{N} \frac{1}{\sigma_i^2} \sum_{i=1}^{N} \frac{x_i^2}{\sigma_i^2} - \left(\sum_{i=1}^{N} \frac{x_i}{\sigma_i^2} \right)^2} \tag{13.25}$$

These equations look horrendous but are actually quite straightforward to use in practice. As the uncertainty in the measurement in the y_i values given in eqn (13.21) is now accounted for in the weighting, the uncertainties associated with m and c under these conditions then become:

$$\sigma_m = \sqrt{\frac{\sum\limits_{i=1}^{N} \frac{1}{\sigma_i^2}}{\sum\limits_{i=1}^{N} \frac{1}{\sigma_i^2} \sum\limits_{i=1}^{N} \frac{x_i^2}{\sigma_i^2} - \left(\sum\limits_{i=1}^{N} \frac{x_i}{\sigma_i^2}\right)^2}} \qquad (13.26)$$

and:

$$\sigma_c = \sqrt{\frac{\sum\limits_{i=1}^{N} \frac{x_i^2}{\sigma_i^2}}{\sum\limits_{i=1}^{N} \frac{1}{\sigma_i^2} \sum\limits_{i=1}^{N} \frac{x_i^2}{\sigma_i^2} - \left(\sum\limits_{i=1}^{N} \frac{x_i}{\sigma_i^2}\right)^2}} \qquad (13.27)$$

Although this whole process can be time-consuming and even tedious if performing multiple linear regressions on a set of data by hand, doing so can provide a degree of insight not available when running a linear regression analysis on a computer using modern data plotting software such as Excel or Origin.

13.4 Propagation of Errors

13.4.1 Uncertainty in a Single Variable

It is often the case that the value of a particular physical quantity is not measured directly in an experiment but instead is calculated from some other measured quantity. For example, in rotational spectroscopy, adjacent pairs of lines in the microwave spectrum of a heteronuclear diatomic molecule such as ^{12}CO are separated by twice the rotational constant B (neglecting the effects of centrifugal distortion) which itself is related to the bond length, r, through the relationship:

$$B = \frac{h^2}{8\pi^2 c \mu r^2} \, cm^{-1} \qquad (13.28)$$

where $\mu = \frac{m_C m_O}{m_C + m_O}$ is the reduced mass and c, the speed of light in cm s^{-1}. In this example, we need to be confident about how the uncertainties associated with our measurement of B transfer to an error associated with our subsequent determination of the bond length, r. The process of transferring error in one quantity to another quantity is known as the propagation of errors.

Let us imagine we have measured the value of some quantity, x, and have an estimate of its uncertainty, σ_x. If x is related to the value of another quantity, y, through some functional relationship $y = f(x)$ then the uncertainty in y, σ_y will be related to the slope of the function, $\dfrac{d}{dx}f$ and the magnitude of σ_x (see Figure 13.7)

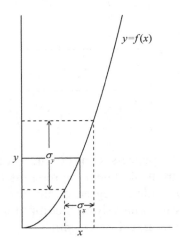

Figure 13.7 For two quantities, x and y related through a function $y = f(x)$, the uncertainty in y, σ_y, will be related to the slope of the function, $\frac{d}{dx}f$ and the magnitude of σ_x.

Assuming σ_x is small enough that the slope is approximately constant across the range $x \pm \sigma_x$ we can make the approximation that:

$$\sigma_y \approx \left| \frac{d}{dx}f \right| \sigma_x \tag{13.29}$$

or, avoiding the use of the modulus:

$$\sigma_y^2 \approx \left(\frac{d}{dx}f \right)^2 \sigma_x^2 \tag{13.30}$$

Clearly, eqns (13.29) and (13.30) are equivalent in the sense that they will yield exactly the same value for σ_y. However, we shall see later, in discussing how errors propagate in functions of more than one independent variable, that the two analogous expressions in that case do not produce identical values for the error propagated through to the dependent variable.

The quantities σ_x and σ_y are known as the **absolute uncertainties** in x and y. The **relative uncertainties** for particular values of x and y are obtained by dividing σ_x and σ_y by x and y, respectively.

Worked Problem 13.3

Q. Two quantities are related through the linear relationship:

$$y = kx$$

where the absolute uncertainty associated with measurements of quantity x is σ_x. Find the absolute and relative uncertainties in y and write down the relationship between the relative uncertainties in x and y.

A. The modulus of the derivative is:

$$\left|\frac{dy}{dx}\right| = |k|$$

and so the absolute uncertainty in y from eqn (13.29) is:

$$\sigma_y = |k|\sigma_x \qquad (13.31)$$

The relative uncertainty in y is:

$$\frac{\sigma_y}{|y|} = \frac{|k|\sigma_x}{|y|} = \frac{|k|\sigma_x}{|k||x|} = \frac{\sigma_x}{|x|} \qquad (13.32)$$

For a linear relationship between x and y, the absolute uncertainty in y is k times the absolute uncertainty in x, whereas the relative uncertainties in x and y are equal.

Worked Problem 13.4

Q. Two quantities are related through the relationship:

$$y = x^n$$

with absolute uncertainty in x, σ_x. Find the absolute and relative uncertainties in y and write down the relationship between the relative uncertainties in x and y.

A. The modulus of the derivative is:

$$\left|\frac{dy}{dx}\right| = |nx^{n-1}|$$

and so the absolute uncertainty in y is:

$$\sigma_y = \left| nx^{n-1} \right| \sigma_x \qquad (13.33)$$

with relative uncertainty:

$$\frac{\sigma_y}{|y|} = \frac{\left| nx^{n-1} \right| \sigma_x}{|x^n|} = \frac{|n| \sigma_x}{|x|} \qquad (13.34)$$

This time the relative uncertainty in y is $|n|$ times the relative uncertainty in x.

Problem 13.5

Bimolecular collision theory provides some insight into the significance of the Arrhenius parameters appearing Arrhenius equation $k = Ae^{-E_a/RT}$. Consider the elementary reaction

$$CH_3 + H_2 \rightarrow CH_4 + H$$

The first step in estimating the pre-exponential factor A for this reaction involves calculating the collision cross-section $\sigma = \pi d^2$ where d is the sum of the hard-sphere radii $d = r_{CH_3} + r_{H_2}$ of the reactants CH_3 and H_2. Given $d = (6.50 \pm 0.14) \times 10^{-10}$ m use eqn (13.34) to calculate the relative and hence absolute uncertainties in the collision cross-section σ.
Note: Try not to confuse the symbol σ used for the collision cross-section with the σ's used to represent the respective uncertainties!

Worked Problem 13.5

Q. Two quantities are related through the relationship:

$$y = \ln x$$

with absolute uncertainty in x, σ_x. Find the absolute and relative uncertainties in y and write down the relationship between the absolute and relative uncertainties in x and y.

A. The modulus of the derivative is:

$$\left|\frac{dy}{dx}\right| = \left|\frac{1}{x}\right|$$

and so the absolute uncertainty in y is:

$$\sigma_y = \frac{\sigma_x}{|x|} \qquad (13.35)$$

which yields the unexpected result that the absolute uncertainty in y is the same as the relative uncertainty in x.

Problem 13.6

Two quantities are related through the relationship:

$$y = e^x$$

with absolute uncertainty in x, σ_x. Find the absolute and relative uncertainties in y and comment on how your result compares to that in Worked Problem 13.5.

Problem 13.7

The rotational constant, B, for ^{12}CO is measured as 1.923 cm^{-1} with an uncertainty of ± 0.005 cm^{-1}. Use eqn (13.28) to calculate the bond length, r, of ^{12}CO assuming $c = 2.998 \times 10^{10}$ cm s^{-1}, $h = 6.626 \times 10^{-34}$ J s and $\mu_{CO} = 1.139 \times 10^{-26}$ kg. Hence use your answer and the result from Worked Problem 13.4 to find the absolute and relative errors in r.

13.4.2 Combining Uncertainties in More Than One Variable

If we consider a functional relationship $y = f(u, v, \ldots)$ in which the dependent variable y depends on two or more independent variables, $u, v, \ldots$, then we need to exercise a little care in deciding which of the propagation of errors formulae in eqns (13.29) and (13.30) we extend to include the additional variables when combining their uncertainties. Extending first the modulus expression in eqn (13.29) yields:

$$\sigma_y = \left|\frac{\partial}{\partial u}f\right|\sigma_u + \left|\frac{\partial}{\partial v}f\right|\sigma_v + \ldots \qquad (13.36)$$

The problem here is that eqn (13.36) will tend to over-estimate σ_y because it assumes that all of the component uncertainties, σ_u, σ_v, ... will contribute to the total uncertainty to their full extent in the same direction, *i.e.* it assumes that they will either all be positive or all be negative. In practice, if the component uncertainties are independent and random, there will be a higher probability that some of the contributing errors will be positive and some negative with the resulting uncertainty then being smaller than that implied by eqn (13.36). In order to acknowledge this implicit overestimation of the total uncertainty, we modify eqn (13.36) to:

$$\sigma_y \leqslant \left|\frac{\partial}{\partial u}f\right|\sigma_u + \left|\frac{\partial}{\partial v}f\right|\sigma_v + \ldots \qquad (13.37)$$

which describes the error in y as being no larger than the sum of the terms on the right-hand side. Thus, if you wish to calculate an upper limit to the uncertainty, eqn (13.37) provides a safe choice.

If instead, we now combine the terms as suggested by extension of eqn (13.30) by squaring them, adding the squares and then taking the square root, we end up with an estimate of σ_y that will always be smaller than that deriving from simple addition of terms. For situations in which the errors in u and v are random and independent (*i.e.* uncorrelated), a more realistic estimate of σ_y will then be obtained by the expression:

$$\sigma_y^2 \approx \left(\frac{\partial}{\partial u}f\right)^2 \sigma_u^2 + \left(\frac{\partial}{\partial v}f\right)^2 \sigma_v^2 + \ldots \qquad (13.38)$$

In the next three Worked Problems, as well as Problem 13.8, we will see how the relative and absolute uncertainties in dependent and independent variables are related for functions involving addition and subtraction, and multiplication and division.

Worked Problem 13.6

Addition and Subtraction
Q. Use eqn (13.38) to find how the absolute uncertainty in y in a linear combination, $y = u \pm v$ is related to those of u and v.

A.
$$y = u \pm v; \quad \frac{\partial y}{\partial u} = 1; \quad \frac{\partial y}{\partial v} = 1$$

and so:

$$\sigma_y^2 = 1^2 \times \sigma_u^2 + 1^2 \times \sigma_v^2 = \sigma_u^2 + \sigma_v^2 \qquad (13.39)$$

with:

$$\sigma_y = \sqrt{\sigma_u^2 + \sigma_v^2}. \qquad (13.40)$$

In other words, the absolute uncertainty in y is the root mean square of the absolute uncertainties in u and v. This result is independent of whether u and v are added or subtracted in the initial function.

Working through the same process, but this time starting with eqn (13.37), yields the upper limit to the uncertainty in y which results from a simple addition in uncertainties in u and v:

$$\sigma_y \leq \sigma_u + \sigma_v \qquad (13.41)$$

Worked Problem 13.7

In Problem 13.5, we used the parameter d, representing the sum of hard-sphere radii $d = r_{CH_3} + r_{H_2}$, to calculate the collision cross-section for the reaction $CH_3 + H_2 \rightarrow CH_4 + H$.

Q. Given $r_{CH_3} = (3.6 \pm 0.1) \times 10^{-10}$ m and $r_{H_2} = (2.9 \pm 0.1) \times 10^{-10}$ m, use first eqn (13.40) and then eqn (13.41) to calculate two different estimates for the absolute uncertainty in d. Comment on the significance of the two outcomes.

A. Using first eqn (13.40), $\sigma_d = \sqrt{(0.1 \times 10^{-10}\ \mathrm{m}) + (0.1 \times 10^{-10}\ \mathrm{m})} = 0.14 \times 10^{-10}$ m. If we use instead eqn (13.41), then the uncertainty in d is now just a simple sum of the two components, $\sigma_d \leqslant 0.1 \times 10^{-10} + 0.1 \times 10^{-10} = 0.2 \times 10^{-10}$ m. The second, larger values provides us with an upper limit to the uncertainty, whereas the first provides us with a more realistic estimate assuming the errors associated with r_{CH_3} and r_{H_2} are independent and random.

Worked Problem 13.8

Multiplication

Q. Use eqn (13.38) to find how the relative uncertainty in y in a product, $y = uv$ is related to those of u and v.

A.
$$y = uv; \quad \frac{\partial y}{\partial u} = v; \quad \frac{\partial y}{\partial v} = u$$

and so:

$$\sigma_y^2 = v^2 \times \sigma_u^2 + u^2 \times \sigma_v^2 = v^2 \sigma_u^2 + u^2 \sigma_v^2$$

We can develop this further by dividing through by y^2:

$$\frac{\sigma_y^2}{y^2} = \frac{v^2 \sigma_u^2}{u^2 v^2} + \frac{u^2 \sigma_v^2}{u^2 v^2} = \frac{\sigma_u^2}{u^2} + \frac{\sigma_v^2}{v^2}$$

$$\Downarrow$$

$$\left(\frac{\sigma_y}{y}\right)^2 = \left(\frac{\sigma_u}{u}\right)^2 + \left(\frac{\sigma_v}{v}\right)^2$$

and:

$$\frac{\sigma_y}{y} = \sqrt{\left(\frac{\sigma_u}{u}\right)^2 + \left(\frac{\sigma_v}{v}\right)^2} \tag{13.42}$$

i.e. the relative uncertainty in y is the root mean square of the relative uncertainties in u and v. This result is also true for division.

Problem 13.8

Division

Prove that eqn (13.42) also applies for division.

Problem 13.9

In an experiment, we allow 0.0200 mol of solid CO_2 (a little less than 1 g) to sublime in an evacuated 10 cm^3 round-bottomed flask at room temperature (298 K). Assuming no measurable uncertainties apply to the volume of the flask, and that the estimated uncertainty associated with our measurement of the amount of CO_2 is $\sigma_n = \pm 0.0002$ mol and of the temperature, $\sigma_T = \pm 0.5$ K, use the ideal gas law to find the pressure exerted by the CO_2 on the flask and the relative and absolute uncertainties associated with that value.

Problem 13.10

Use eqn (13.37) to calculate an *upper limit* to the absolute uncertainty associated with the calculated pressure of CO_2 from Problem 13.9.

Hint: The approach you take here is equivalent to that described in Section 5.3 which discussed the use of differentials in the calculation of propagation of errors. The answer you get should be larger than the value you determined from Problem 13.9.

Summary of Key Points

This chapter describes how simple statistics can be used in the interpretation of experimental results and in the treatment of errors arising from experimental measurement. The chapter is divided into four sections covering types of error, the statistics of repeated measurement, linear regression analysis and the propagation of errors. The key points discussed include:

1. The distinction between the accuracy and precision of an experimental measurement and between systematic and random errors.
2. How the binomial distribution can be used to compute the probabilities associated with experiments yielding one of two possible outcomes.

3. The Gaussian (or normal) distribution commonly used to model the results of repeated experimental measurement.

4. The variance and standard deviation as measures of the uncertainties associated with the 'true' mean value of the Gaussian distribution.

5. The sample variance and sample standard deviation as measures of the uncertainties associated with the *sample* mean.

6. The significance of 68% and 95% confidence intervals.

7. The definition of the standard error of the mean and standard error of the sample mean.

8. Student *t*-factors and the 95% confidence limit.

9. The least-squares method in linear regression analysis

10. The propagation of errors in functions of a single variable

11. The propagation of errors in functions of more than one variable

12. A selection of mathematical and chemically based examples and problems to illustrate practical applications of statistics and error analysis.

References

1. T. J. Scanlon, R. N. Luben, F. L. Scanlon and N. Singleton, *BMJ*, 1993, **307**, 1584.

Further Reading

J. R. Taylor, *An Introduction to Error Analysis*, University Science Books, Sausalito, 2nd edn, 1997.

Appendix

Appendix 1: Worked Answers to Problems

Chapter 1

1.1 (a) (i) 2.554 455 = 2.554 46 to 5 d.p.

                2.554 5 to 4 d.p.

                2.554 5 to 3 d.p.

  (ii) 2.554 455 = 2.55 to 3 sig. fig.

(b) (i) 1.723 205 08 = 1.723 21 to 5 d.p.

             = 1.723 2 to 4 d.p.

             = 1.723 to 3 d.p.

  (ii) 1.723 205 08 = 1.72 to 3 sig. fig.

(c) (i) 3.141 592 653 ... = 3.141 59 to 5 d.p.

             = 3.141 6 to 4 d.p.

             = 3.142 to 3 d.p.

  (ii) 3.141 592 6 ... = 3.14 to 3 sig. fig.

(d) (i) 2.718 281 828 ... = 2.718 28 to 5 d.p.

             = 2.718 3 to 4 d.p.

             = 2.718 to 3 d.p.

  (ii) 2.718 281 2 ... = 2.72 to 3 sig .fig.

1.2 (a) (i) 1.378 423 784 2 rational $= 1.378$ to 4 sig. fig.

(ii) 1.378 423 784 2... irrational $= 1.378$ to 4 sig. fig.

(iii) $1/70 = 0.014\ 285\ 714...$ rational $= 0.014\ 29$ to 4 sig. fig.

(iv) $\pi/4 = 0.785\ 398\ 163...$ irrational $= 0.785\ 4$ to 4 sig. fig.

(v) 0.005068 rational $= 0.005068$ to 4 sig. fig.

(vi) $e/10 = 0.271\ 828\ 182...$ irrational $= 0.271\ 8$ to 4 sig.fig.

(b) 23.3 cm^3 : 3 sig. fig., 1 d.p.; max. titre, 23.35 cm^3;

min. titre, 23.25 cm^3.

1.3 (a) $\dfrac{10^2 \times 10^{-4}}{10^6} = \dfrac{10^{-2}}{10^6} = 10^{-8}$;

(b) $\dfrac{9 \times 2^4 \times 3^{-2}}{4^2} = \dfrac{2^4}{4^2} = \dfrac{2^4}{(2^2)^2} = \dfrac{2^4}{2^4} = 1$

(c) $\left(\dfrac{10}{3^2 + 4^2 + 5^2}\right)^{-1/2} = \left(\dfrac{10}{9 + 16 + 25}\right)^{-1/2}$

$= \left(\dfrac{10}{50}\right)^{-1/2} = \left(\dfrac{1}{5}\right)^{-1/2} = 5^{1/2} = \sqrt{5}$

(d) $\dfrac{(2^4)^3}{4^4} = \dfrac{2^{12}}{4^4} = \dfrac{2^{12}}{(2^2)^4} = \dfrac{2^{12}}{2^8} = 2^4 = 16.$

1.4 (a) $(2.5 \times 10^2 - 0.5 \times 10^2)^2 / 4 \times 10^4$

$= (250 - 50)^2 / 4 \times 10^4 = 200^2 / 40000 = 1$

(b) $\left(\dfrac{1}{2 \times 4}\right)^{1/3} - 4 \times \dfrac{1}{16} = \dfrac{1}{8^{1/3}} - \dfrac{1}{4} = \dfrac{1}{2} - \dfrac{1}{4} = \dfrac{1}{4}.$

1.5 (a) I.E. $= \dfrac{2.179\ \text{aJ}}{0.1602\ \text{aJ eV}^{-1}} = 13.6$ eV to 3 sig. fig.

(b) $\varepsilon_{vib} = 6.626 \times 10^{-34}$ J s $\times 1.2404 \times 10^{14}$ s$^{-1} = 8.22 \times 10^{-20}$ J.

1.6 1×10^6 cm^3=1 m^3: Therefore, $\rho = \dfrac{879 \text{ kg m}^{-1}}{1 \times 10^6 \text{ cm}^3 \text{ m}^{-1}} = 8.79 \times 10^{-4}$ kg cm^{-3} and so 1 cm^3 of benzene weighs 8.79×10^{-4} kg which is equivalent to: $\dfrac{8.79 \times 10^{-4} \text{ kg}}{0.078 \text{ kg mol}^{-1}} = 0.0113$ mol.

1.7 Benzene diameter $= 600$ pm $= 600 \times 10^{-12}$ m; circumference of the earth $= 2\pi \times 6.378 \times 10^6$ m; therefore $n = \dfrac{2\pi \times 6.378 \times 10^6 \text{ m}}{600 \times 10^{-12} \text{ m}} = 6.679 \times 10^{16}$

$\Rightarrow \dfrac{6.679 \times 10^{16}}{6.022 \times 10^{23}} = 1.109 \times 10^{-7}$ moles.

1.109×10^{-7} moles of benzene weighs

$1.109 \times 10^{-7} \times 0.078$ kg $= 8.65 \times 10^{-9}$ kg $= 8.65 \times 10^{-6}$g$= 8.65$ μg

1.8 Volume of cube $= a^3 = 6.7832 \times 10^{-29}$ m^3, so mass of Au $= 19.321 \times 10^3$ kg m^{-3} $\times 6.7832 \times 10^{-29}$ kg $= 1.3106 \times 10^{-24}$ kg $= 1.3106 \times 10^{-21}$ g.
R.M.M ^{197}Au (100%)$=196.97$ g mol^{-1} so 1 molecule of Au $= 3.2708 \times 10^{-22}$g and therefore
$n = \dfrac{1.3106 \times 10^{-21}}{3.2708 \times 10^{-22}} = 4.007$.

1.9 (a) $2 < 6$; $6 > 2$.
   (b) $1.467 < 1.469$; $1.469 > 1.467$.
   (c) $\pi > e$; $e < \pi$.

1.10 (a) $2.4555 < 2.456 < 2.4565$
   (b) $-5.34 > -5.35$; $5.34 < 5.35$
       $5.34 > -5.35$; $-5.34 < 5.35$

1.11 (a) $|4-9|=5$; $|-3-6|=9$; $|9-4|=5$
   (b) $0, \infty, 0, -\infty$

1.12 (a) $S_{100} = \dfrac{100 \times 101}{2} = 5050$
   (b) $S_{68} = \dfrac{-68 \times 69}{2} = -2346$.

1.13 (a) 3 spin states, 2 nuclei $\Rightarrow 3^2=9$ spin states.
        3 spin states, 3 nuclei $\Rightarrow 3^3=27$ spin states.
   (b) Number of spin states associated with n equivalent nuclei with spin $I = (2I+1)^n$.

(c) $^{51}V \Rightarrow I=7/2$ so for a single atom of ^{51}V, there are $(2 \times 7/2)+1=8$ spin states.

1.14 (a) $\dfrac{(u^2+v^2)}{(v-u)} = \dfrac{2(x^2+y^2)}{-2y} = \dfrac{-(x^2+y^2)}{y}$.

(b) $\dfrac{uv}{2u-v} = \dfrac{x^2-y^2}{x+3y}$.

(c) $\dfrac{10^{u+v}}{10^{u-v}} = \dfrac{10^{2x}}{10^{2y}}$.

1.15 (a) (i) $4p-q-(2q+3p)=p-3q$; (ii) $3p^2-p(4p-7)=-p^2+7p$

(b) (i) $(1+x)^2-(1-x)^2=4x$; (ii) $x(2x+1)-(1+x-x^2)=3x^2-1$.

1.16 (a) $\dfrac{p^4q^2}{p^2q^3} = \dfrac{p^2}{q}$; $q\neq0,\ p\neq0$ (b) $\dfrac{p^8q^{-3}}{p^{-5}q^2} = \dfrac{p^{13}}{q^5}$; $q\neq0,p\neq0$

(c) $\dfrac{4x}{6x^2-2x} = \dfrac{4}{6x-2} = \dfrac{2}{3x-1}$; $x\neq\dfrac{1}{3},0$ (d) $\dfrac{3x^2-12xy}{3} = x^2-4xy$.

1.17 (a) (i) $x^2-3x+2=(x-1)(x-2)$ since the solutions to $a^2-3a+2=0$, are $a=1,2$.

(ii) $x^3-7x+6=(x-1)(x-2)(x+3)$ since the solutions to $a^3-7a+6=0$ are $a=1,2-3$.

(b) (i) $\dfrac{x^3-7x+6}{x-2} = \dfrac{(x-1)(x-2)(x+3)}{x-2} = (x-1)(x+3)$.

(ii) $\dfrac{x^2-1}{x-1} = \dfrac{(x+1)(x-1)}{x-1} = x+1$.

1.18 (a) (i) $\dfrac{3x}{4}-\dfrac{x}{2} = \dfrac{3x-2x}{4} = \dfrac{x}{4}$;   (ii) $\dfrac{2}{x}-\dfrac{1}{x^2} = \dfrac{2x-1}{x^2}$;

(iii) $1-\dfrac{1}{x}+\dfrac{2}{x^2} = \dfrac{x^2-x+2}{x^2}$.

(b) (i) $\dfrac{1}{1+x}-\dfrac{1}{1-x} = \dfrac{1-x-1-x}{(1+x)(1-x)} = \dfrac{-2x}{1-x^2} = \dfrac{2x}{x^2-1}$;

(ii) $\dfrac{2x}{x^2+1}-\dfrac{2}{x} = \dfrac{2x^2-2x^2-2}{x(x^2-1)} = \dfrac{-2}{x(x^2+1)}$.

1.19 (a) $\dfrac{RT}{F} \Rightarrow \dfrac{JK^{-1}mol^{-1} \times K}{Cmol^{-1}} = JC^{-1} = CVC^{-1} = V.$

(b) $\dfrac{m_e e^4}{8h^3 c\varepsilon_0^2} \Rightarrow \dfrac{kgC^4}{J^3\, s^3\, ms^{-1}\, J^{-2}C^4 m^{-2}} = \dfrac{kg}{Js^2 m^{-1}} = kg\, J^{-1}ms^{-2}$

$$= m^{-1} \text{ since } 1\, J = 1\, kg\, m^2 s^{-2}.$$

Chapter 2

2.1 (a) Yes. Domain registration numbers.
(b) No. One keeper may own >1 car.
(c) No. More than one element per group.
(d) Yes. Each element only belongs to one group.

2.2 $s = \dfrac{e\, z\, E}{6\pi\eta a} \Rightarrow \dfrac{CVm^{-1}}{kg\, m^{-1}s^{-1}m} = CVm^{-1}kg^{-1}s$

$= Jm^{-1}kg^{-1}s = kgm^2 s^{-2}m^{-1}kg^{-1}\ s = ms^{-1}.$

2.3 (a) $f(x) = \begin{cases} x-1, & x \geqslant 1 \\ -x+1, & x < 1 \end{cases}.$

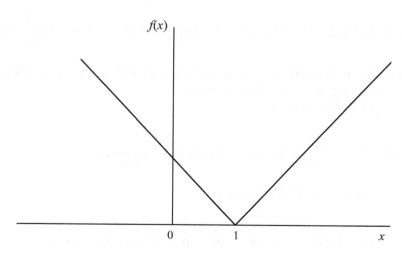

(b)

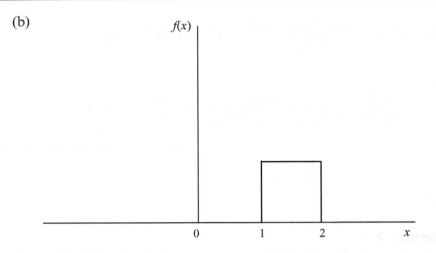

2.4 (a) $n_i = n_0 e^{-(\varepsilon_i - \varepsilon_0)/kT} \Rightarrow \dfrac{n_i}{n_0} = e^{-\left(i + \frac{1}{2} - \frac{1}{2}\right)h\nu/kT} = e^{-ih\nu/kT}$.

(b)

i	1	2	3	4	5
n_i/n_0	1.583×10^{-3}	2.506×10^{-6}	3.966×10^{-9}	6.278×10^{-12}	9.938×10^{-15}

2.5 (a) (i) $\log 4 = \log 2^2 = 2 \log 2$; (ii) $\log 8 = \log 2^3 = 3 \log 2$;

(iii) $\log 6 - \log 3 = \log \dfrac{6}{3} = \log 2$;

(iv) $\ln 8 = 3 \ln 2 = 3 \ln 10\ \log 2$; (v) $\ln \dfrac{1}{2} = \ln 2^{-1} = -\ln 2$.

(b) (i) $\log 2 + \log 3 = \log 2 \times 3 = \log 6$; (ii) $\ln 3 - \ln 6 = \log \dfrac{3}{6} = \log \dfrac{1}{2}$.

2.6 (a) Given that $pH = -\log a_H$ and $\log_{10} y = \log_e y\ \log_{10} e$ it follows that $\log a_H = \ln a_H \log e$ and so
$pH = -\ln a_H \log e$.

(b) $E^\ominus = -\dfrac{RT}{nF} \ln K$ and from (a) $\ln K = \dfrac{1}{\log e} \log K$

so $E^\ominus = -2.303 \dfrac{RT}{nF} \log K$.

(c) $pK = 4.756 = -\log K \Rightarrow K = 10^{-4.756} = 1.754 \times 10^{-5}$.

2.7 (a) $r_{H-H} = 2 \times r_{P-H} \times \sin(\theta/2) = 2 \times 140 \times \sin 61.5° = 246.1$ pm.

(b) $(r_{H-H})_{\text{excited state}} = 195.1$ pm, and so $\Delta r_{H-H} = 51$ pm.

2.8 $\cot\theta = \dfrac{1}{\tan\theta} = \dfrac{\cos\theta}{\sin\theta}$; domain all real numbers except those values of θ for which $\sin\theta = 0$. *i.e.* $\theta = n\pi$, $n = 0$, ± 1, ± 2, $\pm 3 \ldots$

Similarly, $\text{cosec}\,\theta = \dfrac{1}{\sin\theta}$, has the same domain as $\cot\theta$.

2.9 $\cos(-\theta) = \cos(2\pi - \theta) \Rightarrow \cos(-\theta) = \cos 2\pi\,\cos\theta + \sin 2\pi \sin\theta = \cos\theta$

$\tan(-\theta) = \dfrac{\sin(-\theta)}{\cos(-\theta)} = \dfrac{-\sin\theta}{\cos\theta} = -\tan\theta.$

2.10 (a) (i) $\sinh x + \cosh x = \frac{1}{2}e^x - \frac{1}{2}e^{-x} + \frac{1}{2}e^x + \frac{1}{2}e^{-x} = e^x.$

(ii) $\sinh x - \cosh x = \frac{1}{2}e^x - \frac{1}{2}e^{-x} - \frac{1}{2}e^x - \frac{1}{2}e^{-x} = -e^{-x}.$

(b) (i) $\cosh^2 x - \sinh^2 x = \frac{1}{4}e^{2x} + \frac{1}{4} + \frac{1}{4} + \frac{1}{4}e^{-2x} - \frac{1}{4}e^{2x} + \frac{1}{4} + \frac{1}{4} - \frac{1}{4}e^{-2x} = 1.$

(ii) $\cosh^2 x + \sinh^2 x = \frac{1}{4}e^{2x} + \frac{1}{4} + \frac{1}{4} + \frac{1}{4}e^{-2x} + \frac{1}{4}e^{2x} - \frac{1}{4} - \frac{1}{4} + \frac{1}{4}e^{-2x} = \frac{1}{2}e^{2x} + \frac{1}{2}e^{-2x} = \cosh 2x.$

(iii) $\sinh x \cosh x = \left(\frac{1}{2}e^x - \frac{1}{2}e^{-x}\right)\left(\frac{1}{2}e^x + \frac{1}{2}e^{-x}\right)$
$= \frac{1}{4}e^{2x} + \frac{1}{4} - \frac{1}{4} - \frac{1}{4}e^{-2x} = \frac{1}{4}\left(e^{2x} - e^{-2x}\right) = \frac{1}{2}\sinh 2x.$

$\Rightarrow \sinh 2x = 2\sinh x \cosh x.$

2.11 (a) (i) $x^2 + x - 6 = 0 \Rightarrow \dfrac{-1 \pm \sqrt{1^2 - 4.1.-6}}{2} = -\dfrac{1}{2} \pm \dfrac{\sqrt{25}}{2} = 2, -3.$

(ii) $x^2 - 1 = 0 \Rightarrow \dfrac{0 \pm \sqrt{0^2 - 4.1.-1}}{2} = \pm\dfrac{\sqrt{4}}{2} = \pm 1.$

(iii) $x^2 - 2\sqrt{2}x + 2 = 0 \Rightarrow \dfrac{2\sqrt{2} \pm \sqrt{8 - 4.1.2}}{2} = \sqrt{2}$ twice.

(b) (i) $x^2 + x - 6 = (x-2)(x+3)$; (ii) $x^2 - 1 = (x+1)(x-1)$;

(iii) $x^2 - 2\sqrt{2}x + 2 = (x - \sqrt{2})(x - \sqrt{2}).$

2.12 (a) If $R_{3s} = N\left\{27 - \dfrac{18r}{a_0} + 2\dfrac{r^2}{a_0{}^2}\right\}e^{-\frac{r}{3a_0}}$, then $R_{3s, r=0} = 27N$.

$$\lim_{r \to \infty} R_{3s} = N\{27 - \text{v.large} + \text{v.v.large}\} \times \text{v.v.v.small} = 0$$

(b) $\left\{27 - 18\left(\dfrac{r}{a_0}\right) + 2\left(\dfrac{r}{a_0}\right)^2\right\} = 0$ when $\dfrac{r}{a_0} = \dfrac{18}{4} \pm \dfrac{\sqrt{18^2 - 4.27.2}}{4}$

$$= \dfrac{9}{2} \pm \dfrac{\sqrt{108}}{4} = \dfrac{9}{2} \pm \dfrac{5.196}{2}$$

$$\Rightarrow \dfrac{r}{a_0} = 7.098 \text{ and } 1.902 \text{ or } r = 7.098a_0 \text{ and } 1.902a_0$$

(c)

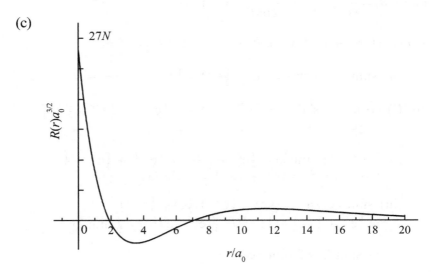

2.13 (a) Carbon dioxide (σ only) $\Rightarrow 1s_C$, $1s_{O(1)}$, $1s_{O(2)}$, $2s_C$, $2s_{O(1)}$, $2s_{O(2)}$, $2p_C$, $2p_{O(1)}$, $2p_{O(2)}$ $\therefore$ 9th degree.

(b) Benzene (π only) $\Rightarrow 6 \times 2p_{C(\pi)}$ $\therefore$ 6th degree.

2.14 (a) $K = hc/e$

(b) $K_w = hb$

(c) $e = 1 - c$

(d) $K = \dfrac{hc}{e} \Rightarrow K = \dfrac{hc}{1-c} \Rightarrow (1-c)K = hc \Rightarrow \dfrac{1-c}{c}$

$\quad = \dfrac{h}{K} \Rightarrow \dfrac{1}{c} - 1 = \dfrac{h}{K} \Rightarrow \dfrac{1}{c} = \dfrac{h}{K} + 1$

and so $c = \dfrac{1}{h/K+1}$.

(e) $\dfrac{1}{h/K+1} + b = 1 + h \Rightarrow b = 1 + h - \dfrac{1}{h/K+1} = 1 + h - \dfrac{K}{h+K}$.

(f) $K_\mathrm{w} = hb = h + h^2 - \dfrac{Kh}{h+K}$.

(g) $K_\mathrm{w} = h\dfrac{K+h}{K+h} + h^2\dfrac{K+h}{K+h} - \dfrac{Kh}{h+K}$

$\quad = \dfrac{1}{K+h}\left(hK + h^2 + h^2K + h^3 - Kh\right)$

$\Rightarrow K_\mathrm{w} = \dfrac{1}{K+h}\left(h^3 + h^2(1+K) + hK - hK\right)$

$\quad = \dfrac{1}{K+h}\left(h^3 + h^2(1+K)\right)$

$\Rightarrow K_\mathrm{w}K + K_\mathrm{w}h = h^3 + h^2(1+K)$

$\Rightarrow h^3 + h^2(1+K) - K_\mathrm{w}K - K_\mathrm{w}h = 0.$

(h) $h^3 + 1.000018h^2 - 10^{-14}h - 1.8 \times 10^{-19}$

(i) $h = 4.243 \times 10^{-9}$; and $c = \dfrac{1}{h/K+1}$, so

$\quad c = \dfrac{1}{(4.243 \times 10^{-9}/1.8 \times 10^{-5})+1} = 0.999764333.$

$\quad e = 1 - c = 1 - 0.999764333 = 2.3567 \times 10^{-4}.$

$\quad c + b = 1 + h \Rightarrow b = 1 + h - c =$

$\quad 1 + 4.243 \times 10^{-9} - 0.999764333 = 2.3567 \times 10^{-4}.$

Chapter 3

3.1 (a) $\lim\limits_{x\to\infty} x^2 e^{-x} = 0$; the e^{-x} term dominates.

(b) $\lim\limits_{x\to\infty} \cos(2x) e^{-x} = 0$; again the e^{-x} term dominates.

3.2 (a) $\lim\limits_{x\to\infty} x^2 e^{-x} = 0$; the x^2 term dominates as $\lim\limits_{x\to 0} e^{-x} = 1$.

(b) $\lim\limits_{x\to 0} \cos(2x) e^{-x} = 1$; as both $\cos(2x)$ and e^{-x} tend to 1 as $x\to 0$.

3.3 (a) $f(x) = \dfrac{2x}{x-4}$ is indeterminate at $x=4$. Thus,

$$\lim_{x\to 4}\frac{2x}{x-4} \equiv \lim_{\delta\to 0}\frac{2(4+\delta)}{4+\delta-4} = \frac{8}{\delta} = \infty.$$

(b) $f(x) = \dfrac{x^2-4}{x+2}$ is indeterminate at $x=-2$. Thus,

$$\lim_{x\to -2}\frac{x^2-4}{x+2} \equiv \lim_{\delta\to 0}\frac{(-2+\delta)^2-4}{-2+\delta+2} = \lim_{\delta\to 0}\frac{4-4\delta+\delta^2-4}{\delta}$$

$$= \lim_{\delta\to 0} -4+\delta = -4.$$

(c) $f(x) = \dfrac{x-1}{x^2-1}$ is indeterminate at $x=1,-1$; Thus

$$\lim_{x\to +1}\frac{x-1}{x^2-1} \equiv \lim_{\delta\to 0}\frac{1+\delta-1}{(1+\delta)^2-1} = \lim_{\delta\to 0}\frac{\delta}{2\delta-\delta^2} = \lim_{\delta\to 0}\frac{1}{2-\delta} = \frac{1}{2}.$$

$$\lim_{x\to -1}\frac{x-1}{x^2-1} \equiv \lim_{\delta\to 0}\frac{-1+\delta-1}{(-1+\delta)^2-1} = \lim_{\delta\to 0}\frac{-2+\delta}{-2\delta+\delta^2}$$

$$= \lim_{\delta\to 0}\frac{-2}{-2\delta} = \infty.$$

(d) $f(x) = 3x^2 - \dfrac{2}{x} - 1$ is indeterminate *at* $x=0$. Thus,

$$\lim_{x\to 0}\left(3x^2 - \frac{2}{x} - 1\right) = -\infty.$$

3.4 (a) $\lim\limits_{x\to\infty}\dfrac{5}{x+1} = 0.$

(b) $\lim\limits_{x\to\infty}\dfrac{3x}{x-4} = \dfrac{3x}{x} = 3.$

(c) $\lim\limits_{x\to\infty}\dfrac{x^2}{x+1} = \dfrac{x^2}{x} = x = \infty.$

(d) $\displaystyle\lim_{x\to\infty}\frac{x+1}{x+2}=\frac{x}{x}=1$

3.5 $\displaystyle\lim_{x\to0}(\ln x-\ln 2x)=\lim_{x\to0}\left(\ln\frac{x}{2x}\right)=\ln\frac{1}{2}.$

3.6 $\displaystyle C_V=3R(ax)^2\left\{\frac{e^{\frac{ax}{2}}}{e^{ax}-1}\right\}^2$

$$\lim_{x\to\infty}3R(ax)^2\left\{\frac{e^{\frac{ax}{2}}}{e^{ax}-1}\right\}^2=3R(ax)^2\left\{e^{-\frac{ax}{2}}\right\}^2=3R(ax)^2e^{-ax}=0$$

since the exponential term will dominate. Therefore $C_V\to 0$ as $T\to 0$ K.

3.7 (a) $\displaystyle\lim_{r\to0}R_{3s}=N\left(\frac{r}{a_0}\right)^2\times1=0.$

(b) $\displaystyle\lim_{r\to\infty}R_{3s}=N\left(\frac{r}{a_0}\right)^2\times0=0.$

3.8 (a) For $k_{-1}\gg k_2[H_2]$, $k_{-1}+k_2[H_2]\approx k_{-1}$ and so

$$\frac{d[N_2]}{dt}=\frac{k_1k_2[H_2][NO]^2}{k_{-1}}.$$

(b) For $k_{-1}\ll k_2[H_2]$, $k_{-1}+k_2[H_2]\approx k_2[H_2]$ and so

$$\frac{d[N_2]}{dt}=\frac{k_1k_2[H_2][NO]^2}{k_2[H_2]}=k_1[NO]^2.$$

Chapter 4

4.1 $\displaystyle\frac{dy}{dx}=\lim_{\Delta x\to0}\left\{\frac{f(x+\Delta x)-f(x)}{\Delta x}\right\}=\lim_{\Delta x\to0}\left\{\frac{3-3}{\Delta x}\right\}=0.$

4.2 (a) $\displaystyle\frac{dy}{dx}=\lim_{\Delta x\to0}\left\{\frac{3(x+\Delta x)^2-3x^2}{\Delta x}\right\}$

$\displaystyle=\lim_{\Delta x\to0}\left\{\frac{3x^2+6x\Delta x+3\Delta x^2-3x^2}{\Delta x}\right\}$

$\displaystyle=\lim_{\Delta x\to0}6x+\Delta x$

$=6x.$

(b) $\dfrac{dy}{dx} = \lim\limits_{\Delta x \to 0} \left\{ \dfrac{\dfrac{1}{(x+\Delta x)^2} - \dfrac{1}{x^2}}{\Delta x} \right\}$

$= \lim\limits_{\Delta x \to 0} \left\{ \dfrac{x^2 - (x+\Delta x)^2}{x^2(x+\Delta x)^2 \Delta x} \right\}$

$= \lim\limits_{\Delta x \to 0} \left\{ \dfrac{x^2 - x^2 - 2x\Delta x - \Delta x^2}{x^2(x+\Delta x)^2 \Delta x} \right\}$

$= \lim\limits_{\Delta x \to 0} \left\{ \dfrac{-2x - \Delta x}{x^2(x+\Delta x)^2} \right\} = -\dfrac{2x}{x^4} - \dfrac{2}{x^3}.$

4.3 (a) $\dfrac{d}{dx} x^{3/4} = \dfrac{3}{4} x^{-1/4}$, (b) $\dfrac{d}{dx} e^{-3x} = -3e^{-3x}$,

(c) $\dfrac{d}{dx} 1/x = -x^{-2}$,   (d) $\dfrac{d}{dx} a \cos ax = -a^2 \sin ax$.

4.4 $\left(\dfrac{d}{dx} + 2 \right) e^{-2x} = \dfrac{d}{dx} e^{-2x} + 2e^{-2x} = -2e^{-2x} + 2e^{-2x} = 0.$

4.5 (a) $\dfrac{d}{dx} (x-1)(x^2+4) = 3x^2 - 2x + 4$,

(b) $\dfrac{d}{dx} \dfrac{x}{x+1} = \dfrac{1}{(x+1)^2}$,

(c) $\dfrac{d}{dx} \sin^2 x = 2 \sin x \cos x = \sin 2x$,

(d) $\dfrac{d}{dx} x \ln x = 1 + \ln x$,

(e) $\dfrac{d}{dx} e^x \sin x = e^x(\cos x + \sin x).$

4.6 $y = e^{x\sin x}$; $u = x \sin x \Rightarrow y = e^u$

$\dfrac{dy}{du} = e^u$; $\dfrac{du}{dx} = x \cos x + \sin x \Rightarrow \dfrac{dy}{dx} = \dfrac{dy}{du} \times \dfrac{du}{dx} = e^u(x \cos x + \sin x)$

$= e^{x\sin x}(x \cos x + \sin x).$

4.7 (a) $y = \ln(2+x^2)$; $u = 2 + x^2 \Rightarrow y = \ln u$

$\dfrac{dy}{du} = \dfrac{1}{u}$; $\dfrac{du}{dx} = 2x \Rightarrow \dfrac{dy}{dx} = \dfrac{1}{u} \times 2x = \dfrac{2x}{2+x^2}.$

(b) $y = 2\sin(x^2 - 1)$; $u = x^2 - 1 \Rightarrow y = 2\sin u$

$$\frac{dy}{du} = 2\cos u; \; \frac{du}{dx} = 2x \Rightarrow \frac{dy}{dx} = 2\cos u \times 2x = 4x\cos(x^2 - 1).$$

4.8 (a) $y = 1/x$; $\dfrac{dy}{dx} = -x^{-2}$; $\dfrac{d^2y}{dx^2} = 2x^{-3}$; $\dfrac{d^3y}{dx^3} = -6x^{-4}$.

(b) $y = N\sin ax$; $\dfrac{dy}{dx} = Na\cos ax$;

$$\frac{d^2y}{dx^2} = -Na^2\sin ax; \; \frac{d^3y}{dx^3} = -Na^3\cos ax.$$

4.9 (a) $\dfrac{d^2}{dx^2}x^3 = 6x$; (b) $\dfrac{d^2}{dx^2}\sin kx = -k^2\sin kx$;

(c) $\dfrac{d^2}{dx^2}\{\sin kx + \cos kx\} = -k^2\{\sin kx + \cos kx\}$

(d) $\dfrac{d}{dx}e^{ax} = ae^{ax}$.

4.10 $\left(\dfrac{d^2}{dx^2} - 2\dfrac{d}{dx} - 3\right)e^{mx} = m^2 e^{mx} - 2me^{mx} - 3e^{mx}$

$$= (m^2 - 2m - 3)e^{mx}.$$

$\hat{A}$ annihilates $f(x)$, when $m^2 - 2m - 3 = 0$;

i.e. when $(m+1)(m-3) = 0 \Rightarrow m = -1, 3$.

4.11 $-\dfrac{h^2}{8\pi^2 m}\dfrac{d^2\psi}{dx^2} = E\psi$; $\psi = \sqrt{\dfrac{2}{L}}\sin\dfrac{\pi x}{L}$

$$\Rightarrow -\frac{h^2}{8\pi^2 m}\frac{d^2}{dx^2}\left(\sqrt{\frac{2}{L}}\sin\frac{\pi x}{L}\right) = E\sqrt{\frac{2}{L}}\sin\frac{\pi x}{L}$$

$$\Rightarrow \frac{h^2}{8\pi^2 m}\sqrt{\frac{2}{L}}\left(\frac{\pi}{L}\right)^2\sin\frac{\pi x}{L} = E\sqrt{\frac{2}{L}}\sin\frac{\pi x}{L} \Rightarrow \frac{h^2}{8\pi^2 m}\left(\frac{\pi}{L}\right)^2 = E$$

$$\Rightarrow E = \frac{h^2}{8mL^2}.$$

4.12 (a) $D(r) = Nr^2 e^{-2r/a_0}$;

$$\frac{dD(r)}{dr} = Nr^2 \times \frac{-2}{a_0} e^{-2r/a_0} + 2Nr \times e^{-2r/a_0}$$

$$= 2Ne^{-2r/a_0}\left(r - \frac{r^2}{a_0}\right).$$

(b) $D(r)$ displays a turning point when

$$\frac{dD(r)}{dr} = 2Ne^{-2r/a_0}\left(r - \frac{r^2}{a_0}\right) = 0.$$

i.e. when $r = \frac{r^2}{a_0} \Rightarrow r = a_0$.   When $r = a_0$, $D(r) = Na_0^2 e^{-2}$.

(c) $\dfrac{d^2 D(r)}{dr^2} = -\dfrac{4}{a_0} Ne^{-2r/a_0}\left(r - \dfrac{r^2}{a_0}\right) + 2Ne^{-2r/a_0}\left(1 - \dfrac{2r}{a_0}\right)$

$$= 2Ne^{-2r/a_0}\left\{1 - \frac{2r}{a_0} - \frac{2r}{a_0} + \frac{2r^2}{a_0^2}\right\}$$

$$\Rightarrow \frac{d^2 D(r)}{dr^2} = 2Ne^{-2r/a_0}\left\{1 - \frac{4r}{a_0} + \frac{2r^2}{a_0^2}\right\}.$$

When $r = a_0$, $\dfrac{d^2 D(r)}{dr^2} = 2Ne^{-2}\{1 - 4 + 2\}$

$$= -2Ne^{-2} \Rightarrow \text{maximum.}$$

(d) Points of inflection occur when

$$\frac{d^2 D(r)}{dr^2} = 2Ne^{-2r/a_0}\left\{1 - \frac{4r}{a_0} + \frac{2r^2}{a_0^2}\right\} = 0;$$

i.e. when $1 - \dfrac{4r}{a_0} + \dfrac{2r^2}{a_0^2} = 0$

$$\Rightarrow \frac{r}{a_0} = \frac{4 \pm \sqrt{16 - 4 \cdot 2 \cdot 1}}{4} = 1 \pm \frac{\sqrt{8}}{4} = 1 \pm \frac{\sqrt{2}}{2} = 1 \pm \frac{1}{\sqrt{2}}.$$

i.e. when $r = \left(1 \pm \dfrac{1}{\sqrt{2}}\right) a_0$.

4.13 $P = \dfrac{nRT}{V}$; $\dfrac{\partial P}{\partial T} = \dfrac{nR}{V}$; $\dfrac{\partial P}{\partial V} = -\dfrac{nRT}{V^2}$.

Chapter 5

5.1 $y=x^{1/3}$; $\dfrac{dy}{dx}=\dfrac{1}{3}x^{-2/3} \Rightarrow dy=\dfrac{1}{3}x^{-2/3}dx$.

(a) $dy=\dfrac{1}{3} \times 27^{-2/3} \times 3 = 0.111$; $\Delta y=30^{1/3}-27^{1/3} = 0.107$

$\Rightarrow dy$ overestimates Δy by 3.6%.

(b) $dy=\dfrac{1}{3} \times 27^{-2/3} \times 0.1 = 0.0037037$; $\Delta y=27.1^{1/3}-27^{1/3}=0.0036991$

$\Rightarrow dy$ overestimates Δy by 0.12%.

5.2 (a) $C_P=\alpha+\beta T+\gamma T^2$
$C_{P,500K}=14.143 \; JK^{-1}mol^{-1}+75.495 \times 10^{-3}JK^{-2}mol^{-1} \times 500K-179.64 \times 10^{-7}JK^{-3}mol^{-1} \times 500^2K^2$
$\Rightarrow C_{P,500K}=47.3995 \; JK^{-1}mol^{-1}$.
Similarly, $C_{P,650K}=55.625 \; JK^{-1}mol^{-1}$

(b) $dC_P=(\beta+2\gamma T) \times dT = \{75.495 \times 10^{-3} \; JK^{-2}mol^{-1} - (2 \times 179.64 \times 10^{-7}JK^{-3}mol^{-1})\} \times 150K$
$dC_P=8.62965 \; JK^{-1}mol^{-1}$
$\Rightarrow C_{P,650K}=(47.3995 + 8.62965) \; JK^{-1}mol^{-1}=56.029 \; JK^{-1}mol^{-1}$

(c) The estimate is 0.73% larger than the actual value.

5.3 $z = xy/w$; $dz=\dfrac{\partial z}{\partial x}dx+\dfrac{\partial z}{\partial y}dy+\dfrac{\partial z}{\partial w}dw=\dfrac{y}{w}dx+\dfrac{x}{w}dy-\dfrac{xy}{w^2}dw$

5.4 (a) $U = f(V,T)$; $dU=\left(\dfrac{\partial U}{\partial V}\right)_T dV+\left(\dfrac{\partial U}{\partial T}\right)_V dT$.

(b) (i) $\pi_T=\left(\dfrac{\partial U}{\partial V}\right)_T$; $C_V=\left(\dfrac{\partial U}{\partial T}\right)_V$.

(ii) $dU=\pi_T dV+C_V dT$
$=(840 \; Jm^{-3} \times -10^{-4}m^3)+(27.32 \; JK^{-1} \times 2K)$
$=54.56J$
Contribution from compression is far smaller than that from the increase in temperature.

5.5 (a) $dV = \dfrac{\partial V}{\partial a}\,da + \dfrac{\partial V}{\partial b}\,db + \dfrac{\partial V}{\partial c}\,dc = bc\cdot da + ac\cdot db + ab\cdot dc$

(b) The relative error in V is $\dfrac{dV}{V}$ and so using the result from (a), we have

$$\frac{dV}{V} = \frac{bc\cdot da + ac\cdot db + ab\cdot dc}{abc}$$

$$\Rightarrow \frac{dV}{V} = \frac{da}{a} + \frac{db}{b} + \frac{dc}{c}$$

And the percentage error in V is:

$$100 \times \left(\frac{da}{a} + \frac{db}{b} + \frac{dc}{c} \right)$$

5.6 (a) $(CaCO_3) \times 4 = Ca_4C_4O_{12}$
$\Rightarrow$ molar mass $= 4 \times 40.08 + 4 \times 12.01 + 12 \times 16.00 = 400.36$ g mol^{-1}.
$= 0.40036$ kg mol^{-1} $= 6.648 \times 10^{-25}$ kg.

(b) $V = abc = 4.94 \times 10^{-10}$m $\times\, 7.94 \times 10^{-10}$m $\times\ 5.72 \times 10^{-10}$m
$= 2.244 \times 10^{-28}$m^3
$\rho = M/V \Rightarrow \rho = \dfrac{6.648 \times 10^{-25}\text{kg}}{2.244 \times 10^{-28}\text{m}^3} = 2.963 \times 10^3$ k gm^{-3}.

(c) $\dfrac{dV}{V} = \dfrac{da}{a} + \dfrac{db}{b} + \dfrac{dc}{c} = \dfrac{0.005 \times 10^{-10}}{4.94 \times 10^{-10}} + \dfrac{0.005 \times 10^{-10}}{7.94 \times 10^{-10}} + \dfrac{0.005 \times 10^{-10}}{5.72 \times 10^{-10}}$

$= 1.012 \times 10^{-3} + 6.28 \times 10^{-4} + 8.74 \times 10^{-4} = 2.51 \times 10^{-3}$
$\Rightarrow 0.25\%$ error.

(d) Greatest unit cell volume $= 4.945 \times 10^{-10} \times 7.945 \times 10^{-10} \times$
5.725×10^{-10}m^3 $= 2.249 \times 10^{-28}$ m^3
$\Rightarrow \rho = 2.956 \times 10^3$ kg m^{-3} $\Rightarrow 0.237\%$ error.
Least unit cell volume $= 4.935 \times 10^{-10} \times 7.935 \times 10^{-10} \times$
5.715×10^{-10} m^3 $= 2.238 \times 10^{-28}$ m^3
$\Rightarrow \rho = 2.971 \times 10^3$ kg m^{-3} $\Rightarrow 0.27\%$ error.

Chapter 6

6.1 (a) $\dfrac{d}{dx}e^{2x} = 2e^{2x} \Rightarrow \displaystyle\int 2e^{2x} = 2\int e^{2x} = e^{2x} + B \Rightarrow \int e^{2x} = \dfrac{1}{2}e^{2x} + C.$

(b) $\dfrac{d}{dx}\dfrac{1}{1+e^x} = \dfrac{d}{dx}(1+e^x)^{-1};$ Let $u = 1+e^x;\ \dfrac{du}{dx} = e^x$

$\dfrac{d}{dx}(1+e^x)^{-1} = -(1+e^x)^{-2} \times e^x = -\dfrac{e^x}{(1+e^x)^2}.$

Thus, it follows that $-\displaystyle\int \dfrac{e^x}{(1+e^x)^2}dx = \dfrac{1}{1+e^x} + B$

and so $\displaystyle\int \dfrac{e^x}{(1+e^x)^2}dx = -\dfrac{1}{1+e^x} + C$

6.2 $\displaystyle\int 9x^2 + 2e^{2x} + \dfrac{1}{x}dx = \int 9x^2 dx + \int 2e^{2x}dx + \int \dfrac{1}{x}dx$

$= 3x^3 + e^{2x} + \ln x + C.$

6.3 (a) $\displaystyle\int xe^{-x}dx = uv - \int v\dfrac{du}{dx}dx$ where $u = x$ and $\dfrac{dv}{dx} = e^{-x}$

$\Rightarrow \displaystyle\int xe^{-x}dx = -xe^{-x} - \int -e^{-x} \times 1 dx = -xe^{-x} - e^{-x}$

$= -e^{-x}(x+1) + C.$

(b) $\displaystyle\int xe^{-x}dx = uv - \int v\dfrac{du}{dx}dx$ where $u = e^{-x}$ and $\dfrac{dv}{dx} = x$

$\Rightarrow \displaystyle\int xe^{-x}dx = e^{-x} \times \dfrac{1}{2}x^2 + \int \dfrac{1}{2}x^2 \times e^{-x}dx.$

This solution requires us to integrate $\displaystyle\int \dfrac{1}{2}x^2 \times e^{-x}dx$ which is arguably more complicated than the original integral. Thus, method (a) would seem the more appropriate.

6.4 $\displaystyle\int xe^{ax^2}dx.$ **For** $u = x^2,\ du = \dfrac{du}{dx}dx = 2xdx \Rightarrow dx = \dfrac{1}{2x}du$

$\Rightarrow \displaystyle\int xe^{au}\dfrac{du}{2x} = \dfrac{1}{2}\int e^{au}du = \dfrac{1}{2a}e^{au} + C \Rightarrow \int xe^{ax^2}dx = \dfrac{1}{2a}e^{ax^2} + C.$

6.5 $\int \dfrac{x}{\left(1-x^2\right)^{1/2}}\,dx$; For $u=1-x^2$, $du=\dfrac{du}{dx}\,dx=-2x\,dx$

$\Rightarrow dx=-\dfrac{1}{2x}\,du$

$\Rightarrow \int \dfrac{-x}{u^{1/2}}\dfrac{1}{2x}\,du - \int \dfrac{1}{2u^{1/2}}\,du = -\dfrac{1}{2}\int u^{-1/2}\,du$

$=-\dfrac{1}{2}2u^{1/2}+C=-u^{1/2}+C=-\left(1-x^2\right)^{1/2}+C.$

6.6 (a) $\int x\left(x^2+4\right)^{1/2}\,dx$; For $u=x^2+4$, $du=\dfrac{du}{dx}\,dx=2x\,dx$

$\Rightarrow dx=\dfrac{1}{2x}\,du$

$\Rightarrow \int xu^{1/2}\dfrac{1}{2x}\,du=\dfrac{1}{2}\int u^{1/2}\,du=\dfrac{1}{2}\dfrac{2}{3}u^{3/2}+C=\dfrac{1}{3}u^{3/2}+C$

$=\dfrac{1}{3}\left(x^2+4\right)^{3/2}+C.$

(b) $\int \dfrac{1}{x\ln x}\,dx$; For $u=\ln x$, $du=\dfrac{du}{dx}\,dx=\dfrac{1}{x}\,dx\Rightarrow dx=x\,du$

$\Rightarrow \int \dfrac{x}{xu}\,du=\int \dfrac{1}{u}\,du=\ln u+C=\ln(\ln x)+C.$

6.7 (a) $\int \dfrac{x}{\left(1-x^2\right)^{1/2}}\,dx$; For $x=\cos u$; $\dfrac{dx}{du}=-\sin u\Rightarrow dx=-\sin u\,du$

$\Rightarrow \int \dfrac{-\cos u\times \sin u}{\left(1-\cos^2 u\right)^{1/2}}\,du=-\int \dfrac{\cos u\times \sin u}{\left(\sin^2 u\right)^{1/2}}\,du$

$=-\int \dfrac{\cos u\times \sin u}{\sin u}\,du=-\int \cos u\,du=-\sin u+C.$

$=-\left(1-\cos^2 u\right)^{1/2}+C=-\left(1-x^2\right)^{1/2}+C.$

(b) $\int \dfrac{\cos x}{\sin x}\,dx$; For $u=\sin x$, $\dfrac{du}{dx}=\cos x\Rightarrow dx=\dfrac{du}{\cos x}$

$\Rightarrow \int \dfrac{\cos x}{u}\dfrac{du}{\cos x}=\int \dfrac{du}{u}=\ln u+C=\ln(\sin x)+C.$

6.8 (a) (i) $\int_1^2 \frac{1}{x^3} dx = \left[-\frac{1}{2} x^{-2} + C \right]_1^2 = -\frac{1}{2} 2^{-2} + \frac{1}{2} 1^{-2} = -\frac{1}{8} + \frac{1}{2} = \frac{3}{8}.$

(ii) $\int_0^2 x(x^2+4)^{1/2} dx = \left[\frac{1}{3} (x^2+4)^{3/2} + C \right]_0^2 = \frac{1}{3} 8^{3/2} - \frac{1}{3} 4^{3/2}$

$$= \frac{1}{3} 8\sqrt{8} - \frac{1}{3} 4\sqrt{4} = \frac{16}{3} \sqrt{2} - \frac{8}{3} = 4.876.$$

(b) $\int_0^2 \frac{x}{(x^2+4)} dx;$ Let $u = x^2 + 4;$ $\frac{du}{dx} = 2x \Rightarrow dx = \frac{du}{2x}$

$\int_0^2 \frac{x}{u} \frac{du}{2x} = \frac{1}{2} \int_0^2 \frac{du}{u} = \frac{1}{2} \left[\ln(x^2+4) \right]_0^2 = \frac{1}{2} \left\{ \ln(2^2+4) - \ln 4 \right\}$

$= \frac{1}{2} (\ln 8 - \ln 4) = \frac{1}{2} \ln \frac{8}{4} = \frac{1}{2} \ln 2.$

6.9 $W = \int_{V_a}^{V_b} P dV;$ $P = nRT/V \Rightarrow$

$W = \int_{V_a}^{V_b} \frac{nRT}{V} dV = nRT \int_{V_a}^{V_b} \frac{dV}{V}$

$= nRT [\ln V]_{V_a}^{V_b} = nRT \{ \ln V_b - \ln V_a \} = nRT \ln \frac{V_b}{V_a}.$

6.10 (a) $\frac{d}{dT} \ln K = \frac{\Delta H^{\ominus}}{RT^2} \Rightarrow \ln K = \int \frac{\Delta H^{\ominus}}{RT^2} dT = \frac{\Delta H^{\ominus}}{R} \int \frac{1}{T^2} dT$

$$= \frac{-\Delta H^{\ominus}}{RT} + C.$$

(b) $\Delta \ln K = \int_{500\,K}^{600\,K} \frac{\Delta H^{\ominus}}{RT^2} dT = \frac{-\Delta H^{\ominus}}{R \times 600} - \frac{-\Delta H^{\ominus}}{R \times 500}$

$= \Delta H^{\ominus} \left\{ \frac{1}{R \times 500} - \frac{1}{R \times 600} \right\} = 42.3 \times 10^3 \, \text{Jmol}^{-1} \times$

$\left\{ \frac{1}{8.314 \, \text{JK}^{-1}\text{mol}^{-1} \times 500 \, \text{K}} - \frac{1}{8.314 \, \text{JK}^{-1}\text{mol}^{-1} \times 600 \, \text{K}} \right\}$

$= 1.696.$

Chapter 7

7.1   (a) $y = x^{-1}$; $\dfrac{dy}{dx} = -x^{-2} = -y^2$ first order differential equation.

$$\frac{d^2y}{dx^2} = 2x^{-3} = 2y^3; \text{ second order.}$$

(b) $y = \cos ax$; $\dfrac{dy}{dx} = -a \sin ax$; $\dfrac{d^2y}{dx^2} = -a^2 \cos ax = -a^2 y$.

(c) $y = Ae^{4x}$; $\dfrac{d}{dx} Ae^{4x} - 4Ae^{4x} = 4Ae^{4x} - 4Ae^{4x} = 0$.

$$\frac{d^2}{dx^2} Ae^{4x} - 5\frac{d}{dx} Ae^{4x} + 4Ae^{4x} = 16Ae^{4x} - 5 \times 4Ae^{4x} + 4Ae^{4x} = 0.$$

7.2 $\dfrac{dy}{dx} = -6y^2 \Rightarrow \dfrac{dy}{y^2} = -6y \Rightarrow \displaystyle\int \dfrac{dy}{y^2} = \int -6dx \Rightarrow \dfrac{-1}{y} = -6x + C$

$\Rightarrow 1 = y(6x + D) \Rightarrow y = \dfrac{1}{6x + D}$.

$y = 1$, $x = 0$, so $1 = \dfrac{1}{D} \Rightarrow D = 1$

Therefore $y = \dfrac{1}{6x + 1}$.

7.3 $\dfrac{dy}{dx} = -\lambda y \Rightarrow \dfrac{dy}{y} = -\lambda dx \Rightarrow \displaystyle\int \dfrac{dy}{y} = \int -\lambda dx \Rightarrow \ln y = -\lambda x + C.$

$y = N, x = 0$, so $\ln N = C \Rightarrow \ln y = -\lambda x + \ln N \Rightarrow \ln \dfrac{y}{N} = -\lambda x$

$\dfrac{y}{N} = e^{-\lambda x} \Rightarrow y = Ne^{-\lambda x}.$

7.4   (a) $\dfrac{d}{dx}\left(\dfrac{1}{3}e^x + Ce^{-2x}\right) + 2\left(\dfrac{1}{3}e^x + Ce^{-2x}\right)$

$= \dfrac{1}{3}e^x - 2Ce^{-2x} + \dfrac{2}{3}e^x + 2Ce^{-2x} = e^x.$

(b) $\dfrac{dy}{dx} + \dfrac{y}{x} = x^2;\ P(x) = \dfrac{1}{x};\ Q(x) = x^2;$

$$R(x) = e^{\int \frac{1}{x}dx} = e^{\ln x + C} = e^C e^{\ln x} = e^C x = Ax.$$

Here $g(x) = \ln x$, therefore, $y = \dfrac{1}{x} \int x \times x^2 dx = \dfrac{1}{x} \times \left\{ \dfrac{x^4}{4} + C \right\} = \dfrac{x^3}{4} + \dfrac{C}{x}$

$y = 0,\ x = 1$; so $0 = \dfrac{1}{4} + C \Rightarrow C = -\dfrac{1}{4}$

$$y = \dfrac{x^3}{4} - \dfrac{1}{4x}.$$

7.5 (a) $P(x) \equiv \lambda_2$ and $Q(x) \equiv \lambda_1 (N_1)_0 e^{-\lambda_1 t}$.

$$R(t) = e^{\int \lambda_2 dt} = e^{\lambda_2 t + C} = A e^{\lambda_2 t}, \text{ where } \lambda_2 t = g(t).$$

(b) $N_2 = e^{-\lambda_2 t} \displaystyle\int e^{\lambda_2 t} \lambda_1 (N_1)_0 e^{-\lambda_1 t} dt = e^{-\lambda_2 t} \displaystyle\int \lambda_1 (N_1)_0 e^{(\lambda_2 - \lambda_1)t} dt$

$$= e^{-\lambda_2 t} \left(\dfrac{\lambda_1 (N_1)_0}{\lambda_2 - \lambda_1} e^{(\lambda_2 - \lambda_1)t} + C \right)$$

$N_2 = e^{-\lambda_2 t} \left(\dfrac{\lambda_1 (N_1)_0}{\lambda_2 - \lambda_1} e^{(\lambda_2 - \lambda_1)t} + C \right) = \dfrac{\lambda_1 (N_1)_0}{\lambda_2 - \lambda_1} e^{(\lambda_2 - \lambda_2 - \lambda_1)t} + C e^{-\lambda_2 t}$

$$= \dfrac{\lambda_1 (N_1)_0}{\lambda_2 - \lambda_1} e^{-\lambda_1 t} + C e^{-\lambda_2 t}.$$

(c) If $N_2 = 0$ at $t = 0$, then $0 = \dfrac{\lambda_1 (N_1)_0}{\lambda_2 - \lambda_1} + C \Rightarrow C = \dfrac{\lambda_1 (N_1)_0}{\lambda_2 - \lambda_1};$

Therefore, $N_2 = \dfrac{\lambda_1 (N_1)_0}{\lambda_2 - \lambda_1} e^{-\lambda_1 t} - \dfrac{\lambda_1 (N_1)_0}{\lambda_2 - \lambda_1} e^{-\lambda_2 t} = \dfrac{\lambda_1 (N_1)_0}{\lambda_2 - \lambda_1} \left(e^{-\lambda_1 t} - e^{-\lambda_2 t} \right).$

(d) $\dfrac{dN_1}{dt} = -\lambda_1 N_1$ has same form as $\dfrac{dy}{dx} = -\lambda y$, for which the solution is $y = N e^{-\lambda x}$ for $y = N, x = 0$.

Thus it follows that the solution to $\dfrac{dN_1}{dt} = -\lambda_1 N_1$, where $N_1 = (N_1)_0,\ t = 0$ is given by $N_1 = (N_1)_0 e^{-\lambda_1 t} = e^{-\lambda_1 t}$, since $(N_1)_0 = 1$ mol.

(e) Since $N_1 + N_2 + N_3 = 1$ mol and $N_1 = e^{-\lambda_1 t}$,

$$N_2 = \frac{\lambda_1}{\lambda_2 - \lambda_1} \left(e^{-\lambda_1 t} - e^{-\lambda_2 t} \right), \text{ it follows that}$$

$$N_3 = 1 - e^{-\lambda_1 t} - \frac{\lambda_1}{\lambda_2 - \lambda_1} \left(e^{-\lambda_1 t} - e^{-\lambda_2 t} \right).$$

$$= 1 - e^{-\lambda_1 t} - \frac{\lambda_1}{\lambda_2 - \lambda_1} e^{-\lambda_1 t} + \frac{\lambda_1}{\lambda_2 - \lambda_1} e^{-\lambda_2 t}$$

$$= 1 - \frac{\lambda_2 - \lambda_1}{\lambda_2 - \lambda_1} e^{-\lambda_1 t} - \frac{\lambda_1}{\lambda_2 - \lambda_1} e^{-\lambda_1 t} + \frac{\lambda_1}{\lambda_2 - \lambda_1} e^{-\lambda_2 t}$$

$$= 1 - \left(\frac{\lambda_2 - \lambda_1}{\lambda_2 - \lambda_1} e^{-\lambda_1 t} + \frac{\lambda_1}{\lambda_2 - \lambda_1} e^{-\lambda_1 t} - \frac{\lambda_1}{\lambda_2 - \lambda_1} e^{-\lambda_2 t} \right)$$

$$= 1 - \left(\frac{(\lambda_2 - \lambda_1 + \lambda_1) e^{-\lambda_1 t} - \lambda_1 e^{-\lambda_2 t}}{\lambda_2 - \lambda_1} \right)$$

$$= 1 - \left(\frac{\lambda_2 e^{-\lambda_1 t} - \lambda_1 e^{-\lambda_2 t}}{\lambda_2 - \lambda_1} \right)$$

$$= 1 + \left(\frac{\lambda_1 e^{-\lambda_2 t} - \lambda_2 e^{-\lambda_1 t}}{\lambda_2 - \lambda_1} \right).$$

(f) $t_{1/2} = \dfrac{\ln 2}{\lambda_i}$;

Thus, $\lambda_1 = \dfrac{\ln 2}{t_{1/2}} = \dfrac{\ln 2}{23.5}$ min$^{-1} = 2.95 \times 10^{-2}$ min^{-1}.

$$\lambda_2 = \frac{\ln 2}{t_{1/2}} = \frac{\ln 2}{2.3} \text{ day}^{-1} = \frac{\ln 2}{3312} \text{ min}^{-1} = 2.093 \times 10^{-4} \text{ min}^{-1}.$$

(g) $N_2 = \dfrac{\lambda_1}{\lambda_2 - \lambda_1} \left(e^{-\lambda_1 t} - e^{-\lambda_2 t} \right)$ - maximum when $\dfrac{dN_2}{dt} = 0$.

$$\frac{dN_2}{dt} = \frac{\lambda_1}{\lambda_2 - \lambda_1} \left(-\lambda_1 e^{-\lambda_1 t} - -\lambda_2 e^{-\lambda_2 t} \right)$$

$$= \frac{\lambda_1}{\lambda_2 - \lambda_1} \left(\lambda_2 e^{-\lambda_2 t} - \lambda_1 e^{-\lambda_1 t} \right) = 0$$

when $\lambda_2 e^{-\lambda_2 t} = \lambda_1 e^{-\lambda_1 t}$.

Taking logs, gives

$$\ln \lambda_2 - \lambda_2 t = \ln \lambda_1 - \lambda_1 t$$

$$\Rightarrow \ln \frac{\lambda_2}{\lambda_1} = \lambda_2 t - \lambda_1 t$$

$$\Rightarrow t = \frac{1}{\lambda_2 - \lambda_1} \ln \frac{\lambda_2}{\lambda_1}$$

$$= \frac{1}{0.0002092 - 0.0295} \ln \frac{0.0002092}{0.0295} = 168.96 \ \text{min}.$$

Thus, $N_{2, \ \text{max}} = \dfrac{0.0295}{0.0002092 - 0.0295} \times$

$$\left(e^{-0.0295 \times 168.96} - e^{-0.0002092 \times 168.96} \right)$$

$$= 0.965 \, \text{mol}.$$

7.6 (a)

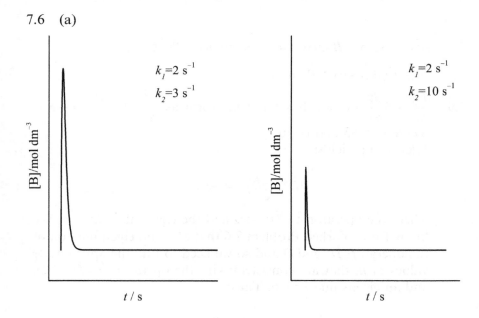

(b) As k_2 increases relative to k_1, so the maximum concentration of the intermediate decreases.

7.7   (a) $\dfrac{d^2y}{dx^2} = n^2y \Rightarrow y = Ae^{nx} + Be^{-nx}$.

(b) $\sinh nx = \dfrac{1}{2}(e^{nx} - e^{-nx}) \Rightarrow e^{nx} - e^{-nx} = 2\sinh nx$;

$\cosh nx = \dfrac{1}{2}(e^{nx} + e^{-nx}) \Rightarrow e^{nx} + e^{-nx} = 2\cosh nx$

$\Rightarrow 2e^{nx} = 2\sinh nx + 2\cosh nx$

$\Rightarrow e^{nx} = \sinh nx + \cosh nx$

Similarly $2e^{-nx} = 2\cosh nx - 2\sinh nx$

$\Rightarrow e^{-nx} = \cosh nx - \sinh x$

$y = A\sinh nx + A\cosh nx + B\cosh nx - B\sinh nx$

$y = (A - B)\sinh nx + (A + B)\cosh nx$.

(c) $0 = (A - B)\sinh 0 + (A + B)\cosh 0 = A + B \Rightarrow B = -A$

$y = (A - -A)\sinh nx + (A + -A)\cosh nx = 2A\sinh nx$.

7.8 $\hat{D}^2(A\cos nx + B\sin nx) = -An^2\cos nx - Bn^2\sin nx$

$= -n^2(A\cos nx + B\sin nx)$.

7.9 $\dfrac{d^2y}{dx^2} - 5\dfrac{dy}{dx} + 6y = 0$ has the same form as $\dfrac{d^2y}{dx^2} + c_1\dfrac{dy}{dx} = -c_2y$,

where $c_1 = -5$ and $c_2 = 6$.

Factorising yields:

$$(\hat{D}^2 - 5\hat{D})y = -6y$$

where the operator $\hat{A} = \hat{D}^2 - 5\hat{D}$ and the eigenvalue, $\lambda = -6$. We know from Worked Problem 7.6 that e^{nx} is an eigenfunction of the operators $\hat{D}^2$ and $\hat{D}$ and so we need to find the appropriate values of n, that are consistent with the operator, $\hat{A} = \hat{D}^2 - 5\hat{D}$, and an eigenvalue $\lambda = -6$. Thus:

$$\hat{A}e^{nx} = (n^2 - 5n)e^{nx} = -6e^{nx}$$
$$\Rightarrow (n^2 - 5n) = -6$$
$$\Rightarrow n^2 - 5n + 6 = 0.$$

Factorising yields:

$(n-2)(n-3)=0$, and so $n=2, 3$.

Thus the general solution is $y=Be^{3x}+Ce^{2x}$. Imposing the boundary condition, $y=0$, $x=0 \Rightarrow 0=B+C$ and $dy/dx = 1$ when $x = 0$

$\Rightarrow \dfrac{dy}{dx} = 3Be^{3x}+2Ce^{2x}=1$, when $x = 0$. Thus:

$3B+2C=1$. But $B=-C$, and so $-3C+2C=1 \Rightarrow C=-1$ and so $B=1$. Therefore the solution is:

$$y=e^{3x}-e^{2x}.$$

Chapter 8

8.1 (a) $\displaystyle\lim_{r \to \infty} \dfrac{1}{2^r}=0$; converges;

    (b) $\displaystyle\lim_{n \to \infty} \dfrac{n-1}{2n} = \dfrac{n}{2n} = \dfrac{1}{2}$, converges;

    (c) $\displaystyle\lim_{r \to \infty} \cos r\pi$ – oscillates between ± 1.

8.2 $n=6$; $r=0, 1, 2, 3, 4, 5, 6$; $^6C_r = \dfrac{6!}{(6-r)!r!}$

$\Rightarrow \dfrac{720}{720 \times 1}=1$; $\dfrac{720}{120 \times 1}=6$; $\dfrac{720}{24 \times 2}=15$;

$\dfrac{720}{6 \times 6}=20$; $\dfrac{720}{2 \times 24}=15$; $\dfrac{720}{1 \times 120}=6$; $\dfrac{720}{1 \times 720}=1$;

8.3 Geometric series, 1, 2, 4, 8, ...2^r, $\Rightarrow a=1$, $ax=2$, $ax^2=4$, $\Rightarrow a=1$, $x=2$

Using eqn (8.20) $S_n=a\left(\dfrac{1-x^n}{1-x}\right) = \dfrac{1-2^n}{1-2} = \dfrac{1-2^n}{-1} = 2^n-1.$

8.4 (a) $S=1+2x+3x^2+4x^3+\ldots,\ u_r=rx^{r-1}$;

$$\frac{u_{r+1}}{u_r}=\frac{(r+1)x^r}{rx^{r-1}}=\left(\frac{r+1}{r}\right)x$$

$$\therefore \lim_{r\to\infty}\left|\frac{u_{r+1}}{u_r}\right|=\lim_{r\to\infty}\left|\left(\frac{r+1}{r}\right)x\right|=|x|\ \text{converges if } |x|<1;\ i.e.\ \text{for}$$
$$-1<x<1.$$

(b) $S=1-x+\dfrac{x^2}{2!}-\dfrac{x^3}{3!}+\dfrac{x^4}{4!}-\cdots+(-1)^{r-1}\dfrac{x^{r-1}}{(r-1)!}+\cdots$

$$\frac{u_{r+1}}{u_r}=\frac{(-1)^r\frac{x^r}{r!}}{(-1)^{r-1}\frac{x^{r-1}}{(r-1)!}}=\frac{(-1)(r-1)!x^r}{r!x^{r-1}}=\frac{-x}{r}$$

$$\lim_{r\to\infty}\left|\frac{u_{r+1}}{u_r}\right|=\left|\frac{-x}{r}\right|=0\ \therefore\text{converges for all } x.$$

(c) $S=1+\dfrac{x^2}{2}-\dfrac{x^4}{4}+\cdots+(-1)^{r-1}\dfrac{x^{2r-2}}{2r-2}$

$$\frac{u_{r+1}}{u_r}=\frac{(-1)^r\frac{x^{2r}}{2r}}{(-1)^{r-1}\frac{x^{2r-2}}{2r-2}}=\frac{-x^2(2r-2)}{2r}=-\left(1-\frac{1}{r}\right)x^2=\frac{x^2}{r}-x^2$$

$$\lim_{r\to\infty}\left|\frac{u_{r+1}}{u_r}\right|=\left|\frac{x^2}{r}-x^2\right|=\left|-x^2\right|\ \therefore\text{converges for } |-x^2|<1;$$
$$i.e.\ \text{when}\ -1<x<1.$$

8.5 (a)

$f(x)=e^{-x}$	$f^{(1)}(x)=-e^{-x}$	$f^{(2)}(x)=e^{-x}$	$f^{(3)}(x)=-e^{-x}$	$f^{(4)}(x)=e^{-x}$	$f^{(5)}(x)=-e^{-x}$
$f(0)=1$	$f^{(1)}(0)=-1$	$f^{(2)}(0)=1$	$f^{(3)}(0)=-1$	$f^{(4)}(0)=1$	$f^{(5)}(0)=-1$

therefore $f(x)=1-x+\dfrac{x^2}{2!}-\dfrac{x^3}{3!}+\ldots\quad u_n=\dfrac{(-1)^{n-1}x^{n-1}}{(n-1)!}.$

(b)

$f(x)=\cos x$	$f^{(1)}(x)=-\sin x$	$f^{(2)}(x)=-\cos x$	$f^{(3)}(x)=\sin x$	$f^{(4)}(x)=\cos x$	$f^{(5)}(x)=-\sin x$
$f(0)=1$	$f^{(1)}(0)=0$	$f^{(2)}(0)=-1$	$f^{(3)}(0)=0$	$f^{(4)}(0)=1$	$f^{(5)}(0)=0$

$$f(x) = 1 - \frac{x^2}{2!} + \frac{x^4}{4!} - \frac{x^6}{6!} + \dots \quad u_n = \frac{(-1)^{n-1} x^{2n-2}}{(2n-2)!}.$$

(c)

$f(x)=(1-x)^{-1}$	$f^{(1)}(x)=(1-x)^{-2}$	$f^{(2)}(x)=1 \cdot 2(1-x)^{-3}$	$f^{(3)}(x)=1 \cdot 2 \cdot 3(1-x)^{-4}$
$f(0)=1$	$f^{(1)}(0)=1$	$f^{(2)}(0)=2$	$f^{(3)}(0)=6$

$$f(x) = 1 + x + \frac{2! x^2}{2!} + \frac{3! x^3}{3!} - \dots$$

$$\Rightarrow f(x) = 1 + x + x^2 + x^3 + \dots \quad u_n = x^{n-1}.$$

8.6 (a)

$f(x)=(1-x)^{-1}$	$f^{(1)}(x)=(1-x)^{-2}$	$f^{(2)}(x)=1 \cdot 2(1-x)^{-3}$	$f^{(3)}(x)=1 \cdot 2 \cdot 3(1-x)^{-4}$
$f(-1)=\frac{1}{2}$	$f^{(1)}(-1)=\frac{1}{4}$	$f^{(2)}(-1)=\frac{2}{8}$	$f^{(3)}(-1)=\frac{6}{16}$

$$f(x) = \frac{1}{2} + \frac{1}{4}(x+1) + \frac{2(x+1)^2}{8 \times 2!} + \frac{6(x+1)^3}{16 \times 3!} + \dots$$

$$f(x) = \frac{1}{2} + \frac{(x+1)}{4} + \frac{(x+1)^2}{8} + \frac{(x+1)^3}{16} + \dots, u_n = \frac{(x+1)^{n-1}}{2^n}.$$

(b)

$f(x)=\sin x$	$f^{(1)}(x)=\cos x$	$f^{(2)}(x)=-\sin x$	$f^{(3)}(x)=-\cos x$	$f^{(4)}(x)=\sin x$	$f^{(5)}(x)=\cos x$
$f(\frac{\pi}{2})=1$	$f^{(1)}(\frac{\pi}{2})=0$	$f^{(2)}(\frac{\pi}{2})=-1$	$f^{(3)}(\frac{\pi}{2})=0$	$f^{(4)}(\frac{\pi}{2})=1$	$f^{(5)}(\frac{\pi}{2})=0$

$$f(x) = 1 - \frac{1}{2!}\left(x - \frac{\pi}{2}\right)^2 + \frac{1}{4!}\left(x - \frac{\pi}{2}\right)^4 - \frac{1}{6!}\left(x - \frac{\pi}{2}\right)^6 \dots,$$

$$u_n = \frac{(-1)^{n-1}}{(2n-2)!}\left(x - \frac{\pi}{2}\right)^{2n-2}.$$

(c)

$f(x)=\ln x$	$f^{(1)}(x)=1/x$	$f^{(2)}(x)=-1/x^2$	$f^{(3)}(x)=2/x^3$	$f^{(4)}(x)=-6/x^4$	$f^{(5)}(x)=24/x^5$
$f(1)=0$	$f^{(1)}(1)=1$	$f^{(2)}(1)=-1$	$f^{(3)}(1)=2$	$f^{(4)}(1)=-6$	$f^{(5)}(1)=24$

$$f(x) = 0 + (x-1) - \frac{1}{2!}(x-1)^2 + \frac{2}{3!}(x-1)^3 - \frac{6}{4!}(x-1)^4 \cdots$$

$$f(x) = (x-1) - \frac{(x-1)^2}{2} + \frac{(x-1)^3}{3} - \frac{(x-1)^4}{4} \cdots ,$$

$$u_n = (-1)^{n-1} \frac{(x-1)^n}{n}.$$

8.7 (a) (i) $E(R) = D_e \left\{ 1 - e^{-\alpha(R-R_e)} \right\}^2$;

Let $u = 1 - e^{-\alpha(R-R_e)}$

$$\Rightarrow \frac{du}{dR} = \alpha e^{-\alpha(R-R_e)}.$$

$$E(R) = D_e u^2 \Rightarrow \frac{dE(R)}{du} = 2D_e u = 2D_e \left\{ 1 - e^{-\alpha(R-R_e)} \right\}$$

$$\Rightarrow \frac{dE(R)}{dR} = \frac{dE(R)}{du} \times \frac{du}{dR}$$

$$= 2D_e \left\{ 1 - e^{-\alpha(R-R_e)} \right\} \times \alpha e^{-\alpha(R-R_e)}$$

$$= E^{(1)}(R).$$

(ii) $E^{(1)}(R) = \underbrace{2D_e \left\{ 1 - e^{-\alpha(R-R_e)} \right\}}_{u} \times \underbrace{\alpha e^{-\alpha(R-R_e)}}_{v}$;

using the product rule,

$$E^{(2)}(R) = 2D_e \left\{ -e^{-\alpha(R-R_e)} \right\} \times -\alpha^2 e^{-\alpha(R-R_e)} + \alpha e^{-\alpha(R-R_e)}$$

$$\times 2\alpha D_e e^{-\alpha(R-R_e)}$$

$$= -2\alpha^2 D_e e^{-\alpha(R-R_e)} + 2\alpha^2 D_e e^{-2\alpha(R-R_e)} + 2\alpha^2 e^{-2\alpha(R-R_e)}$$

$$\Rightarrow E^{(2)}(R) = 4\alpha^2 D_e e^{-2\alpha(R-R_e)} - 2\alpha^2 D_e e^{-\alpha(R-R_e)}$$

$$\Rightarrow E^{(2)}(R) = 2\alpha^2 D_e \left\{ 2e^{-2\alpha(R-R_e)} - e^{-\alpha(R-R_e)} \right\}.$$

(b) $E^{(1)}(R_e) = 2D_e \left\{ 1 - e^{-\alpha(R_e-R_e)} \right\} \times \alpha e^{-\alpha(R_e-R_e)} = 2D_e \{1-1\} \times \alpha = 0$,

therefore, a maximum or minimum.

$$E^{(2)}(R_e) = 2\alpha^2 D_e \left\{ 2e^{-2\alpha(R_e - R_e)} - e^{-\alpha(R_e - R_e)} \right\} = 2\alpha^2 D_e \{2 - 1\} = 2\alpha^2 D_e;$$

positive, therefore a minimum.

(c) $E(R) = \alpha^2 D_e (R - R_e)^2 \Rightarrow$ Let $u = R - R_e \Rightarrow \dfrac{du}{dR} = 1$.

$$E(R) = \alpha^2 D_e u^2 \Rightarrow \frac{dE(R)}{du} = 2\alpha^2 D_e u = 2\alpha^2 D_e (R - R_e).$$

$$\Rightarrow \frac{dE(R)}{dR} = \frac{dE(R)}{du} \times \frac{du}{dR} = 2\alpha^2 D_e (R - R_e) \times 1 = 2\alpha^2 D_e (R - R_e).$$

Thus $F = -\dfrac{dE(R)}{dR} = -2\alpha^2 D_e (R - R_e).$

(d) If $k = 2\alpha^2 D_e$ and $x = R - R_e$, then $F = -kx \Rightarrow$ the expression for F obtained in (c) has the same form as the restoring force acting on a simple harmonic oscillator.

8.8 (a) $\sinh x = \dfrac{1}{2}(e^x - e^{-x})$

$$= \frac{1}{2}\left\{ \left(1 + x + \frac{x^2}{2!} + \frac{x^3}{3!} + \cdots + \frac{x^{n-1}}{(n-1)!} + \cdots\right) - \right.$$
$$\left. \left(1 - x + \frac{x^2}{2!} - \frac{x^3}{3!} + \frac{x^4}{4!} - \cdots (-1)^{n-1}\frac{x^{n-1}}{(n-1)!} + \cdots\right) \right\}$$

$$= \frac{1}{2}\left\{ 2x + \frac{2x^3}{3!} + \cdots + \frac{2x^{2n-1}}{(2n-1)!} + \cdots \right\}$$

$$= x + \frac{x^3}{3!} + \frac{x^5}{5!} + \cdots + \frac{x^{2n-1}}{(2n-1)!} + \cdots$$

(b) $f(x) = \dfrac{e^{-x}}{(1-x)} = e^{-x} \times \dfrac{1}{(1-x)}$

$$= \left\{ 1 - x + \frac{x^2}{2!} - \frac{x^3}{3!} + \frac{x^4}{4!} - \cdots (-1)^{n-1}\frac{x^{n-1}}{(n-1)!} + \cdots \right\}$$
$$\times \left\{ 1 + x + x^2 + x^3 + \cdots x^{n-1} + \cdots \right\}$$

$$= 1 + \frac{x^2}{2!} + \left\{ \frac{x^3}{2!} - \frac{x^3}{3!} \right\} + \left\{ \frac{x^4}{2!} - \frac{x^4}{3!} + \frac{x^4}{4!} \right\} + \cdots$$

$$= 1 + \frac{x^2}{2} + \frac{x^3}{3} + \cdots$$

Maclaurin power series expansion of e^{-x} converges for all x, and of $\dfrac{1}{(1-x)}$ converges for $-1<x<1$, so the interval of convergence of the Maclaurin series of $f(x)=\dfrac{e^{-x}}{(1-x)}$ is $-1<x<1$.

8.9   (a) $e^X = 1 + X + \dfrac{X^2}{2!} + \dfrac{X^3}{3!} + \dfrac{X^4}{4!} + \cdots \dfrac{X^{n-1}}{(n-1)!} + \cdots\,;$

if $X = ax$, then

$$e^{ax} = 1 + ax + \frac{(ax)^2}{2!} + \frac{(ax)^3}{3!} + \frac{(ax)^4}{4!} + \cdots \frac{(ax)^{n-1}}{(n-1)!} + \cdots$$

(b) (i) $\sin 2x = 2 \sin x \cos x = 2\left\{ x - \dfrac{x^3}{3!} + \dfrac{x^5}{5!} - \cdots \right\}$

$$\times \left\{ 1 - \frac{x^2}{2!} + \frac{x^4}{4!} - \cdots \right\}$$

$$\sin 2x = 2x - \frac{8x^3}{6} + \frac{32x^5}{120} - \cdots = 2x - \frac{8x^3}{3!} + \frac{32x^5}{5!} - \cdots$$

(ii) $\sin X = X - \dfrac{X^3}{3!} + \dfrac{X^5}{5!} - \cdots\,;$ If $X = 2x$, then

$$\sin 2x = 2x - \frac{(2x)^3}{3!} + \frac{(2x)^5}{5!} - \cdots$$

$$\Rightarrow \sin 2x = 2x - \frac{8x^3}{3!} + \frac{32x^5}{5!} - \cdots$$

8.10 $\quad C_V = 3R\left(\dfrac{hv}{kT}\right)^2\left\{\dfrac{e^{\frac{hv}{2kT}}}{e^{\frac{hv}{kT}}-1}\right\}^2$

$e^{\frac{hv}{2kT}} = 1 + \dfrac{hv}{2k}\cdot\dfrac{1}{T} + \dfrac{\left(\frac{hv}{2k}\right)^2\cdot\frac{1}{T^2}}{2!} + \cdots \approx 1, \text{ for large } T.$

$e^{\frac{hv}{kT}} - 1 = 1 + \dfrac{hv}{k}\cdot\dfrac{1}{T} + \dfrac{\left(\frac{hv}{k}\right)^2\cdot\frac{1}{T^2}}{2!} + \cdots - 1$

$\qquad\qquad = \dfrac{hv}{k}\cdot\dfrac{1}{T} + \dfrac{\left(\frac{hv}{k}\right)^2\cdot\frac{1}{T^2}}{2!} + \cdots$

$\qquad\qquad \approx \dfrac{hv}{k}\cdot\dfrac{1}{T}, \text{ for large } T.$

Therefore, $\displaystyle\lim_{\text{Large T}} C_V = 3R\left(\dfrac{hv}{kT}\right)^2\left\{\dfrac{1}{\frac{hv}{kT}}\right\}^2$

$\qquad\qquad\qquad = 3R\left(\dfrac{hv}{kT}\right)^2\left\{\dfrac{kT}{hv}\right\}^2 = 3R.$

8.11 (a) $\sin\dfrac{n\pi x}{L} \approx \dfrac{n\pi x}{L}$ for small x.

At $x = 0$, $\sin\dfrac{n\pi x}{L} = 0 \Rightarrow \psi = 0.$

(b) At $x = L$, $\psi = \sqrt{\dfrac{2}{L}}\sin n\pi = 0,$

when $n = 1,2,3\ldots$

(c) When x is small, $\psi = \sqrt{\dfrac{2}{L}}\dfrac{n\pi}{L}x.$

Chapter 9

9.1 (a)

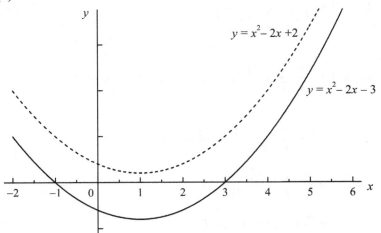

(b) (i) $x^2 - 2x - 3 = 0$ $\Rightarrow$

$$x = \frac{2 \pm \sqrt{4 - -(4 \times 1 \times 3)}}{2} = 1 \pm \frac{\sqrt{16}}{2} = 1 \pm 2 = 3, -1.$$

(ii) $x^2 - 2x + 2 = 0$ $\Rightarrow$

$$x = \frac{2 \pm \sqrt{4 - (4 \times 1 \times 2)}}{2}$$

$$= 1 \pm \frac{\sqrt{-4}}{2} = 1 \pm \frac{\sqrt{4}\sqrt{-1}}{2} = 1 \pm \frac{2i}{2} = 1 \pm i.$$

Part (i) has two real roots corresponding to where the curve cuts the x-axis. For part (ii), there are no real roots and so the curve does not cut the x-axis.

9.2 (a) $i^3 = i \times i^2 = i \times -1 = -i$
   (b) $i^4 = i \times i^3 = i \times -i = 1$
   (c) $i^5 = i \times i^4 = i \times 1 = i.$

9.3 (a) $z_1 + z_2 = (2 + 3i) + (-1 + i) = 1 + 4i$
   $z_1 + z_2 - 2z_3 = (1 + 4i) - 2(3 - 2i) = -5 + 8i.$
   (b) $z_1 z_2 = (2 + 3i)(-1 + i) = -2 + 2i - 3i + 3i^2 = -2 - i - 3$
   $= -5 - i.$

$$z_3^2 = (3-2i)(3-2i) = 9 - 6i - 6i + 4i^2$$
$$= 9 - 12i - 4 = 5 - 12i.$$

$$z_1z_2 + z_3^2 = (-5-i) + (5-12i) = -13i.$$

9.4 (a) $z = (-1 - 2i) + (2 + 7i) = 1 + 5i$; $z^* = 1 - 5i$.
  (b) $z = (3 - i) - (4 - 2i) = -1 + i$; $z^* = -1 - i$.
  (c) $z = i(1 + 3i) = i - 3 = -3 + i$; $z^* = -3 - i$.
  (d) $z = (1 + 3i)(3 + 2i) = 3 + 2i + 9i - 6 = -3 + 11i$;
  $z^* = -3 - 11i$.

9.5 (a) $\dfrac{1}{i} = \dfrac{1}{i} \cdot \dfrac{-i}{-i} = \dfrac{-i}{1} = -i.$

  (b) $\dfrac{1-i}{2-i} = \dfrac{1-i}{2-i} \cdot \dfrac{2+i}{2+i} = \dfrac{2+i-2i+1}{4+2i-2i+1} = \dfrac{3-i}{5} = \dfrac{3}{5} - \dfrac{i}{5}.$

  (c) $\dfrac{i(2+i)}{(1-2i)(2-i)} = \dfrac{-1+2i}{2-i-4i-2} = \dfrac{-1+2i}{-5i} = \dfrac{-1+2i}{-5i} \cdot \dfrac{5i}{5i}$

  $= \dfrac{-5i-10}{25} = \dfrac{-2}{5} - \dfrac{i}{5}.$

9.6

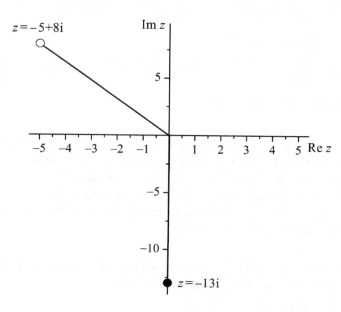

9.7 (a) $z = -1 - 2i; r = \sqrt{1^2 + 2^2} = \sqrt{5}; \theta = \tan^{-1}(-2/-1) = \tan^{-1} 2$
$= 63.43°, -116.56°$. z lies in the 3rd quadrant and so $\theta = -116.56°$.

(b) $z = 2i; r = \sqrt{2^2} = 2; \theta = \tan^{-1}(2/0)$ – undefined at $\theta = 90°$, $-90°$.

In this instance, $\theta = 90°$.

9.8 (a) $(i\theta)^2 = -\theta^2$
(b) $(i\theta)^3 = -i\theta^3$
(c) $(i\theta)^4 = \theta^4$
(d) $(i\theta)^5 = i\theta^5$

$$z = r \left\{ 1 + i\theta + \frac{(i\theta)^2}{2!} + \frac{(i\theta)^3}{3!} + \frac{(i\theta)^4}{4!} + \frac{(i\theta)^5}{5!} + \cdots \right\}$$

$$= r \left\{ 1 + i\theta - \frac{\theta^2}{2!} - \frac{i\theta^3}{3!} + \frac{\theta^4}{4!} + \frac{i\theta^5}{5!} + \cdots \right\} \text{ – same as eqn (9.16).}$$

9.9 (a) $z_1 z_2 = r_1 e^{i\theta_1} \cdot r_2 e^{i\theta_2} = r_1 r_1 e^{i(\theta_1 + \theta_2)}$; modulus $= r_1 r_2$, argument $= \theta_1 + \theta_2$.

(b) $\dfrac{z_1}{z_2} = \dfrac{r_1 e^{i\theta_1}}{r_2 e^{i\theta_2}} = \dfrac{r_1}{r_2} e^{i\theta_1} \cdot e^{-i\theta_2} = \dfrac{r_1}{r_2} e^{i(\theta_1 - \theta_2)}$; modulus $= \dfrac{r_1}{r_2}$,

argument $= \theta_1 - \theta_2$.

(c) $\dfrac{z_1^2}{z_2^4} = \dfrac{r_1^2 e^{2i\theta_1}}{r_2^4 e^{4i\theta_2}} = \dfrac{r_1^2}{r_2^4} e^{2i\theta_1} e^{-4i\theta_2} = \dfrac{r_1^2}{r_2^4} e^{i(2\theta_1 - 4\theta_2)}$; modulus $= \dfrac{z_1^2}{z_2^4}$,

argument $= 2\theta_1 - 4\theta_2$.

9.10 $z = -1 - i, r = \sqrt{1^2 + 1^2} = \sqrt{2}; \theta = \tan^{-1}(-1/-1) = \tan^{-1} 1 = \pi/4, -3\pi/4$ but in third quadrant and so $\theta = -3\pi/4$.
Thus $z = \sqrt{2} e^{-\frac{3\pi i}{4}} \Rightarrow z^2 = 2 e^{-\frac{6\pi i}{4}} = 2 e^{-\frac{3\pi i}{2}}$; modulus$=2$,
argument$= -3\pi/2 = \pi/2$. $z^{-4} = \frac{1}{4} e^{3\pi i}$; modulus$=1/4$,
argument$= 3\pi = \pm \pi$, but from the definition of the argument $\theta = \pi$ is the only acceptable result.

9.11 (a) $(\cos \theta + i \sin \theta)^{-1} = \cos -\theta + i \sin -\theta = \cos \theta - i \sin \theta$.
(b) $(\cos \theta + i \sin \theta)^{1/2} = \cos \theta/2 + i \sin \theta/2$.

(c) $z^n = \underbrace{r^n \cos n\theta}_{\text{real}} + \underbrace{i r^n \sin n\theta}_{\text{imaginary}}$

(d) $z = -1 - i$; $r = \sqrt{2}$; $\theta = -3\pi/4$

$$\Rightarrow z = \sqrt{2}\{\cos(3\pi/4) - i\sin(3\pi/4)\}$$

$$z^3 = \sqrt{2}^3\{\cos(9\pi/4) - i\sin(9\pi/4)\}$$
$$= \sqrt{2}^3\sqrt{2}^{-1} - i\sqrt{2}^3\sqrt{2}^{-1} = 2 - 2i.$$

$$z^{-2} = \sqrt{2}^{-2}\{\cos(-3\pi/2) - i\sin(-3\pi/2)\}$$

$$= \frac{1}{2}\{\cos(3\pi/2) + i\sin(3\pi/2)\} = \frac{1}{2} \times -i = -\frac{i}{2}.$$

9.12 $e^{i2m\pi} = \cos 2m\pi + i\sin 2m\pi = 1$; $e^{i(\theta + 2m\pi)} = e^{i\theta}e^{i2m\pi} = e^{i\theta}$.

9.13 (a) $e^{-i\theta} = \cos-\theta + i\sin-\theta = \cos\theta - i\sin\theta$.
(b) $e^{-i\theta} = \cos\theta - i\sin\theta$ and $e^{i\theta} = \cos\theta + i\sin\theta$.
(i) Adding the two expressions yields:
$e^{i\theta} + e^{-i\theta} = 2\cos\theta + i\sin\theta - i\sin\theta = 2\cos\theta$

$$\Rightarrow \cos\theta = \frac{1}{2}\left(e^{i\theta} + e^{-i\theta}\right).$$

(ii) Similarly, subtracting the two expressions yields:
$e^{i\theta} - e^{-i\theta} = \cos\theta - \cos\theta + i\sin\theta + i\sin\theta = 2i\sin\theta$

$$\Rightarrow \sin\theta = \frac{1}{2i}\left(e^{i\theta} - e^{-i\theta}\right).$$

9.14 $y = A\cos kt + B\sin kt$

$$y = A \times \frac{1}{2}\left(e^{ikt} + e^{-ikt}\right) + B \times \frac{1}{2i}\left(e^{ikt} - e^{-ikt}\right)$$

$$= \frac{A}{2}e^{ikt} + \frac{A}{2}e^{-ikt} + \frac{B}{2i}e^{ikt} - \frac{B}{2i}e^{-ikt}$$

$$= \left(\frac{A}{2} + \frac{B}{2i}\right)e^{ikt} + \left(\frac{A}{2} - \frac{B}{2i}\right)e^{-ikt}.$$

9.15 (a) $\psi_1 = \underbrace{N_1e^{-r/2a_0}r\sin\theta}_{\text{real}}\ \underbrace{e^{i\phi}}_{\text{imaginary}}$,

$\psi_0 = N_2e^{-r/2a_0}r\cos\theta$, real; no imaginary part

$\psi_{-1} = \underbrace{N_1e^{-r/2a_0}r\sin\theta}_{\text{real}}\ \underbrace{e^{-i\phi}}_{\text{imaginary}}$.

(b) (i) $\psi_1 + \psi_{-1} = N_1e^{-r/2a_0}r\sin\theta e^{i\phi} + N_1e^{-r/2a_0}r\sin\theta e^{-i\phi}$

$= N_1e^{-r/2a_0}r\sin\theta\left(e^{i\phi} + e^{-i\phi}\right)$.

but $e^{i\phi} + e^{-i\phi} = 2\cos\phi$ and so:

$\dfrac{1}{\sqrt{2}}(\psi_1 + \psi_{-1}) = \dfrac{2}{\sqrt{2}}N_1e^{-r/2a_0}r\sin\theta \times \cos\phi$

$= \sqrt{2}N_1e^{-r/2a_0}r\sin\theta\cos\phi$.

(ii) $\psi_1 - \psi_{-1} = N_1e^{-r/2a_0}r\sin\theta e^{i\phi} - N_1e^{-r/2a_0}r\sin\theta e^{-i\phi}$

$= N_1e^{-r/2a_0}r\sin\theta\left(e^{i\phi} - e^{-i\phi}\right)$

but $e^{i\phi} - e^{-i\phi} = 2i\sin\phi$ and so:

$\dfrac{-i}{\sqrt{2}}(\psi_1 - \psi_{-1}) = \dfrac{-i}{\sqrt{2}}N_1e^{-r/2a_0}r\sin\theta \times 2i\sin\phi$

$= \sqrt{2}N_1e^{-r/2a_0}r\sin\theta\sin\phi$.

(c) $\psi_0 = N_2e^{-r/2a_0}r\cos\theta$; $z = r\cos\theta \Rightarrow \psi_0 = N_2e^{-r/2a_0}z$. Thus we can relabel ψ_0 as ψ_z.

$\dfrac{1}{\sqrt{2}}(\psi_1 + \psi_{-1}) = \sqrt{2}N_1e^{-r/2a_0}r\sin\theta\cos\phi$;

$x = r\sin\theta\cos\phi \Rightarrow \psi_x = \sqrt{2}N_1e^{-r/2a_0}x$.

$\dfrac{-i}{\sqrt{2}}(\psi_1 - \psi_{-1}) = \sqrt{2}N_1e^{-r/2a_0}r\sin\theta\sin\phi$;

$y = r\sin\theta\sin\phi \Rightarrow \psi_y = \sqrt{2}N_1e^{-r/2a_0}y$.

9.16 (a) $F(hkl) = \sum_{j}^{cell} f_j e^{2\pi i[hx_j + ky_j + lz_j]} \Rightarrow$

$$F(hkl) = f_{Na} e^{2\pi i[h0 + k0 + l0]} + f_{Na} e^{2\pi i\left[h\frac{1}{2} + k\frac{1}{2} + l\frac{1}{2}\right]}$$

$$= f_{Na} + f_{Na} e^{\pi i[h+k+l]}.$$

Euler's formula: $e^{i\theta} = \cos\theta + i\sin\theta$;

$$\Rightarrow F(hkl) = f_{Na} + f_{Na} e^{\pi i[h+k+l]}$$

$$= f_{Na} + f_{Na}\{\cos(h+k+l)\pi + i\sin(h+k+l)\pi\}.$$

(b) $F(hkl) = f_{Na} + f_{Na}\{ \underbrace{\cos(h+k+l)\pi}_{\substack{=1,\text{ when } h+k+l,\text{ even}\\ =-1 \text{ when } h+k+l,\text{ odd}}} + \underbrace{i\sin(h+k+l)\pi}_{=0,\text{ for } h+k+l,\text{ even, odd}} \}$

$\therefore F(hkl) = 2f_{Na}$, for n even; $=0$ for n odd.

9.17 For the complex number i, $r = 1$ and $\theta = \dfrac{\pi}{2}$; hence

$$i^{1/3} = e^{i(\pi/2 + 2m\pi) \times 1/3} = e^{i(\pi/6 + 2m\pi/3)}$$

$$= \cos\left(\frac{\pi}{6} + \frac{2m\pi}{3}\right) + i\sin\left(\frac{\pi}{6} + \frac{2m\pi}{3}\right)$$

For $m = 1$,

$$= \cos\left(\frac{5\pi}{6}\right) + i\sin\left(\frac{5\pi}{6}\right) = -\frac{\sqrt{3}}{2} + \frac{1}{2}i;$$

For $m = 2$,

$$= \cos\left(\frac{3\pi}{2}\right) + i\sin\left(\frac{3\pi}{2}\right) = -i;$$

For $m = 3$,

$$= \cos\left(\frac{13\pi}{6}\right) + i\sin\left(\frac{13\pi}{6}\right) = \frac{\sqrt{3}}{2} + \frac{1}{2}i.$$

For $m = 4$,

$$= \cos\left(\frac{17\pi}{6}\right) + i\sin\left(\frac{17\pi}{6}\right) = -\frac{\sqrt{3}}{2} + \frac{1}{2}i,$$

and so on. We see that taking $m \geqslant 4$ merely repicates the roots already found, and so the three cube roots of i are $-\dfrac{\sqrt{3}}{2} + \dfrac{1}{2}i$, $-i$ and $\dfrac{\sqrt{3}}{2} + \dfrac{1}{2}i$.

Chapter 10

10.1 $2x + y = 5$

$\frac{1}{2}x + 8y = 9$

$$x = \dfrac{\begin{vmatrix} 5 & 9 \\ 1 & 8 \end{vmatrix}}{\begin{vmatrix} 2 & 1 \\ \frac{1}{2} & 8 \end{vmatrix}} = \dfrac{40 - 9}{16 - \frac{1}{2}} = \dfrac{31}{15\frac{1}{2}} = 2;$$

$$y = \dfrac{\begin{vmatrix} 9 & 5 \\ \frac{1}{2} & 2 \end{vmatrix}}{\begin{vmatrix} 2 & 1 \\ \frac{1}{2} & 8 \end{vmatrix}} = \dfrac{18 - 2\frac{1}{2}}{16 - \frac{1}{2}} = \dfrac{15\frac{1}{2}}{15\frac{1}{2}} = 1.$$

10.2 (a) $a_{11} \equiv 1$, $a_{12} \equiv -\dfrac{1}{RT_1}$, $b_1 \equiv \ln k_1$; $a_{21} \equiv 1$,

$a_{22} \equiv -\dfrac{1}{RT_2}$, $b_2 \equiv \ln k_2$; $x = \ln A, y = E_a$.

(b) $a_{11} \equiv 100$, $a_{12} \equiv 1$, $b_1 \equiv 212$; $a_{21} \equiv 0$, $a_{22} \equiv 1$, $b_2 \equiv 32$;

$$a = \dfrac{\begin{vmatrix} 212 & 32 \\ 1 & 1 \end{vmatrix}}{\begin{vmatrix} 100 & 1 \\ 0 & 1 \end{vmatrix}} = \dfrac{212 - 32}{100} = \dfrac{180}{100} = \dfrac{9}{5} \text{ and}$$

$$b = \dfrac{\begin{vmatrix} 32 & 212 \\ 0 & 100 \end{vmatrix}}{\begin{vmatrix} 100 & 1 \\ 0 & 1 \end{vmatrix}} = \dfrac{3200}{100} = 32$$

and so the formula relating T to t is

$T = \frac{9}{5}t + 32.$

10.3 (a) $\begin{vmatrix} 1 & -1 & 2 \\ 0 & 3 & 0 \\ 2 & -2 & -2 \end{vmatrix} = 1\begin{vmatrix} 0 & 0 \\ 2 & -2 \end{vmatrix} + 3\begin{vmatrix} 1 & 2 \\ 2 & -2 \end{vmatrix} + 2\begin{vmatrix} 1 & 2 \\ 0 & 0 \end{vmatrix}$

$$= 0 + 3(-2-4) + 0 = -18.$$

(b) $\begin{vmatrix} 1 & -1 & 2 \\ 0 & 3 & 0 \\ 2 & -2 & -2 \end{vmatrix} = 0\begin{vmatrix} -1 & 2 \\ -2 & -2 \end{vmatrix} + 3\begin{vmatrix} 1 & 2 \\ 2 & -2 \end{vmatrix} + 0\begin{vmatrix} 1 & -1 \\ 2 & -2 \end{vmatrix}$

$$= 0 + 3(-2-4) + 0 = -18.$$

10.4 (a) $\begin{vmatrix} 1 & 0 & -2 \\ 2 & 8 & 4 \\ 3 & 2 & 2 \end{vmatrix}$

$$A_{33} = (-1)^6 \begin{vmatrix} 1 & 0 \\ 2 & 8 \end{vmatrix} = 8;$$

$$A_{22} = (-1)^4 \begin{vmatrix} 1 & -2 \\ 3 & 2 \end{vmatrix} = 2 - -6 = 8;$$

$$A_{32} = (-1)^5 \begin{vmatrix} 1 & -2 \\ 2 & 4 \end{vmatrix} = -1(4 - -4) = -8;$$

$$A_{23} = (-1)^5 \begin{vmatrix} 1 & 0 \\ 3 & 2 \end{vmatrix} = -1 \times 2 = -2.$$

(b) $\begin{vmatrix} \cos\theta & -\sin\theta & 0 \\ \sin\theta & \cos\theta & 0 \\ 0 & 0 & 1 \end{vmatrix}$

$$A_{12} = (-1)^3 \begin{vmatrix} \sin\theta & 0 \\ 0 & 1 \end{vmatrix} = -\sin\theta;$$

$$A_{21} = (-1)^3 \begin{vmatrix} -\sin\theta & 0 \\ 0 & 1 \end{vmatrix} = -1 \times -\sin\theta = \sin\theta.$$

10.5 (a)

(i) $\begin{vmatrix} 1 & 2 & 3 \\ 0 & 8 & 2 \\ -2 & 4 & 2 \end{vmatrix} = 0\begin{vmatrix} 2 & 3 \\ 4 & 2 \end{vmatrix} + 8\begin{vmatrix} 1 & 3 \\ -2 & 2 \end{vmatrix} - 2\begin{vmatrix} 1 & 2 \\ -2 & 4 \end{vmatrix}$

$= 0 + 8(2--6) - 2(4--4) = 64 - 16 = 48.$

(ii) $\begin{vmatrix} 1 & 2 & 3 \\ 0 & 8 & 2 \\ -2 & 4 & 2 \end{vmatrix}$ Subtracting twice col 1 from col 2

$= \begin{vmatrix} 1 & 0 & 3 \\ 0 & 8 & 2 \\ -2 & 8 & 2 \end{vmatrix}$ and then 3 times col 1

from col 3 $= \begin{vmatrix} 1 & 0 & 0 \\ 0 & 8 & 2 \\ -2 & 8 & 8 \end{vmatrix} = 1\begin{vmatrix} 8 & 2 \\ 8 & 8 \end{vmatrix}$

$= 64 - 16 = 48.$

(iii) Starting from $\begin{vmatrix} 1 & 0 & 0 \\ 0 & 8 & 2 \\ -2 & 8 & 8 \end{vmatrix}$,

subtract $\frac{1}{4} \times$ col 2 from col 3

$= \begin{vmatrix} 1 & 0 & 0 \\ 0 & 8 & 0 \\ -2 & 8 & 6 \end{vmatrix} = 1 \times 8 \times 6 = 48.$

(b)

(i) $\begin{vmatrix} 1 & 0 & -2 \\ 2 & 8 & 4 \\ 3 & 2 & 2 \end{vmatrix} = -2\begin{vmatrix} 0 & -2 \\ 2 & 2 \end{vmatrix} + 8\begin{vmatrix} 1 & -2 \\ 3 & 2 \end{vmatrix} - 4\begin{vmatrix} 1 & 0 \\ 3 & 2 \end{vmatrix}$

$= -2(0--4) + 8(2--6) - 4(2-0)$

$= -8 + 64 - 8 = 48.$

(ii) $\begin{vmatrix} 1 & 0 & -2 \\ 2 & 8 & 4 \\ 3 & 2 & 2 \end{vmatrix}$ Add twice col 1 to col 3

$$= \begin{vmatrix} 1 & 0 & 0 \\ 2 & 8 & 8 \\ 3 & 2 & 8 \end{vmatrix} = 1 \begin{vmatrix} 8 & 8 \\ 2 & 8 \end{vmatrix} = 64 - 16 = 48.$$

(iii) Starting from $\begin{vmatrix} 1 & 0 & 0 \\ 2 & 8 & 8 \\ 3 & 2 & 8 \end{vmatrix}$, subtract col 2 from col 3

$$= \begin{vmatrix} 1 & 0 & 0 \\ 2 & 8 & 0 \\ 3 & 2 & 6 \end{vmatrix} = 1 \begin{vmatrix} 8 & 0 \\ 2 & 6 \end{vmatrix} = 1 \times 8 \times 6 = 48.$$

10.6 $\begin{vmatrix} \alpha - \varepsilon & \beta & 0 \\ \beta & \alpha - \varepsilon & \beta \\ 0 & \beta & \alpha - \varepsilon \end{vmatrix} = 0$

(a) $\begin{vmatrix} \dfrac{\alpha - \varepsilon}{\beta} & 1 & 0 \\ 1 & \dfrac{\alpha - \varepsilon}{\beta} & 1 \\ 0 & 1 & \dfrac{\alpha - \varepsilon}{\beta} \end{vmatrix} \beta^3 = 0$

$\Rightarrow \begin{vmatrix} \dfrac{\alpha - \varepsilon}{\beta} & 1 & 0 \\ 1 & \dfrac{\alpha - \varepsilon}{\beta} & 1 \\ 0 & 1 & \dfrac{\alpha - \varepsilon}{\beta} \end{vmatrix} = 0$

$$(b) \begin{vmatrix} x & 1 & 0 \\ 1 & x & 1 \\ 0 & 1 & x \end{vmatrix} = x \begin{vmatrix} x & 1 \\ 1 & x \end{vmatrix} - 1 \begin{vmatrix} 1 & 1 \\ 0 & x \end{vmatrix}$$

$$= x(x^2 - 1) - (x - 0)$$

$$= x^3 - x - x = x^3 - 2x$$

$$\Rightarrow \begin{vmatrix} x & 1 & 0 \\ 1 & x & 1 \\ 0 & 1 & x \end{vmatrix} = x^3 - 2x = 0.$$

(c) $x^3 - 2x = 0 \Rightarrow x(x^2 - 2) = 0$, when $x = 0, \pm\sqrt{2}$.

(d) $x = (\alpha - \varepsilon)/\beta \Rightarrow \varepsilon = \alpha - \beta x \Rightarrow \varepsilon = \alpha, \alpha \mp \sqrt{2}\beta$.

Chapter 11

11.1 $b = \begin{pmatrix} 1 & 1 & 1 \\ 2 & -2 & 2 \end{pmatrix}$, $c = \begin{pmatrix} 3 & -1 \\ 1 & -3 \end{pmatrix}$, $d = \begin{pmatrix} 1 \\ 0 \end{pmatrix}$ and $e = (0 \ -i \ 1 \ i)$

(a) b, rectangular; c, square; d, column; e, row.
(b) $b_{11} = 1, b_{12} = 1, b_{13} = 1, b_{21} = 2, b_{22} = -2, b_{23} = 2$
    $c_{11} = 3, c_{12} = -1, c_{21} = 1, c_{22} = -3$.
    $d_{11} = 1, d_{21} = 0$.
    $e_{11} = 0, e_{12} = -i, e_{13} = 1, e_{14} = i$.
(c) b, 2×3; c, 2×2; d, 2×1; e, 1×4.

11.2 (a) $2B = 2 \begin{pmatrix} 4 & 5 \\ 1 & 6 \\ -4 & 3 \end{pmatrix} = \begin{pmatrix} 8 & 10 \\ 2 & 12 \\ -8 & 6 \end{pmatrix}$

(b) $2C = 2 \begin{pmatrix} 2 & \frac{5}{2} \\ \frac{1}{2} & 3 \\ -2 & \frac{3}{2} \end{pmatrix} = \begin{pmatrix} 4 & 5 \\ 1 & 6 \\ -4 & 3 \end{pmatrix}$.

11.3 (a) $\mathbf{A}+\mathbf{B}=\begin{pmatrix} 1 & i \\ -i & 1 \end{pmatrix}+\begin{pmatrix} 1 & -i \\ i & 1 \end{pmatrix}=\begin{pmatrix} 2 & 0 \\ 0 & 2 \end{pmatrix}=2\mathbf{D}.$

(b) $\mathbf{A}-\mathbf{B}=\begin{pmatrix} 1 & i \\ -i & 1 \end{pmatrix}-\begin{pmatrix} 1 & -i \\ i & 1 \end{pmatrix}=\begin{pmatrix} 0 & 2i \\ -2i & 0 \end{pmatrix}=2i\mathbf{C}.$

(c) $\mathbf{R}+\mathbf{S}=\begin{pmatrix} \cos\theta & \sin\theta \\ -\sin\theta & \cos\theta \end{pmatrix}+\begin{pmatrix} \cos\theta & -\sin\theta \\ \sin\theta & \cos\theta \end{pmatrix}$

$\qquad=\begin{pmatrix} 2\cos\theta & 0 \\ 0 & 2\cos\theta \end{pmatrix}=2\cos\theta\mathbf{D}.$

(d) $\mathbf{R}-\mathbf{S}=\begin{pmatrix} \cos\theta & \sin\theta \\ -\sin\theta & \cos\theta \end{pmatrix}-\begin{pmatrix} \cos\theta & -\sin\theta \\ \sin\theta & \cos\theta \end{pmatrix}$

$\qquad=\begin{pmatrix} 0 & 2\sin\theta \\ -2\sin\theta & 0 \end{pmatrix}=2\sin\theta\mathbf{C}.$

11.4 $\mathbf{AB}=\begin{pmatrix} 1 & 2 \\ 2 & 1 \end{pmatrix}\begin{pmatrix} 1 & -1 \\ -1 & 2 \end{pmatrix}=\begin{pmatrix} -1 & 3 \\ 1 & 0 \end{pmatrix}, 2\times 2;$

$\mathbf{BA}=\begin{pmatrix} 1 & -1 \\ -1 & 2 \end{pmatrix}\begin{pmatrix} 1 & 2 \\ 2 & 1 \end{pmatrix}=\begin{pmatrix} -1 & 1 \\ 3 & 0 \end{pmatrix}, 2\times 2$

$\mathbf{AC}=\begin{pmatrix} 1 & 2 \\ 2 & 1 \end{pmatrix}\begin{pmatrix} -1 & 1 \\ -1 & 1 \end{pmatrix}=\begin{pmatrix} -3 & 3 \\ -3 & 3 \end{pmatrix}, 2\times 2;$

$\mathbf{BC}=\begin{pmatrix} 1 & -1 \\ -1 & 2 \end{pmatrix}\begin{pmatrix} -1 & 1 \\ -1 & 1 \end{pmatrix}=\begin{pmatrix} 0 & 0 \\ -1 & 1 \end{pmatrix}, 2\times 2.$

$\mathbf{DE}=\begin{pmatrix} 1 & 2 \end{pmatrix}\begin{pmatrix} 3 \\ -1 \end{pmatrix}=1, 1\times 1;$

$\mathbf{ED}=\begin{pmatrix} 3 \\ -1 \end{pmatrix}\begin{pmatrix} 1 & 2 \end{pmatrix}=\begin{pmatrix} 3 & 6 \\ -1 & -2 \end{pmatrix}, 2\times 2.$

$$\mathbf{DA} = (1 \quad 2)\begin{pmatrix} 1 & 2 \\ 2 & 1 \end{pmatrix} = (5 \quad 4), 1 \times 2; \quad \mathbf{AD} \text{ not defined.}$$

EA not defined.

$$\mathbf{AE} = \begin{pmatrix} 1 & 2 \\ 2 & 1 \end{pmatrix}\begin{pmatrix} 3 \\ -1 \end{pmatrix} = \begin{pmatrix} 1 \\ 5 \end{pmatrix}, 2 \times 1.$$

$$\Rightarrow \mathbf{AB} - \mathbf{BA} = \begin{pmatrix} -1 & 3 \\ 1 & 0 \end{pmatrix} - \begin{pmatrix} -1 & 1 \\ 3 & 0 \end{pmatrix} = \begin{pmatrix} 0 & 2 \\ -2 & 0 \end{pmatrix}, 2 \times 2.$$

$$(\mathbf{AB})\mathbf{C} = \begin{pmatrix} -1 & 3 \\ 1 & 0 \end{pmatrix}\begin{pmatrix} -1 & 1 \\ -1 & 1 \end{pmatrix} = \begin{pmatrix} -2 & 2 \\ -1 & 1 \end{pmatrix}, 2 \times 2;$$

$$\mathbf{A}(\mathbf{BC}) = \begin{pmatrix} 1 & 2 \\ 2 & 1 \end{pmatrix}\begin{pmatrix} 0 & 0 \\ -1 & 1 \end{pmatrix} = \begin{pmatrix} -2 & 2 \\ -1 & 1 \end{pmatrix}, 2 \times 2.$$

$$\mathbf{A}(\mathbf{B} + \mathbf{C}) = \begin{pmatrix} 1 & 2 \\ 2 & 1 \end{pmatrix}\begin{pmatrix} 0 & 0 \\ -2 & 3 \end{pmatrix} = \begin{pmatrix} -4 & 6 \\ -2 & 3 \end{pmatrix}, 2 \times 2.$$

$$\mathbf{AB} + \mathbf{AC} = \begin{pmatrix} -1 & 3 \\ 1 & 0 \end{pmatrix} + \begin{pmatrix} -3 & 3 \\ -3 & 3 \end{pmatrix} = \begin{pmatrix} -4 & 6 \\ -2 & 3 \end{pmatrix}, 2 \times 2.$$

11.5 (a) Reflection in the line $y = x$ will result in the x and y values interchanging. We can represent this coordinate transformation as:

$$\begin{pmatrix} x' \\ y' \end{pmatrix} = \begin{pmatrix} y \\ x \end{pmatrix} = \begin{pmatrix} d_{11} & d_{12} \\ d_{21} & d_{22} \end{pmatrix}\begin{pmatrix} x \\ y \end{pmatrix}.$$
$$\mathbf{r}' \quad = \qquad\qquad \mathbf{D} \qquad\quad \mathbf{r}$$

Multiplying out gives:

$$d_{11}x + d_{12}y = y \Rightarrow d_{11} = 0, d_{12} = 1$$

$$d_{21}x + d_{22}y = x \Rightarrow d_{21} = 1, d_{22} = 0$$

$$\Rightarrow \mathbf{D} = \begin{pmatrix} 0 & 1 \\ 1 & 0 \end{pmatrix}.$$

(b)   (i) $\mathbf{E} = \mathbf{CD} = \begin{pmatrix} -1 & 0 \\ 0 & 1 \end{pmatrix} \begin{pmatrix} 0 & 1 \\ 1 & 0 \end{pmatrix} = \begin{pmatrix} 0 & -1 \\ 1 & 0 \end{pmatrix}.$

(ii) $\mathbf{F} = \mathbf{DC} = \begin{pmatrix} 0 & 1 \\ 1 & 0 \end{pmatrix} \begin{pmatrix} -1 & 0 \\ 0 & 1 \end{pmatrix} = \begin{pmatrix} 0 & 1 \\ -1 & 0 \end{pmatrix}.$

11.6 (a) $\mathbf{A}^{\mathrm{T}} = \begin{pmatrix} \cos\theta & \sin\theta \\ -\sin\theta & \cos\theta \end{pmatrix}$

(b) $\mathbf{C}^{\mathrm{T}} = \begin{pmatrix} -1 & -1 \\ 1 & 1 \end{pmatrix}$

(c) $\mathbf{D}^{\mathrm{T}} = \begin{pmatrix} 1 & 1 \\ 3 & 2 \\ 4 & 1 \end{pmatrix}.$

11.7 (a) $\mathbf{X}_{nm}\mathbf{X}_{mn}{}^{\mathrm{T}}$ yields $n \times n$; $\mathbf{X}_{mn}{}^{\mathrm{T}}\mathbf{X}_{nm}$ yields $m \times m$.

(b) $\mathbf{BB}^{\mathrm{T}} = \begin{pmatrix} 1 & 1 & 2 \\ 1 & 2 & 1 \end{pmatrix} \begin{pmatrix} 1 & 1 \\ 1 & 2 \\ 2 & 1 \end{pmatrix} = \begin{pmatrix} 6 & 5 \\ 5 & 6 \end{pmatrix};$

$\mathbf{B}^{\mathrm{T}}\mathbf{B} = \begin{pmatrix} 1 & 1 \\ 1 & 2 \\ 2 & 1 \end{pmatrix} \begin{pmatrix} 1 & 1 & 2 \\ 1 & 2 & 1 \end{pmatrix} = \begin{pmatrix} 2 & 3 & 3 \\ 3 & 5 & 4 \\ 3 & 4 & 5 \end{pmatrix}.$

11.8 $\mathbf{A} = \begin{pmatrix} 1+i & i \\ -i & 1 \end{pmatrix};$ $\mathbf{A}^{*} = \begin{pmatrix} 1-i & -i \\ i & 1 \end{pmatrix};$

$\mathbf{A}^{\dagger} = \begin{pmatrix} 1-i & i \\ -i & 1 \end{pmatrix}.$

11.9 (a) $\mathbf{AB} = \begin{pmatrix} 1 & 1-i \\ 1+i & -1 \end{pmatrix} \begin{pmatrix} 1 & 1+i \\ 1+i & 0 \end{pmatrix} = \begin{pmatrix} 3 & 1+i \\ 0 & 2i \end{pmatrix};$

$(\mathbf{AB})^{*} = \begin{pmatrix} 3 & 1-i \\ 0 & -2i \end{pmatrix};$

$$\mathbf{A}^*\mathbf{B}^* = \begin{pmatrix} 1 & 1+i \\ 1-i & -1 \end{pmatrix} \begin{pmatrix} 1 & 1-i \\ 1-i & 0 \end{pmatrix} = \begin{pmatrix} 3 & 1-i \\ 0 & -2i \end{pmatrix}.$$

(b) $(\mathbf{AB})^\dagger = \begin{pmatrix} 3 & 0 \\ 1-i & -2i \end{pmatrix};$

$$\mathbf{B}^\dagger\mathbf{A}^\dagger = \begin{pmatrix} 1 & 1-i \\ 1-i & 0 \end{pmatrix} \begin{pmatrix} 1 & 1-i \\ 1+i & -1 \end{pmatrix} = \begin{pmatrix} 3 & 0 \\ 1-i & -2i \end{pmatrix}.$$

11.10 (a) $\mathbf{AB} = \begin{pmatrix} 1 & -1 \\ 0 & 3 \end{pmatrix} \begin{pmatrix} 0 & 1 \\ 1 & -2 \end{pmatrix} = \begin{pmatrix} -1 & 3 \\ 3 & -6 \end{pmatrix}$

$\Rightarrow \mathrm{tr}(\mathbf{AB}) = -1-6 = -7.$

$$\mathbf{BA} = \begin{pmatrix} 0 & 1 \\ 1 & -2 \end{pmatrix} \begin{pmatrix} 1 & -1 \\ 0 & 3 \end{pmatrix} = \begin{pmatrix} 0 & 3 \\ 1 & -7 \end{pmatrix}$$

$\Rightarrow \mathrm{tr}(\mathbf{BA}) = 0-7 = -7.$

(b) $\mathbf{ABC} = \begin{pmatrix} -1 & 3 \\ 3 & -6 \end{pmatrix} \begin{pmatrix} -1 & 1 \\ 1 & 0 \end{pmatrix} = \begin{pmatrix} 4 & -1 \\ -9 & 3 \end{pmatrix}$

$\Rightarrow \mathrm{tr}(\mathbf{ABC}) = 4+3 = 7.$

$$\mathbf{CAB} = \begin{pmatrix} -1 & 1 \\ 1 & 0 \end{pmatrix} \begin{pmatrix} -1 & 3 \\ 3 & -6 \end{pmatrix} = \begin{pmatrix} 4 & -9 \\ -1 & 3 \end{pmatrix}$$

$\Rightarrow \mathrm{tr}(\mathbf{CAB}) = 4+3 = 7.$

$$\mathbf{BCA} = \begin{pmatrix} 0 & 1 \\ 1 & -2 \end{pmatrix} \begin{pmatrix} -1 & 4 \\ 1 & -1 \end{pmatrix} = \begin{pmatrix} 1 & -1 \\ -3 & 6 \end{pmatrix}$$

$\Rightarrow \mathrm{tr}(\mathbf{BCA}) = 1+6 = 7.$

(c) $\mathbf{D}^\mathrm{T}\mathbf{D} = \begin{pmatrix} 1 & 1 \\ -1 & -2 \\ 0 & 0 \end{pmatrix} \begin{pmatrix} 1 & -1 & 0 \\ 1 & -2 & 0 \end{pmatrix} = \begin{pmatrix} 2 & -3 & 0 \\ -3 & 5 & 0 \\ 0 & 0 & 0 \end{pmatrix}$

$\Rightarrow \mathbf{tr}(\mathbf{D}^\mathrm{T}\mathbf{D}) = 2+5+0 = 7.$

$$\mathbf{DD}^{\mathrm{T}} = \begin{pmatrix} 1 & -1 & 0 \\ 1 & -2 & 0 \end{pmatrix} \begin{pmatrix} 1 & 1 \\ -1 & -2 \\ 0 & 0 \end{pmatrix} = \begin{pmatrix} 2 & 3 \\ 3 & 5 \end{pmatrix}$$

$$\Rightarrow \mathrm{tr}\left(\mathbf{DD}^{\mathrm{T}}\right) = 2 + 5 = 7.$$

11.11 (a) $\begin{pmatrix} 0 & 0 & 0 \\ 0 & 0 & 0 \end{pmatrix} \begin{pmatrix} 1 & 3 \\ 2 & 2 \\ 0 & 1 \end{pmatrix} = \begin{pmatrix} 0 & 0 \\ 0 & 0 \end{pmatrix}.$

  (b) $\begin{pmatrix} 1 & 2 & 3 \\ 4 & 5 & 6 \end{pmatrix} \begin{pmatrix} 0 & 0 & 0 \\ 0 & 0 & 0 \end{pmatrix}$ is undefined.

  (c) $\begin{pmatrix} 1 & 0 & 0 \\ 0 & 1 & 0 \\ 0 & 0 & 1 \end{pmatrix} \begin{pmatrix} 1 & 3 \\ 2 & 2 \\ 0 & 1 \end{pmatrix} = \begin{pmatrix} 1 & 3 \\ 2 & 2 \\ 0 & 1 \end{pmatrix}.$

  (d) $\begin{pmatrix} 1 & 2 & 3 \\ 4 & 5 & 6 \end{pmatrix} \begin{pmatrix} 1 & 0 & 0 \\ 0 & 1 & 0 \\ 0 & 0 & 1 \end{pmatrix} = \begin{pmatrix} 1 & 2 & 3 \\ 4 & 5 & 6 \end{pmatrix}.$

  (e) $\begin{pmatrix} 1 & 3 \\ 2 & 2 \\ 0 & 1 \end{pmatrix} \begin{pmatrix} 1 & 0 & 0 \\ 0 & 1 & 0 \\ 0 & 0 & 1 \end{pmatrix}$ is undefined.

11.12 (a)(i) $\det \mathbf{A} = \begin{vmatrix} \cos\theta & -\sin\theta \\ \sin\theta & \cos\theta \end{vmatrix} = \cos^2\theta + \sin^2\theta = 1;$

  $$\mathbf{B} = \begin{pmatrix} \cos\theta & -\sin\theta \\ \sin\theta & \cos\theta \end{pmatrix}$$

  (ii) $\mathbf{B}^{\mathrm{T}}\mathbf{A} = \begin{pmatrix} \cos\theta & \sin\theta \\ -\sin\theta & \cos\theta \end{pmatrix} \begin{pmatrix} \cos\theta & -\sin\theta \\ \sin\theta & \cos\theta \end{pmatrix} = \begin{pmatrix} 1 & 0 \\ 0 & 1 \end{pmatrix}.$

  $$\mathbf{E}_2 \det \mathbf{A} = \begin{pmatrix} 1 & 0 \\ 0 & 1 \end{pmatrix} \times 1 = \begin{pmatrix} 1 & 0 \\ 0 & 1 \end{pmatrix}$$

  Thus $\mathbf{E}_2 \det \mathbf{A} = \mathbf{B}^{\mathrm{T}}\mathbf{A}.$

(b) $G_{11} = \begin{pmatrix} -1 & 2 \\ 1 & 1 \end{pmatrix} = -3; \quad G_{12} = -\begin{pmatrix} 2 & 2 \\ 1 & 1 \end{pmatrix} = 0;$

$G_{13} = \begin{pmatrix} 2 & -1 \\ 1 & 1 \end{pmatrix} = 3;$

$G_{21} = -\begin{pmatrix} -1 & 1 \\ 1 & 1 \end{pmatrix} = 2; \quad G_{22} = \begin{pmatrix} 0 & 1 \\ 1 & 1 \end{pmatrix} = -1;$

$G_{23} = -\begin{pmatrix} 0 & -1 \\ 1 & 1 \end{pmatrix} = -1;$

$G_{31} = \begin{pmatrix} -1 & 1 \\ -1 & 2 \end{pmatrix} = -1; \quad G_{32} = -\begin{pmatrix} 0 & 1 \\ 2 & 2 \end{pmatrix} = 2;$

$G_{33} = \begin{pmatrix} 0 & -1 \\ 2 & -1 \end{pmatrix} = 2;$

$$\mathbf{H} = \begin{pmatrix} -3 & 0 & 3 \\ 2 & -1 & -1 \\ -1 & 2 & 2 \end{pmatrix}.$$

$$\Rightarrow \mathbf{H}^T \mathbf{G} = \begin{pmatrix} -3 & 2 & -1 \\ 0 & -1 & 2 \\ 3 & -1 & 2 \end{pmatrix} \begin{pmatrix} 0 & -1 & 1 \\ 2 & -1 & 2 \\ 1 & 1 & 1 \end{pmatrix} = \begin{pmatrix} 3 & 0 & 0 \\ 0 & 3 & 0 \\ 0 & 0 & 3 \end{pmatrix}.$$

$$\det \mathbf{G} = 1 \begin{vmatrix} 2 & 2 \\ 1 & 1 \end{vmatrix} + 1 \begin{vmatrix} 2 & -1 \\ 1 & 1 \end{vmatrix} = 0 + 3 = 3;$$

Therefore, $\mathbf{E}_3 \det \mathbf{G} = \begin{pmatrix} 1 & 0 & 0 \\ 0 & 1 & 0 \\ 0 & 0 & 1 \end{pmatrix} \times 3 = \begin{pmatrix} 3 & 0 & 0 \\ 0 & 3 & 0 \\ 0 & 0 & 3 \end{pmatrix}.$

11.13 (a) $\mathbf{A} = \begin{pmatrix} \frac{1}{\sqrt{2}} & k \\ \frac{1}{\sqrt{2}} & -\frac{1}{\sqrt{2}} \end{pmatrix}$;

$\det \mathbf{A} = -\frac{1}{2} - \frac{k}{\sqrt{2}} = \pm 1.$

$\Rightarrow -\frac{k}{\sqrt{2}} = \pm 1 + \frac{1}{2} \Rightarrow \frac{k}{\sqrt{2}} = \mp 1 - \frac{1}{2} = -\frac{3}{2}, \frac{1}{2}$

$\Rightarrow k = -\frac{3\sqrt{2}}{2}, \frac{\sqrt{2}}{2} = -\frac{3}{\sqrt{2}}, \frac{1}{\sqrt{2}}.$

Check : $\mathbf{A}^T \mathbf{A} = \mathbf{A} \mathbf{A}^T = \mathbf{E}_n \Rightarrow$

$$\mathbf{A}^T \mathbf{A} = \begin{pmatrix} \frac{1}{\sqrt{2}} & \frac{1}{\sqrt{2}} \\ k & -\frac{1}{\sqrt{2}} \end{pmatrix} \begin{pmatrix} \frac{1}{\sqrt{2}} & k \\ \frac{1}{\sqrt{2}} & -\frac{1}{\sqrt{2}} \end{pmatrix}$$

$$= \begin{pmatrix} 1 & \frac{k}{\sqrt{2}} - \frac{1}{2} \\ \frac{k}{\sqrt{2}} - \frac{1}{2} & k^2 + \frac{1}{2} \end{pmatrix}$$

$$\mathbf{A} \mathbf{A}^T = \begin{pmatrix} \frac{1}{\sqrt{2}} & k \\ \frac{1}{\sqrt{2}} & -\frac{1}{\sqrt{2}} \end{pmatrix} \begin{pmatrix} \frac{1}{\sqrt{2}} & \frac{1}{\sqrt{2}} \\ k & -\frac{1}{\sqrt{2}} \end{pmatrix}$$

$$= \begin{pmatrix} \frac{1}{2} + k^2 & \frac{1}{2} - \frac{k}{\sqrt{2}} \\ \frac{1}{2} - \frac{k}{\sqrt{2}} & 1 \end{pmatrix}$$

For $k = -\frac{3}{\sqrt{2}}$, $\mathbf{A}^T \mathbf{A} = \begin{pmatrix} 1 & -2 \\ -2 & 5 \end{pmatrix}$;

$$\mathbf{A} \mathbf{A}^T = \begin{pmatrix} 5 & 2 \\ 2 & 1 \end{pmatrix} \Rightarrow \mathbf{A}^T \mathbf{A} \neq \mathbf{A} \mathbf{A}^T \neq \mathbf{E}_n$$

For $k = \frac{1}{\sqrt{2}}$, $\mathbf{A}^T \mathbf{A} = \begin{pmatrix} 1 & 0 \\ 0 & 1 \end{pmatrix}$;

$$\mathbf{A} \mathbf{A}^T = \begin{pmatrix} 1 & 0 \\ 0 & 1 \end{pmatrix} \Rightarrow \mathbf{A}^T \mathbf{A} = \mathbf{A} \mathbf{A}^T = \mathbf{E}_n.$$

The only valid solution is $k = \frac{1}{\sqrt{2}}$.

(b) $\mathbf{R} = \begin{pmatrix} \cos\theta & \sin\theta & 0 \\ \sin\theta & \cos\theta & 0 \\ 0 & 0 & 1 \end{pmatrix}$;

$\det \mathbf{R} = \cos^2\theta - \sin^2\theta = \pm 1$

But $\cos 2\theta = \cos^2\theta - \sin^2\theta$ and so $\cos 2\theta = \pm 1$

$\Rightarrow 2\theta = \cos^{-1}(\pm 1) = 0, \pm n\pi$

$\Rightarrow \theta = 0, \pm \dfrac{n\pi}{2}, n = 1,2,3\ldots$

$\Rightarrow \mathbf{R} = \begin{pmatrix} 1 & 0 & 0 \\ 0 & 1 & 0 \\ 0 & 0 & 1 \end{pmatrix}$, for $\theta = 0$;

$\mathbf{R} = \begin{pmatrix} 0 & 1 & 0 \\ 1 & 0 & 0 \\ 0 & 0 & 1 \end{pmatrix}$, for $\theta = \dfrac{\pi}{2}$;

$\mathbf{R} = \begin{pmatrix} -1 & 0 & 0 \\ 0 & -1 & 0 \\ 0 & 0 & 1 \end{pmatrix}$, for $\theta = \pi$;

$\mathbf{R} = \begin{pmatrix} 0 & -1 & 0 \\ -1 & 0 & 0 \\ 0 & 0 & 1 \end{pmatrix}$, for $\theta = \dfrac{3\pi}{2}, -\dfrac{\pi}{2}$;

In each case $\mathbf{R}^{\mathrm{T}} = \mathbf{R}$

$$\mathbf{RR}^T = \mathbf{R}^T\mathbf{R} = \begin{pmatrix} 1 & 0 & 0 \\ 0 & 1 & 0 \\ 0 & 0 & 1 \end{pmatrix} \begin{pmatrix} 1 & 0 & 0 \\ 0 & 1 & 0 \\ 0 & 0 & 1 \end{pmatrix}$$

$$= \begin{pmatrix} 1 & 0 & 0 \\ 0 & 1 & 0 \\ 0 & 0 & 1 \end{pmatrix} = \mathbf{E}_3$$

$$\mathbf{RR}^T = \mathbf{R}^T\mathbf{R} = \begin{pmatrix} 0 & 1 & 0 \\ 1 & 0 & 0 \\ 0 & 0 & 1 \end{pmatrix} \begin{pmatrix} 0 & 1 & 0 \\ 1 & 0 & 0 \\ 0 & 0 & 1 \end{pmatrix}$$

$$= \begin{pmatrix} 1 & 0 & 0 \\ 0 & 1 & 0 \\ 0 & 0 & 1 \end{pmatrix} = \mathbf{E}_3$$

$$\mathbf{RR}^T = \mathbf{R}^T\mathbf{R} = \begin{pmatrix} -1 & 0 & 0 \\ 0 & -1 & 0 \\ 0 & 0 & 1 \end{pmatrix} \begin{pmatrix} -1 & 0 & 0 \\ 0 & -1 & 0 \\ 0 & 0 & 1 \end{pmatrix}$$

$$= \begin{pmatrix} 1 & 0 & 0 \\ 0 & 1 & 0 \\ 0 & 0 & 1 \end{pmatrix} = \mathbf{E}_3$$

$$\mathbf{RR}^T = \mathbf{R}^T\mathbf{R} = \begin{pmatrix} 0 & -1 & 0 \\ -1 & 0 & 0 \\ 0 & 0 & 1 \end{pmatrix} \begin{pmatrix} 0 & -1 & 0 \\ -1 & 0 & 0 \\ 0 & 0 & 1 \end{pmatrix}$$

$$= \begin{pmatrix} 1 & 0 & 0 \\ 0 & 1 & 0 \\ 0 & 0 & 1 \end{pmatrix} = \mathbf{E}_3$$

For $\theta = 0, \pm \dfrac{n\pi}{2}, n = 1,2,3\ldots, \mathbf{R}^T\mathbf{R} = \mathbf{RR}^T = \mathbf{E}_n$.

(c) det $\mathbf{A} = 1+k \Rightarrow 1+k = \pm 1 \Rightarrow k = 0, -2$

For $k = 0$,

$$\mathbf{A}\mathbf{A}^{\mathrm{T}} = \begin{pmatrix} 1 & 0 \\ -1 & 1 \end{pmatrix} \begin{pmatrix} 1 & -1 \\ 0 & 1 \end{pmatrix} = \begin{pmatrix} 1 & -1 \\ -1 & 2 \end{pmatrix}$$

$$\mathbf{A}^{\mathrm{T}}\mathbf{A} = \begin{pmatrix} 1 & -1 \\ 0 & 1 \end{pmatrix} \begin{pmatrix} 1 & 0 \\ -1 & 1 \end{pmatrix} = \begin{pmatrix} 2 & -1 \\ -1 & 1 \end{pmatrix}$$

For $k = -2$

$$\mathbf{A}\mathbf{A}^{\mathrm{T}} = \begin{pmatrix} 1 & -2 \\ -1 & 1 \end{pmatrix} \begin{pmatrix} 1 & -1 \\ -2 & 1 \end{pmatrix} = \begin{pmatrix} 5 & -3 \\ -3 & 2 \end{pmatrix}$$

$$\mathbf{A}^{\mathrm{T}}\mathbf{A} = \begin{pmatrix} 1 & -1 \\ -2 & 1 \end{pmatrix} \begin{pmatrix} 1 & -2 \\ -1 & 1 \end{pmatrix} = \begin{pmatrix} 2 & -3 \\ -3 & 5 \end{pmatrix}$$

$\therefore$ Not orthogonal.

11.14 (a) $\mathbf{A} = \begin{pmatrix} 0 & 3+i \\ 3-i & 1 \end{pmatrix}$, $\mathbf{A}^{\dagger} = \begin{pmatrix} 0 & 3+i \\ 3-i & 1 \end{pmatrix}$,

$\therefore \mathbf{A} = \mathbf{A}^{\dagger}$.

(b) $\mathbf{A}\mathbf{x} = \begin{pmatrix} 0 & 3+i \\ 3-i & 1 \end{pmatrix} \begin{pmatrix} 1 \\ i \end{pmatrix} = \begin{pmatrix} 3i-1 \\ 3 \end{pmatrix}$;

$\mathbf{x}^{\dagger}\mathbf{A}\mathbf{x} = (1-i) \begin{pmatrix} 3i-1 \\ 3 \end{pmatrix} = 3i - 1 - 3i = -1$.

11.15 (a) $\mathbf{A} = \begin{pmatrix} 1 & i \\ -i & 0 \end{pmatrix}$; $\mathbf{A}^{\dagger} = \begin{pmatrix} 1 & i \\ -i & 0 \end{pmatrix}$ $\therefore$ Hermitian.

$$\mathbf{A}\mathbf{A}^{\dagger} = \begin{pmatrix} 1 & i \\ -i & 0 \end{pmatrix} \begin{pmatrix} 1 & i \\ -i & 0 \end{pmatrix} = \begin{pmatrix} 2 & i \\ -i & 1 \end{pmatrix} \therefore \text{Not unitary.}$$

(b) $\mathbf{B} = \dfrac{1}{\sqrt{2}} \begin{pmatrix} 1 & i \\ -i & -1 \end{pmatrix}$; $\mathbf{B}^\dagger = \dfrac{1}{\sqrt{2}} \begin{pmatrix} 1 & i \\ -i & -1 \end{pmatrix}$ $\therefore$ Hermitian.

$$\mathbf{BB}^\dagger = \frac{1}{2} \begin{pmatrix} 1 & i \\ -i & -1 \end{pmatrix} \begin{pmatrix} 1 & i \\ -i & -1 \end{pmatrix} = \frac{1}{2} \begin{pmatrix} 2 & 0 \\ 0 & 2 \end{pmatrix} = \begin{pmatrix} 1 & 0 \\ 0 & 1 \end{pmatrix}$$

$\therefore$ Unitary.

(c) $\mathbf{C} = \begin{pmatrix} 0 & -1 \\ 1 & 0 \end{pmatrix}$; $\mathbf{C}^T = \begin{pmatrix} 0 & 1 \\ -1 & 0 \end{pmatrix}$ $\therefore$ Not symmetric.

$$\mathbf{CC}^T = \begin{pmatrix} 0 & -1 \\ 1 & 0 \end{pmatrix} \begin{pmatrix} 0 & 1 \\ -1 & 0 \end{pmatrix} = \begin{pmatrix} 1 & 0 \\ 0 & 1 \end{pmatrix}$$ $\therefore$ Orthogonal.

(d) $\mathbf{D} = \begin{pmatrix} 1 & -1 \\ -1 & 0 \end{pmatrix}$; $\mathbf{D}^T = \begin{pmatrix} 1 & -1 \\ -1 & 0 \end{pmatrix}$ $\therefore$ Symmetric.

$$\mathbf{DD}^T = \begin{pmatrix} 1 & -1 \\ -1 & 0 \end{pmatrix} \begin{pmatrix} 1 & -1 \\ -1 & 0 \end{pmatrix} = \begin{pmatrix} 2 & -1 \\ -1 & 1 \end{pmatrix}$$

$\therefore$ Not othogonal.

11.16 $\det \mathbf{A} = \begin{vmatrix} 1 & -1 & 1 \\ -1 & -1 & 1 \\ 1 & 1 & 1 \end{vmatrix}$

$\Rightarrow$ Add column (col) 2 to col 3

$\Rightarrow \begin{vmatrix} 1 & -1 & 0 \\ -1 & -1 & 0 \\ 1 & 1 & 2 \end{vmatrix}$ and then add col 1 to col 2.

$\Rightarrow \det \mathbf{A} = \begin{vmatrix} 1 & 0 & 0 \\ -1 & -2 & 0 \\ 1 & 2 & 2 \end{vmatrix} = 1 \begin{vmatrix} -2 & 0 \\ 2 & 2 \end{vmatrix} = -4.$

The matrix of cofactors of $\mathbf{A}$ is $\mathbf{B} = \begin{pmatrix} -2 & 2 & 0 \\ 2 & 0 & -2 \\ 0 & -2 & -2 \end{pmatrix};$

$$\mathbf{B}^{\mathrm{T}} = \begin{pmatrix} -2 & 2 & 0 \\ 2 & 0 & -2 \\ 0 & -2 & -2 \end{pmatrix}$$

Therefore $\mathbf{A}^{-1} = \dfrac{1}{\det \mathbf{A}} \mathbf{B}^{\mathrm{T}} = -\dfrac{1}{4} \begin{pmatrix} -2 & 2 & 0 \\ 2 & 0 & -2 \\ 0 & -2 & -2 \end{pmatrix}$

$$= \begin{pmatrix} \frac{1}{2} & -\frac{1}{2} & 0 \\ -\frac{1}{2} & 0 & \frac{1}{2} \\ 0 & \frac{1}{2} & \frac{1}{2} \end{pmatrix}.$$

Check: $\mathbf{A}\mathbf{A}^{-1} = \begin{pmatrix} 1 & -1 & 0 \\ -1 & -1 & 1 \\ 1 & 1 & 1 \end{pmatrix} \begin{pmatrix} \frac{1}{2} & -\frac{1}{2} & 0 \\ -\frac{1}{2} & 0 & \frac{1}{2} \\ 0 & \frac{1}{2} & \frac{1}{2} \end{pmatrix}$

$$= \begin{pmatrix} 1 & 0 & 0 \\ 0 & 1 & 0 \\ 0 & 0 & 1 \end{pmatrix}.$$

11.17 $x + 2y + 3z = 1$

$8y + 2z = 1$

$-2x + 4y + 2z = 2$

$$\underset{\mathbf{A}}{\begin{pmatrix} 1 & 2 & 3 \\ 0 & 8 & 2 \\ -2 & 4 & 2 \end{pmatrix}} \underset{\mathbf{x}}{\begin{pmatrix} x \\ y \\ z \end{pmatrix}} = \underset{\mathbf{b}}{\begin{pmatrix} 1 \\ 1 \\ 2 \end{pmatrix}}$$

Matrix of cofactors $\mathbf{B} = \begin{pmatrix} 8 & -4 & 16 \\ 8 & 8 & -8 \\ -20 & -2 & 8 \end{pmatrix}$;

$$\mathbf{B}^{\mathrm{T}} = \begin{pmatrix} 8 & 8 & -20 \\ -4 & 8 & -2 \\ 16 & -8 & 8 \end{pmatrix};$$

$$\det \mathbf{A} = \begin{vmatrix} 8 & 2 \\ 4 & 2 \end{vmatrix} - 2 \begin{vmatrix} 0 & 2 \\ -2 & 2 \end{vmatrix} + 3 \begin{vmatrix} 0 & 8 \\ -2 & 4 \end{vmatrix} = 8 - 8 + 48 = 48;$$

$$\mathbf{A}^{-1} = \frac{1}{\det \mathbf{A}} \mathbf{B}^{\mathrm{T}} = \frac{1}{48} \begin{pmatrix} 8 & 8 & -20 \\ -4 & 8 & -2 \\ 16 & -8 & 8 \end{pmatrix}$$

$$= \begin{pmatrix} \frac{1}{6} & \frac{1}{6} & -\frac{5}{12} \\ -\frac{1}{12} & \frac{1}{6} & -\frac{1}{24} \\ \frac{1}{3} & -\frac{1}{6} & \frac{1}{6} \end{pmatrix};$$

$$\begin{pmatrix} x \\ y \\ z \end{pmatrix} = \begin{pmatrix} \frac{1}{6} & \frac{1}{6} & -\frac{5}{12} \\ -\frac{1}{12} & \frac{1}{6} & -\frac{1}{24} \\ \frac{1}{3} & -\frac{1}{6} & \frac{1}{6} \end{pmatrix} \begin{pmatrix} 1 \\ 1 \\ 2 \end{pmatrix} = \begin{pmatrix} -\frac{1}{2} \\ 0 \\ \frac{1}{2} \end{pmatrix}$$

$$\Rightarrow x = -\tfrac{1}{2},\ y = 0,\ z = \tfrac{1}{2}.$$

11.18 $\begin{pmatrix} (\alpha - \varepsilon) & \beta & 0 \\ \beta & (\alpha - \varepsilon) & \beta \\ 0 & \beta & (\alpha - \varepsilon) \end{pmatrix} \begin{pmatrix} c_1 \\ c_2 \\ c_3 \end{pmatrix} = \mathbf{0}$

(a) For $\varepsilon = \alpha$, $\begin{pmatrix} 0 & \beta & 0 \\ \beta & 0 & \beta \\ 0 & \beta & 0 \end{pmatrix} \begin{pmatrix} c_1 \\ c_2 \\ c_3 \end{pmatrix} = \begin{pmatrix} c_2\beta \\ c_1\beta + c_3\beta \\ c_2\beta \end{pmatrix} = \begin{pmatrix} 0 \\ 0 \\ 0 \end{pmatrix}$

$\Rightarrow c_2\beta = 0;\ c_1\beta + c_3\beta = 0 \Rightarrow c_2 = 0, c_3 = -c_1$

(b) For $\varepsilon = \alpha + \sqrt{2}\beta$,

$$\begin{pmatrix} -\sqrt{2}\beta & \beta & 0 \\ \beta & -\sqrt{2}\beta & \beta \\ 0 & \beta & -\sqrt{2}\beta \end{pmatrix} \begin{pmatrix} c_1 \\ c_2 \\ c_3 \end{pmatrix}$$

$$= \begin{pmatrix} -\sqrt{2}\beta c_1 + c_2\beta \\ c_1\beta - \sqrt{2}\beta c_2 + c_3\beta \\ c_2\beta - \sqrt{2}\beta c_3 \end{pmatrix} = \begin{pmatrix} 0 \\ 0 \\ 0 \end{pmatrix}$$

$$\Rightarrow -\sqrt{2}\beta c_1 + c_2\beta = 0; \ c_1\beta - \sqrt{2}\beta c_2 + c_3\beta = 0; \ c_2\beta - \sqrt{2}\beta c_3 = 0 \Rightarrow$$

$$\Rightarrow c_2 = \sqrt{2}c_1, \ c_3 = c_1.$$

(c) For $\varepsilon = \alpha - \sqrt{2}\beta$,

$$\begin{pmatrix} \sqrt{2}\beta & \beta & 0 \\ \beta & \sqrt{2}\beta & \beta \\ 0 & \beta & \sqrt{2}\beta \end{pmatrix} \begin{pmatrix} c_1 \\ c_2 \\ c_3 \end{pmatrix} = \begin{pmatrix} \sqrt{2}\beta c_1 + c_2\beta \\ c_1\beta + \sqrt{2}\beta c_2 + c_3\beta \\ c_2\beta + \sqrt{2}\beta c_3 \end{pmatrix} = \begin{pmatrix} 0 \\ 0 \\ 0 \end{pmatrix}$$

$$\Rightarrow \sqrt{2}\beta c_1 + c_2\beta = 0; \ c_1\beta + \sqrt{2}\beta c_2 + c_3\beta = 0; \ c_2\beta + \sqrt{2}\beta c_3 = 0 \Rightarrow$$

$$\Rightarrow c_2 = -\sqrt{2}c_1; \ c_3 = c_1.$$

(d) For $\varepsilon = \alpha$, $\mathbf{c} = c_1 \begin{pmatrix} 1 \\ 0 \\ -1 \end{pmatrix}$; For $\varepsilon = \alpha + \sqrt{2}\beta$,

$$\mathbf{c} = c_1 \begin{pmatrix} 1 \\ \sqrt{2} \\ 1 \end{pmatrix}; \text{ For } \varepsilon = \alpha - \sqrt{2}\beta, \ \mathbf{c} = c_1 \begin{pmatrix} 1 \\ -\sqrt{2} \\ 1 \end{pmatrix}$$

11.19 Identity is 1.

$$\left.\begin{array}{l} -1 \times -1 = 1 \\ -1 \times i = -i \\ -1 \times -i = i \\ i \times i = -1 \\ i \times -i = 1 \\ 1 \times -1 = -1 \end{array}\right\} \text{Product of any two yields another member of the group.}$$

Inverse of 1 is 1.
Inverse of -1 is -1.
Inverse of i is $-i$.
Inverse of $-i$ is i.
Multiplication is associative $(-1 \times i) \times -i = -1 \times (i \times -i) = -1$.
$\therefore$ the set G_1 forms a group.

11.20 (a)

	A	B	C	D
A	A	B	C	D
B	B	A	D	C
C	C	D	B	A
D	D	C	A	B

(b)

	A	B	C	D
A	A	B	C	D
B	B	A	D	C
C	C	D	A	B
D	D	C	B	A

11.21 One three-fold axis of rotation, three two-fold axes of rotation, one mirror plane containing the plane of the molecule and three mirror planes perpendicular to the plane of the molecule.

Chapter 12

12.1 (a)

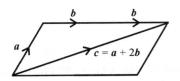

(b)

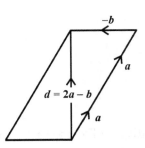

12.2 (a) $3a - 2b = 3(\hat{i} + \hat{j} - 2\hat{k}) - 2(\hat{i} + \hat{k})$
$$= \hat{i} + 3\hat{j} - 8\hat{k}.$$

(b) $-2a - b = -2(\hat{i} + \hat{j} - 2\hat{k}) - (\hat{i} + \hat{k})$
$$= -3\hat{i} - 2\hat{j} + 3\hat{k}.$$

(c) $a + b - c - d = (\hat{i} + \hat{j} - 2\hat{k}) + (\hat{i} + \hat{k})$
$$- (\hat{i} + \hat{j} + \hat{k}) - (\hat{i} - 2\hat{k})$$
$$= 0\hat{i} + 0\hat{j} + 0\hat{k} = 0.$$

(d) $|a - d| = |(\hat{i} + \hat{j} - 2\hat{k}) - (\hat{i} - 2\hat{k})| = |\hat{j}| = 1.$

(e) $\dfrac{a + c}{|a + c|} = \dfrac{(\hat{i} + \hat{j} - 2\hat{k}) + (\hat{i} + \hat{j} + \hat{k})}{|(\hat{i} + \hat{j} - 2\hat{k}) + (\hat{i} + \hat{j} + \hat{k})|}$

$$= \frac{2\hat{i} + 2\hat{j} - \hat{k}}{\sqrt{2^2 + 2^2 + 1^2}} = \frac{2\hat{i} + 2\hat{j} - \hat{k}}{3}$$

$$= \frac{2}{3}\hat{i} + \frac{2}{3}\hat{j} - \frac{1}{3}\hat{k}.$$

(f) $\left|\frac{2}{3}\hat{i}+\frac{2}{3}\hat{j}-\frac{1}{3}\hat{k}\right| = \sqrt{\left(\frac{2}{3}\right)^2 + \left(\frac{2}{3}\right)^2 + \left(\frac{1}{3}\right)^2} = 1.$

(g) $|\boldsymbol{a}| - |\boldsymbol{c}| = \left|(\hat{i}+\hat{j}-2\hat{k})\right| - \left|(\hat{i}+\hat{j}+\hat{k})\right|$
$$= \sqrt{1+1+4} - \sqrt{1+1+1} = \sqrt{6} - \sqrt{3}.$$

12.3 (a) $\hat{i},\ -\hat{j},\ -\hat{i},\ \hat{j}.$

(b) $R\hat{i},\ -R\hat{j},\ -R\hat{i},\ R\hat{j}.$

(c) Shortest : $e.g.\ \boldsymbol{r} = R\hat{i} - R\hat{j};$
$$|\boldsymbol{r}| = \left|R\hat{i} - R\hat{j}\right| = \sqrt{R^2 + R^2} = \sqrt{2}R.$$

Longest : $e.g.\ \boldsymbol{p} = -R\hat{i} + -R\hat{i} = -2R\hat{i};\ |\boldsymbol{p}| = \left|-2R\hat{i}\right| = 2R.$

12.4 (a) $\boldsymbol{a}\cdot\boldsymbol{c} = (2\hat{i}+3\hat{k})\cdot(\hat{i}-2\hat{j}+\hat{k})$
$$= 2\hat{i}\cdot\hat{i} + 3\hat{k}\cdot\hat{k} = 5.$$

(b) $\boldsymbol{a}\cdot(\boldsymbol{b}-2\boldsymbol{c}) = (2\hat{i}+3\hat{k})\cdot(-\hat{i}+5\hat{j}-\hat{k})$
$$= -2\hat{i}\cdot\hat{i} - 3\hat{k}\cdot\hat{k} = -5.$$

(c) $\boldsymbol{a}\cdot(\boldsymbol{b}+\boldsymbol{a}) = (2\hat{i}+3\hat{k})\cdot(3\hat{i}+\hat{j}+4\hat{k})$
$$= 6\hat{i}\cdot\hat{i} + 12\hat{k}\cdot\hat{k} = 18.$$

(d) $\boldsymbol{b}\cdot\boldsymbol{c} = (\hat{i}+\hat{j}+\hat{k})\cdot(\hat{i}-2\hat{j}+\hat{k})$
$$= \hat{i}\cdot\hat{i} - 2\hat{j}\cdot\hat{j} + \hat{k}\cdot\hat{k} = 0.$$

12.5 (a) (i) $\boldsymbol{a}\cdot(\boldsymbol{b}-2\boldsymbol{c}) = -5 = |\boldsymbol{a}|.|(\boldsymbol{b}-2\boldsymbol{c})|\cos\theta$

$$|\boldsymbol{a}| = \sqrt{2^2 + 3^2} = \sqrt{13}.$$

$$|(\boldsymbol{b}-2\boldsymbol{c})| = \sqrt{1^2+5^2+1^2} = \sqrt{27}$$

$$\Rightarrow \frac{-5}{\sqrt{13}\sqrt{27}} = \cos\theta \Rightarrow \theta = 105.48°.$$

(ii) $\boldsymbol{b}\cdot\boldsymbol{c} = 0 = |\boldsymbol{b}|.|\boldsymbol{c}|\cos\theta \Rightarrow \theta = 90°.$

(b) $\boldsymbol{d}\cdot\boldsymbol{e} = (3\hat{\boldsymbol{i}}-2\hat{\boldsymbol{j}}-\hat{\boldsymbol{k}})\cdot(\hat{\boldsymbol{i}}+\lambda\hat{\boldsymbol{j}}+2\hat{\boldsymbol{k}})$

$\qquad = 3\hat{\boldsymbol{i}}\cdot\hat{\boldsymbol{i}}-2\lambda\hat{\boldsymbol{j}}\cdot\hat{\boldsymbol{j}}-2\hat{\boldsymbol{k}}\cdot\hat{\boldsymbol{k}} = 1-2\lambda.$

$|\boldsymbol{d}| = \sqrt{3^2+2^2+1^2} = \sqrt{14}.$

$|\boldsymbol{e}| = \sqrt{1^2+\lambda^2+2^2} = \sqrt{5+\lambda^2}.$

$\therefore 1-2\lambda = \sqrt{14}\sqrt{5+\lambda^2}\cos90° = 0$

$\Rightarrow 1-2\lambda = 0$

$1 = 2\lambda$

$\lambda = \frac{1}{2}.$

12.6 $\boldsymbol{c} = \boldsymbol{a}+\boldsymbol{b} \Rightarrow \boldsymbol{c}\cdot\boldsymbol{c} = (\boldsymbol{a}+\boldsymbol{b})\cdot(\boldsymbol{a}+\boldsymbol{b})$

$$= \underbrace{\boldsymbol{a}\cdot\boldsymbol{a}}_{a^2} + \underbrace{2\boldsymbol{a}\cdot\boldsymbol{b}}_{2ab\,\cos\theta} + \underbrace{\boldsymbol{b}\cdot\boldsymbol{b}}_{b^2}$$

We need to exercise some care here because the quantity $\boldsymbol{a}\cdot\boldsymbol{b}$ will yield an angle $\theta = 180°-C$, which in this example is an acute angle, rather than the obtuse angle required. Consequently, we must substitute $180°-C$ for θ which gives:

$$\therefore \boldsymbol{c}\cdot\boldsymbol{c} = a^2+b^2+2ab\,\cos(180°-C) = a^2+b^2-2ab\cos C.$$

12.7 (a) $\boldsymbol{r}_1 = a\hat{\boldsymbol{i}}-a\hat{\boldsymbol{j}}-a\hat{\boldsymbol{k}};\ \boldsymbol{r}_2 = -a\hat{\boldsymbol{i}}+a\hat{\boldsymbol{j}}-a\hat{\boldsymbol{k}};$

$\qquad \boldsymbol{r}_3 = a\hat{\boldsymbol{i}}+a\hat{\boldsymbol{j}}+a\hat{\boldsymbol{k}};\ \boldsymbol{r}_4 = -a\hat{\boldsymbol{i}}-a\hat{\boldsymbol{j}}+a\hat{\boldsymbol{k}}.$

(b) $|\boldsymbol{r}_3| = \sqrt{3a^2} = \sqrt{3}a \Rightarrow R = \sqrt{3}a \Rightarrow a = \dfrac{R}{\sqrt{3}}.$

(c) $r_3 - r_2 = (a\hat{i} + a\hat{j} + a\hat{k}) - (-a\hat{i} + a\hat{j} - a\hat{k}) = 2a\hat{i} + 2a\hat{k}$;

$$|r_3 - r_2| = \sqrt{(2a)^2 + (2a)^2} = \sqrt{8a^2} = \sqrt{8}a = \frac{\sqrt{8}R}{\sqrt{3}} = \frac{2\sqrt{2}R}{\sqrt{3}}.$$

12.8 (a) $\boldsymbol{a} \times \boldsymbol{c} = (2\hat{i} + 3\hat{k}) \times (\hat{i} - 2\hat{j} + \hat{k})$

$$= (-4\hat{i} \times \hat{j}) + (2\hat{i} \times \hat{k}) + (3\hat{k} \times \hat{i}) - (6\hat{k} \times \hat{j})$$

$$= -4\hat{k} - 2\hat{j} + 3\hat{j} + 6\hat{i} = 6\hat{i} + \hat{j} - 4\hat{k}.$$

(b) $\boldsymbol{c} \times \boldsymbol{a} = (\hat{i} - 2\hat{j} + \hat{k}) \times (2\hat{i} + 3\hat{k})$

$$= (3\hat{i} \times \hat{k}) + (-4\hat{j} \times \hat{i}) - (6\hat{j} \times \hat{k}) + (2\hat{k} \times \hat{i})$$

$$= -3\hat{j} + 4\hat{k} - 6\hat{i} + 2\hat{j} = -6\hat{i} - \hat{j} + 4\hat{k}.$$

(c) $|\boldsymbol{c} \times \boldsymbol{a}| = \sqrt{36 + 1 + 16} = \sqrt{53}.$

(d) $(\hat{i} \times \hat{j}) \times \hat{j} = \hat{k} \times \hat{j} = -\hat{i}.$

(e) $\hat{i} \times (\hat{j} \times \hat{j}) = \hat{i} \times \boldsymbol{0} = \boldsymbol{0}.$

12.9 (a) $\boldsymbol{a} \cdot \boldsymbol{b} = (a_1\hat{i} + a_2\hat{j} + a_3\hat{k}) \cdot (b_1\hat{i} + b_2\hat{j} + b_3\hat{k})$

$$= a_1 b_1 \hat{i} \cdot \hat{i} + a_2 b_2 \hat{j} \cdot \hat{j} + a_3 b_3 \hat{k} \cdot \hat{k}$$

$$= a_1 b_1 + a_2 b_2 + a_3 b_3.$$

(b) $\boldsymbol{b} \times \boldsymbol{c} = (b_1\hat{i} + b_2\hat{j} + b_3\hat{k}) \times (c_1\hat{i} + c_2\hat{j} + c_3\hat{k})$

$$= b_1 c_2 \hat{i} \times \hat{j} + b_1 c_3 \hat{i} \times \hat{k} + b_2 c_1 \hat{j} \times \hat{i}$$

$$+ b_2 c_3 \hat{j} \times \hat{k} + b_3 c_1 \hat{k} \times \hat{i} + b_3 c_2 \hat{k} \times \hat{j}$$

$$= b_1 c_2 \hat{k} - b_1 c_3 \hat{j} - b_2 c_1 \hat{k} + b_2 c_3 \hat{i} + b_3 c_1 \hat{j} - b_3 c_2 \hat{i}$$

$$= (b_2 c_3 - b_3 c_2)\hat{i} - (b_1 c_3 - b_3 c_1)\hat{j} + (b_1 c_2 - b_2 c_1)\hat{k}.$$

$$12.10 \text{ (a) } \boldsymbol{a} \times \boldsymbol{b} = \begin{vmatrix} \hat{\boldsymbol{i}} & \hat{\boldsymbol{j}} & \hat{\boldsymbol{k}} \\ 1 & 1 & 1 \\ 1 & -1 & 1 \end{vmatrix}$$

$$= \hat{\boldsymbol{i}} \begin{vmatrix} 1 & 1 \\ -1 & 1 \end{vmatrix} - \hat{\boldsymbol{j}} \begin{vmatrix} 1 & 1 \\ 1 & 1 \end{vmatrix} + \hat{\boldsymbol{k}} \begin{vmatrix} 1 & 1 \\ 1 & -1 \end{vmatrix}$$

$$= 2\hat{\boldsymbol{i}} - 2\hat{\boldsymbol{k}}.$$

(b) $|\boldsymbol{a} \times \boldsymbol{b}| = |2\hat{\boldsymbol{i}} - 2\hat{\boldsymbol{k}}| = \sqrt{2^2 + 2^2} = \sqrt{8}.$

$$\therefore \text{Unit vector} = \frac{2}{\sqrt{8}}\hat{\boldsymbol{i}} - \frac{2}{\sqrt{8}}\hat{\boldsymbol{k}} = \frac{1}{\sqrt{2}}\hat{\boldsymbol{i}} - \frac{1}{\sqrt{2}}\hat{\boldsymbol{k}}.$$

12.11 For $\boldsymbol{a} = a_2\hat{\boldsymbol{j}} + a_3\hat{\boldsymbol{k}}, \; \boldsymbol{b} = b_1\hat{\boldsymbol{i}}, \; \boldsymbol{c} = c_2\hat{\boldsymbol{j}};$

$$\boldsymbol{a} \cdot (\boldsymbol{b} \times \boldsymbol{c}) = \begin{vmatrix} 0 & a_2 & a_3 \\ b_1 & 0 & 0 \\ 0 & c_2 & 0 \end{vmatrix} = a_3 \begin{vmatrix} b_1 & 0 \\ 0 & c_2 \end{vmatrix} = a_3 b_1 c_2.$$

12.12 (a) $\boldsymbol{b} = b_1\hat{\boldsymbol{i}}$ and $|\boldsymbol{b}| = 0.600$ nm; $\boldsymbol{c} = c_2\hat{\boldsymbol{j}}$ and $|\boldsymbol{c}| = 0.866$ nm

$\therefore b_1 = 0.600$ nm and $c_2 = 0.866$ nm.

(b) $\boldsymbol{a} \cdot \boldsymbol{c} = (a_2\hat{\boldsymbol{j}} + a_3\hat{\boldsymbol{k}}) \cdot (c_2\hat{\boldsymbol{j}}) = a_2 c_2 = 0.824 \times 0.866 \times \cos\beta \text{ nm}^2$

$\Rightarrow a_2 \times 0.866\text{nm} = 0.824 \times 0.866 \times \cos\beta \text{ nm}^2$

$\Rightarrow a_2 = 0.824 \text{ nm} \times \cos122.9° = -0.448 \text{ nm}.$

(c) $|\boldsymbol{a}| = \sqrt{a_2^2 + a_3^2} = \sqrt{0.448^2 + a_3^2} = 0.824 \text{ nm}.$

$\Rightarrow 0.679 = 0.448^2 + a_3^2$

$\therefore a_3^2 = 0.679 - 0.2003 = 0.479 \text{ nm}^2.$

$\Rightarrow a_3 = \pm 0.692 \text{ nm}.$

(d) Volume of the unit cell is given by:
$\boldsymbol{a} \cdot (\boldsymbol{b} \times \boldsymbol{c}) = a_3 b_1 c_2 = 0.692 \times 0.600 \times 0.866 = 0.36\text{nm}^3.$

Chapter 13

13.2 $P(5/6) = \dfrac{6!}{5!(6-5)!} \left(\dfrac{1}{2}\right)^5 \left(\dfrac{1}{2}\right)^1 = 6 \times \left(\dfrac{1}{2}\right)^6 = \dfrac{3}{32} = 0.09375$

13.3 For $\Delta H^\ominus$;
   (a) Mean $= 142.1429$ kJ mol^{-1}.
   (b) The sample standard deviation $= 11.4681$ kJ mol^{-1}.
   (c) The standard error of the sample mean $= 3.0650$ kJ mol^{-1} and so we report the mean enthalpy change as $\overline{\Delta H^\ominus} = 142.1429 \pm 3.0650$ kJ mol^{-1}.
   (d) The 95% confidence interval $= 142.1429 \pm 6.5591$ kJ mol^{-1}.

For $\Delta G^\ominus$:
   (a) Mean $= 20.4571$ kJ mol^{-1}.
   (b) The sample standard deviation $= 0.8410$ kJ mol^{-1}.
   (c) The standard error of the sample mean $= 0.2248$ kJ mol^{-1} and so we report the mean Gibbs free energy as $\overline{\Delta G^\ominus} = 20.4571 \pm 0.2248$ kJ mol^{-1}.
   (d) The 95% confidence interval $= 20.4571 \pm 0.4811$ kJ mol^{-1}.

13.4 The Arrhenius equation can be rearranged to give a form which is suitable for linear regression by taking natural logs both sides to yield

$$\ln k = \ln A - \frac{E_a}{RT}$$

A plot of $\ln k$ against $1/T$ will then yield a straight line with a gradient of $-E_a/R$ and a y-axis intercept at $\ln A$.

$1/T$	0.00131	0.00128	0.00127	0.00126	0.00123	0.00122	0.00121
$\ln k$	-9.3847	-8.3307	-8.3266	-7.8727	-7.0109	-6.5713	-6.3654

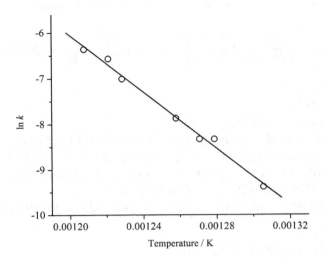

For the gradient, $-E_a/R$, we evaluate each of the terms that appear in eqn (13.17) where we associate the x_i values with $1/T$ and the y_i values with $\ln k$.

$$N=7, \quad \sum_{i=1}^{N} x_i = 0.008770, \quad \sum_{i=1}^{N} y_i = -53.8623,$$

$$\sum_{i=1}^{N} x_i y_i = -0.06771, \quad \sum_{i=1}^{N} x_i^2 = 1.0995 \times 10^{-5},$$

$$\left(\sum_{i=1}^{N} x_i\right)^2 = 7.6913 \times 10^{-5}$$

Inserting these values into eqn (13.17) yields:

$$-\frac{E_a}{R} = \frac{(7 \times -0.06771) - (0.008770 \times -53.8623)}{(7 \times 1.0995 \times 10^{-5}) - 7.6913 \times 10^{-5}} = \frac{-1.5974 \times 10^{-3}}{5.2 \times 10^{-8}}$$

$$= -30964.895 \text{ K mol}$$

which, when multiplied by the gas constant R, gives a value for E_a of 257 457.9 J mol^{-1} or 257.46 kJ mol^{-1}.

For the intercept, $\ln A$:

$$\ln(A/s^{-1}) = \frac{(-53.86233 \times 1.0995 \times 10^{-5}) - (0.00877 \times -0.06771)}{(7 \times 1.0995 \times 10^{-5}) - 7.6913 \times 10^{-5}}$$

$$= \frac{1.6004 \times 10^{-6}}{5.2 \times 10^{-8}} = 31.099$$

which then yields a value for the pre-exponential factor of $3.21 \times 10^{13} s^{-1}$.

The uncertainty in the $\ln k$ values is given by eqn (13.21), for which in this case the term:

$$\sum_{i=1}^{N} (y_i - mx_i - c)^2 = \sum_{i=1}^{N} \left(\ln k_i + \frac{E_a}{R} \frac{1}{T_i} - \ln A \right)^2 = 0.06681$$

and so:

$$\sigma_{\ln k} = \sqrt{\frac{1}{5} \times 0.06681} = 0.1156$$

We can now use this to find the uncertainties associated with $-E_a/R$ and $\ln A$:

$$\sigma_{-E_a/R} = 0.11556 \sqrt{\frac{7}{5.2 \times 10^{-8}}} = 1341.27 \text{ K}^{-1}$$

and:

$$\sigma_{\ln A} = 0.11556 \sqrt{\frac{1.0995 \times 10^{-5}}{5.2 \times 10^{-8}}} = 1.681.$$

13.5 The collision cross-section $\sigma = \pi d^2 = \pi \times (6.5 \times 10^{-10} \text{ m})^2 = 1.327 \times 10^{-18} \text{ m}^2$. The relative error in its calculation (from eqn 13.34) is given by:

$$\frac{\sigma_\sigma}{|\sigma|} = \frac{2 \times \sigma_d}{|d|} = \frac{2 \times 0.14 \times 10^{-10} \text{ m}}{6.5 \times 10^{-10} \text{ m}} = 0.043,$$

i.e. twice the relative error in the parameter d. Multiplying the relative error by the calculated value for the cross-section σ yields a value for the absolute uncertainty σ_σ:

$$\sigma_\sigma = 0.043 \times 1.327 \times 10^{-18} = 5.707 \times 10^{-20} \, \text{m}^2$$

We might then record the calculated cross-section as
$\sigma_\sigma = (1.327 \pm 0.057) \times 10^{-18} \, \text{m}^2$.

13.6 The modulus of the derivative of $y = e^x$ is:

$$\left| \frac{dy}{dx} \right| = e^x$$

and so the absolute uncertainty in y is $\sigma_y = e^x \sigma_x$ with relative uncertainty:

$$\frac{\sigma_y}{|y|} = \frac{e^x \sigma_x}{e^x} = \sigma_x$$

i.e. the relative uncertainty in y is the same as the absolute uncertainty in x, the opposite result to that found for $y = \ln x$.

13.7 Rearranging eqn (13.28) yields:

$$r = \sqrt{\frac{h}{8\pi^2 \mu c B}}$$

$$= \sqrt{\frac{6.626 \times 10^{-34} \, \text{Js}}{8\pi^2 \times 1.139 \times 10^{-26} \, \text{kg} \times 2.998 \times 10^{10} \, \text{cms}^{-1} \times 1.923 \, \text{cm}^{-1}}}$$

$$= 1.1305 \times 10^{-10} \, \text{m}$$

i.e. $r_{CO} = 0.113$ nm.

The absolute error in r, σ_r, can be deduced from the absolute error in B, σ_B, by noting that the two are related through the general functional form $r = k \times B^{-\frac{1}{2}}$ where

$$k = \sqrt{\frac{h}{8\pi^2 \mu c}}.$$

The derivative of this function is:

$$\frac{dr}{dB} = -\frac{1}{2} k B^{-\frac{3}{2}}$$

but its modulus is:

$$\left|\frac{dr}{dB}\right| = \frac{1}{2}kB^{-\frac{3}{2}}$$

and so:

$$\sigma_r = \frac{1}{2} kB^{-\frac{3}{2}}\sigma_B$$

The relative errors are then related as follows:

$$\frac{\sigma_r}{r} = \frac{\frac{1}{2}kB^{-\frac{3}{2}}}{kB^{-\frac{1}{2}}}\sigma_B = \frac{1}{2}\frac{\sigma_B}{B}$$

which is the result suggested by eqn (13.34). Thus if $\sigma_B=0.005$ cm^{-1}, then the relative error will be: $\dfrac{\sigma_B}{B} = \dfrac{0.005}{1.923} = 2.6 \times 10^{-3}$ or 0.26%. We can now calculate the relative error in r from:

$$\frac{\sigma_r}{r} = \frac{1}{2}\frac{\sigma_B}{B} = \frac{2.6 \times 10^{-3}}{2} = 1.3 \times 10^{-3} \text{ or } 0.13\%.$$

A 0.13% error in the bond length of 1.1305×10^{-10} m is 1.4697×10^{-13} m and so we can report the bond length as:

$$r = 0.11305 \pm 0.00015 \text{ nm}$$

13.8 For $y=\dfrac{u}{v}$, $\dfrac{\partial y}{\partial u} = \dfrac{1}{v}$; $\dfrac{\partial y}{\partial v} = \dfrac{-u}{v^2}$

Thus $\sigma_y^2 = \dfrac{1}{v^2}\sigma_u^2 + \dfrac{u^2}{v^4}\sigma_v^2$

As with multiplication, dividing through by y^2 gives:

$$\frac{\sigma_y^2}{y^2} = \frac{v^2\sigma_u^2}{u^2v^2} + \frac{v^2u^2\sigma_v^2}{u^2v^4} = \frac{\sigma_u^2}{u^2} + \frac{\sigma_v^2}{v^2}$$

$$\Downarrow$$

$$\left(\frac{\sigma_y}{y}\right)^2 = \left(\frac{\sigma_u}{u}\right)^2 + \left(\frac{\sigma_v}{v}\right)^2$$

and:

$$\frac{\sigma_y}{y} = \sqrt{\left(\frac{\sigma_u}{u}\right)^2 + \left(\frac{\sigma_v}{v}\right)^2} \text{ as required.}$$

13.9 Assuming the ideal gas law $P = \dfrac{nRT}{V}$, the total pressure exerted

will be: $P = \dfrac{0.02\,\text{mol} \times 8.31451\,\text{J K}^{-1}\,\text{mol}^{-1} \times 298\,\text{K}}{1 \times 10^{-5}\,\text{m}^3} = 4\,955\,544.8$

Pa. With 1 bar = 100 kPa, the sublimed CO_2 will be exerting a pressure of nearly 50 bar! As the ideal gas law is in the form of a product with respect to the independent variables, n and T, the relative uncertainty associated with P will be given by eqn (13.42):

$$\frac{\sigma_p}{p} = \sqrt{\left(\frac{\sigma_n}{n}\right)^2 + \left(\frac{\sigma_T}{T}\right)^2}$$

$$= \sqrt{\left(\frac{0.0002}{0.02}\right)^2 + \left(\frac{0.5}{298}\right)^2}$$

$$= \sqrt{1 \times 10^{-4} + 2.82 \times 10^{-6}} = 0.0101.$$

Multiplying this by P gives a value for the absolute uncertainty of: $\sigma_P = 0.0101 \times 4955544.8 = 50249.3$ Pa.

Therefore, we might report the pressure as approximately 4955 $\pm$ 50 kPa.

13.10 Applying eqn (13.37) to the ideal gas law equation, $P = \dfrac{nRT}{V}$,

yields: $\sigma_P \leqslant \dfrac{\partial P}{\partial n}\sigma_n + \dfrac{\partial P}{\partial T}\sigma_T.$

Partial differentiation of the ideal gas law equation with respect to n and T yields: $\dfrac{\partial P}{\partial n} = \dfrac{RT}{V}$ and $\dfrac{\partial P}{\partial T} = \dfrac{nR}{V}.$

Thus:

$$\sigma_P \leqslant \frac{RT}{V}\sigma_n + \frac{nR}{V}\sigma_T$$

$$= \frac{8.31451\,\text{J K}^{-1}\,\text{mol}^{-1} \times 298\,\text{K}}{1 \times 10^{-5}\,\text{m}^3} \times 0.0002\,\text{mol}$$

$$+ \frac{0.02\,\text{mol} \times 8.31451\,\text{J K}^{-1}\,\text{mol}^{-1}}{1 \times 10^{-5}\,\text{m}^3} \times 0.5\,\text{K}$$

$$= 49551.4 + 8314.5 = 57865.9\,\text{Pa}$$

Note that the computed value for σ_P using this approach is significantly larger than that determined from Problem 13.9.

Symbols

$>$		greater than
$\geqslant$		greater than or equal to
$\gg$		much greater than
$<$		less than
$\leqslant$		less than or equal to
$\ll$		much less than
/ or $\div$		division
$\neq$		not equal to
$\cong$ or $\approx$		approximately equal to
$\Rightarrow$		implies
$\propto$		proportionality
$=$		equality
∞		infinity
Σ		summation sign
Π		product sign
!		factorial
{ }		braces
[ ]		brackets
( )		parentheses

Subject Index

References to tables and charts are in **bold** type